Fires
and
Human Behaviour

Fires
and
Human Behaviour

Edited by

David Canter
Reader in Psychology
University of Surrey

JOHN WILEY & SONS
Chichester · New York · Brisbane · Toronto

Copyright © 1980 by John Wiley & Sons, Ltd.

British Library Cataloguing in Publication Data:

Fires and human behaviour.
  1. Fires—Psychological aspects
  I. Canter, David Victor
  155.9'35   BF789.F5   79-41489

ISBN 0 471 27709 6

Typeset by Activity, Teffont, Salisbury, Wilts.
and printed in United States of America

Contents

Contents

Notes on Contributors

IAN APPLETON BSc (Physics) PhD, *Building Research Establishment, Fire Research Station, Borehamwood, Hertfordshire*
After a background in computing and social work Ian Appleton joined the Fire Research Station in 1974 and in 1976 became Head of the Operational Research Section. Their present activities include behaviour of people, attitudes and perception of risks, various costing exercises, demographic studies, statistical analyses and the cost effectiveness of regulations.

LEONARD BICKMAN, *Loyola University of Chicago, 6525 North Sheridan Road, Chicago, Illinois*
Dr Bickman is a full professor in the Department of Psychology at Loyola University and Director of the Westinghouse Evaluation Institute, Evanston, Illinois. He has directed a large number of research projects and published over 60 research papers. He is president-elect of the Society for the Psychological Study of Social Issues and the American Psychology-Law Society. He is also current editor of the *Applied Social Psychology Annual.*

JOHN BREAUX DPhil, *Department of Psychology, University of Surrey, Guildford, Surrey*
Dr Breaux is Senior Research Fellow with the Fire Research Unit at the University of Surrey. His articles and interests centre on the modelling and analysis of behaviour.

LAURA BAKER BUCHBINDER, *United States Fire Administration, Federal Emergency Management Agency, Washington DC 20472*
Laura Buchbinder is the Director, Public Education Assistance Program which establishes a delivery system for fire safety education information in the states and cities.

DAVID CANTER PhD, *Department of Psychology, University of Surrey, Guildford, Surrey*
Dr Canter is Reader in Psychology with the University of Surrey and Director of the Masters' course in Environmental Psychology. He has been leading a research unit studying human behaviour in fires for a number of years. Besides his research on fires he has published widely in relation to environmental evaluation and the design of therapeutic environments.

DAVID CRONRATH, *University of California, Berkeley*
David Cronrath received a BArch from Pennsylvania State University and a MArch from Berkeley. He is currently the Director of the Architecture Workshop in Philadelphia, a community design service centre dealing primarily with inner city housing issues.

PERRY EDELMAN, *7026 N. Sheridan Road, Chicago, Illinois*
Perry Edelman received his BS degree in Psychology from the State University of New York, Stony Brook and is now completing his PhD degree at Loyola University, Chicago. He teaches part-time and serves as a research analyst for the Chicago Area Agency on Aging. He has also completed a major evaluation project for the Illinois Department of Correction and for two years made a special study of human behaviour in fires on a federal grant.

COLIN H. GREEN, *Faculty of Environmental Studies, Department of Architecture, Duncan of Jordanstone College of Art and University of Dundee, Perth Road, Dundee, Scotland*
Colin Green trained originally as an architect and switched into systems analysis via a MArch at SUNYAB. He became involved in 'Acceptable Risk' and 'The Value of Life' when researching the policy aspects of building regulation whilst Resident Fellow at the Center for Advanced Study, University of Illinois, Urbana.

GILDA MOSS HABER PhD, *University of the District of Columbia, Mount Vernon Square Campus, 929 E. Street N. W. Room 703 Washington DC 20004*
Dr Haber's teaching experience at the University of the District of Columbia includes such topics as organization, work, stratification and medical sociology. She has published various articles in medical sociology and is at present working on two books, one on feudalism in bureaucracy and the other on territoriality in the college classroom.

JANET HALL, *34a Kings Parade, Blackwater, Surrey*
Janet Hall heads a consultancy which advises on matters concerning industrial design, ergonomics and training. They concentrate on social-orientated research and development projects including environmental and artefact design for the disabled, and lately have been concerned with operational and human factor fire precaution problems in hospitals. This work is now at the development stage and will include determining an effective national standard training programme.

ELICIA J. HERZ, *Northwestern University, Evanston, Illinois*
Elicia Herz is a research consultant for the Center for Urban Affairs at Evanston, Illinois. She received her MA in psychology from Loyola University of Chicago and is currently working on her PhD dissertation. For two years she was a research assistant at Loyola University of Chicago on a federally funded grant to study human behaviour and fire.

BRIAN JONES, *National Research Council of Canada, Division of Building Research, Ottawa, K1A OR6*
Brian Jones is a Research Officer with the Division of Building Research, National Research Council of Canada. His research interests in the behaviour and fire field include collective behaviour and decision-making under emergency conditions, the perception and assessment of risk, and the organizational aspects of evacuation planning.

DITSA KAFRY PhD, *University of California, Berkeley*
Dr Kafry received her BA and MA in psychology fron the Hebrew University, Jerusalem, and her PhD in psychology from the University of California, Berkeley. Her interest and research work focus on exploratory and motivational processes of children and adults.

LARS LERUP, *University of California, Berkeley*
Lars Lerup holds a civil engineering diploma from Sweden and degrees in architecture and urban design from Berkeley and Harvard. He teaches environmental design at Berkeley, where he also conducts research in the area of man/environment/ relations. He is currently writing a book on modernism and architecture.

BERNARD M. LEVIN PhD, *Center for Fire Research, National Bureau of Standards, Washington DC 20234*
Dr Levin was trained in psychometrics at the University of North Carolina. He has held a variety of technical and administrative assignments at the United States National Bureau of Standards (NBS) since 1958 in the fields of psychology and operation research. He is currently in the NBS Center for Fire Research where he is involved in the arson problem and the relationship between knowledge about behaviour in fire and fire regulations.

JOHN K. C. LIU, *University of California, Berkeley*
John Liu received a BArch from Cooper Union and a MArch from University of Washington. He is currently a PhD candidate at Berkeley, writing a dissertation on the relation between family life and the physical setting in Chinese cities.

ERIC W. MARCHANT PhD, *Department of Fire Safety Engineering, University of Edinburgh, The King's Buildings, Edinburgh Scotland*
Dr Marchant studied architecture at Nottingham, qualifying in 1962. After postgraduate studies and research in Building Science at Liverpool University, he was appointed as a lecturer in the Department of Civil Engineering at Edinburgh University. His interest in fire problems began in 1969 and he transferred to the Department of Fire Safety Engineering at its inception in 1974.

JAKE L. PAULS, *National Research Council Canada, Division of Building Research, Ottawa, K1A OR6*
Jake Pauls is a Research Officer with the Division of Building Research, Ottawa. He has directed major field studies of evacuation and general movement on office buildings and large buildings for public assembly. Complementary technology-transfer activities have entailed considerable contact with building design, management and regulatory personnel.

JONATHAN D. SIME, *Department of Psychology, University of Surrey, Guildford, Surrey*
Jonathan Sime has a BA in psychology from London University and works as Research Fellow with the Fire Research Unit at the University of Surrey where he has been since completing his MSc in environmental psychology in 1975 at that university. His previous research experience has been in the areas of landscape evaluation and hospital design.

RICHARD STROTHER, *United States Fire Administration, Federal Emergency Management Agency, Washington DC 20472*
Richard Strother is the Associate Administrator of the United States Fire Administration. He is responsible for establishing programmes to change fire-related behaviour including public fire education, arson prevention, emergency medical services and fire protection planning.

ROBERT G. VREELAND PhD, *Department of Psychology, Davie Hall 013-A, University of North Carolina, Chapel Hill, NC 27514*
Dr Vreeland received his PhD in experimental psychology from the University of North Carolina in 1975. He has been engaged in research and therapy with the developmentally disabled, and most recently has been involved in a project on firesetting sponsored by the National Bureau of Standards.

JOHN R. WILSON and CLAIRE WHITTINGTON, *University of Technology, Loughborough, Leics*
John Wilson and Claire Whittington are both Senior Research Officers at the Institute for Consumer Ergonomics, University of Technology, Loughborough. For a number of years, they both worked as members of a research team which carried out a series of projects connected with problems of domestic safety. Claire Whittington is currently working on problems of rehabilitation in the Steel Industry, and John Wilson is involved in a number of research areas which include the behaviour of elderly pedestrians and the design of space in the home.

PETER G. WOOD, *University of Surrey, Guildford, Surrey*
Peter Wood graduated from Manchester University with Hons BSc in psychology in 1962. At first he was a research assistant at Loughborough University and then worked for the Ford Motor Company for three years, in both cases working on the safety and ergonomics of motor vehicles. Work on behaviour in fires was conducted on his return to Loughborough as Research Fellow and his current research interest is in the field of consumer safety. He is at present Lecturer in Behavioural Science, Department of Home Economics, University of Surrey.

List of Figures

List of Tables

List of Tables

Foreword

Over the past three or four years a small number of applied social scientists and government advisers have begun to consider radically new ways of reducing the number and effects of dangerous fires in buildings. This has grown out of an examination both of the human causes of fires and of what people do when they experience a fire. This examination has led to reconsidering whether the causes of many unwanted fires are 'mechanical failures', or if it is always appropriate to cope with potential fire emergencies by the provision of 'hardware' such as special fire escapes, elaborate alarm bells, and the like. The re-examination of fire precautions has pointed to a need for more effective education and training for emergencies and the more sensitive design of buildings for human use in an emergency.

That such a change in direction has been indicated, albeit by a small group of people, at a time of severe economic restraint is fortuitous. It is in such times that circumstances force policy-makers and administrators, at all levels, to reconsider their traditional practices and to explore, however tentatively, other options. The possibility that a social science emphasis for fire prevention and safety may prove less expensive than an exclusively engineering approach, makes the human option all the more worthy of investigation.

Yet such considerations have many consequences, for building design and management, as well as for the processes of legislation and control; for large governmental and commercial enterprises as well as for individuals and families at home. It is therefore important to make available to as wide an audience as possible the scientific bases for the social and psychological approaches to fires. This book is the first attempt to bring together the full range of recent research on fires and human behaviour.

The contributors to this book include most of the small number of people around the world who have been actively involved in this area of research, who were able to meet at the University of Surrey in March 1977 to compare strategies, tactics, and results. It was partly as a consequence of the awareness at that meeting that there are remarkable consistencies in the results, and large areas of agreement in their interpretation amongst independent research groups in many different places, that the desire to produce this book emerged.

I regard it as a privilege to have been part of such a community of researchers. As editor of this book I have been struck by the enthusiasm and commitment of the contributors. I am deeply grateful for the way they have cooperated with me.

The contributors come from a great variety of backgrounds which is also reflected in the very different styles of their presentations. One contributor, however, must be singled out for mention in this foreword. Charles Clisby took the opportunity of his early retirement from a senior post in the London Fire Brigade to write for this book. Then, whilst I was still editing his manuscript, there came the sad news of his untimely death. His contribution to this book is unique, providing a viewpoint on fires in stark contrast to that from the other authors. He was always outspokenly in favour of the objectives of the research reported in this book and gave direct and substantial help to the research efforts of my colleagues and myself. I hope that this book can be taken as an acknowledgement of our debt to D. A. C. O. Clisby and his colleagues in fire services throughout the world.

Surrey, August, 1979   *David Canter*

CHAPTER 1

Fires and Human Behaviour —
an Introduction

DAVID CANTER
Department of Psychology, University of Surrey

PEOPLE AND FIRE

Fire has always both fascinated and repelled. Its complex power has stimulated a profound interest and desire to learn to harness its force for the benefit of mankind. The destructive potential of fire has frequently led to it being seen as the symbol of evil.

However it is perceived, fire is an integral part of human experience of the physical world, as demonstrated by the considerable resources which are expended on the control and reduction of the malicious potential of conflagrations. Yet it is only in recent years that the psychological and social aspects of destructive fires have been systematically explored. In the past fire has been dealt with solely as a problem of engineering. This book brings together, for the first time, the results of recent research which highlights the human aspects of dangerous fires.

Although systematic study of fire and human behaviour has occurred only in the past few years, it is possible to trace the use of fire to the earliest origins of mankind. Indeed, the controlled use of fire may be regarded as an important stage in the development from primitive man to his more civilized descendants, dating back to our human ancestors of at least half a million years ago. At the excavation of Choukoutien, near Peking, where some of the earliest human remains have been found, it was discovered that amongst the human fossils there were also large quantities of woodash and charcoal. There seems little doubt that those distant yet intelligent ancestors of early man not only tended fires in their caves, but also transported fire materials with them whenever they moved.

Hundreds of thousands of years later, in neolithic times, the control of fire developed to the extent that instead of transporting the results of chance conflagrations, people were able to create their own light and flame by the use of flints or wood drills. However, it was not until 1827, when the English chemist John Walker invented the friction match, that a source for creating fires easily became widely available. Later developments in the nineteenth century introduced gas and then electricity throughout the rapidly growing cities, increasing further the possibilities for harnessing fire.

1

The controlled utilization of fire has gone hand in hand with advance in technology. Modern civilization would not have been possible without the use of fire. Very few cultures, if any, have been denied access to spark and flame. But with the growing use of fire has come an increase in its associated risks and hazards. The reduction of the unwanted occurrence of fire may well be an important milestone in the advancement of civilization.

Before the turn of the twentieth century many major cities of the world had been destroyed, many times over, by the effects of fires which had blazed out of control. The greater part of London has been destroyed by fire on at least three occasions in recorded history, apart from the destruction caused by wars. In North America, over the 35 years from 1835 to 1871, New York City, Pittsburgh, San Francisco, Atlanta, Portland and Chicago all had major fires which destroyed great areas of the city. During the same time period St Louis had two major fires; in 1849, fifteen blocks of the city were burned, and in 1851 more than 2500 buildings were burned.

Advance in the control of destructive fires was further necessitated by the use of fire as a weapon in times of war. One useful consequence of the extensive havoc produced by incendiary bombing was the creation of more effective ways of coping with the destructive effects of fire.

There is a long history of some form of precaution, public control and fire-fighting force, used together as a means of combating the evil effects of chance conflagrations. Fire brigades, in fact, existed in Ancient Rome; the custom of ringing a bell so that all heating fires and candles could be extinguished for the night can be traced to the reign of Alfred the Great and has found a place in our language as the curfew *(couvre-feu)*.

However, in Britain it was not until the Second World War that there was a consolidation of all local fire brigades into a national fire fighting service. This gave an orientation to the organization of British fire brigades which has been followed by other Western countries which did not suffer the effects of large-scale bombing.

THE NEED TO KNOW ABOUT HUMAN RESPONSE TO FIRE

It is possible to trace many thousands of years of effective use of fire for human benefit and many hundreds of years of organized methods of controlling its unwanted effects. But whilst centuries of experience have provided some knowledge of how to control the physical effects of fire, there is still very little understanding of its psychological and social concomitants.

There is one major reason for the scarcity of systematic knowledge about the human, behavioural aspects of fire. This is that parallel to the growth of sophistication in fire fighting has occurred an increase in the utilization of engineering (or 'hardware') approaches to the reduction of unwanted fires and the control of those which do take place. These technical developments have been spurred on since the Second World War by the continuing occurrence of major destructive fires. For example a fire in a nursing home in Yokohama in 1955 killed 100 people. In 1961 a fire in Brazil killed 323 people. In 1974 the same country had a fire in Sao Paulo which killed 227 people. Even more recently, in 1977, the fire at the Beverly Hills Supper Club in Kentucky killed 164 people.

Each of these major conflagrations attracted worldwide public attention and usually resulted in new legislation about the provision of escape routes, fire-resistant materials, structural design of buildings and the many other engineering aspects reviewed in such texts as Lie (1972), Marchant (1972) and Langdon-Thomas (1972). This present book is witness to the fact that such provisions are frequently insufficient and in many cases inappropriate. The human aspects of the causes and developments of fire must be understood if its disastrous effects are to be minimized.

'HARDWARE' IS NOT ENOUGH

Fire and its adverse effects cannot be mastered by engineering means alone. As Appleton shows (in Chapter 2 of this volume) many of the fatalities, and the damage caused by fires, can be attributed to the great number of smaller-scale fires. The world's newspapers do not report these day-to-day fires so often, but it is likely that the fatalities which they cause can be traced more to human error than to engineering failure. The statistics are difficult to come by, but it may well be that the gigantic conflagrations of the past century have been greatly reduced in frequency through the development of fire protection and fire management policies. Furthermore the major fire hazards which existed in old cities due to high density housing built out of highly combustible materials have been recognized by official bodies throughout the world, even if there is some difficulty in many areas in putting this knowledge to good effect. Nonetheless there remain the human causes of fire, as well as the effect of human error, and the human agency which magnifies the damaging effects of any fire which does occur.

With hindsight, a case may be made even of the major fires throughout history, that they were either caused directly by human malice or malpractice, or the severity of their effects was increased by the human response to them. As our control of fire develops and the uses to which we put it varies, it is becoming increasingly clear that to control this demoniac workhorse we must learn more about the people who work and deal with it, especially on those occasions when it escapes.

RECURRENT THEMES

In subsequent chapters a variety of research results and perspectives on fire and human behaviour are presented. Yet despite the diversity of studies, conducted across two continents, there are a number of themes which recur throughout the book. These themes provide an initial basis for understanding the human aspects of dangerous fires in buildings. In effect, each of these themes deals with a distinct class of human activity.

(1) The human component in the cause of fires.
(2) The patterns of early response activity which occur when a fire ignites within a building.

(3) The processes of escape and evacuation from fires once they have been identified, and evacuation procedures put in motion.
(4) The frequency, impact and effect of fires upon individuals and society.

Each of these themes are considered below.

Human causes of fire

Study of the *causes* of fire, is surprisingly, scarce. The engineering tradition of dealing with the management of effects rather than the identification of causes seems to have had an influence here. It is also the case that given the fortunate rarity of fires it is much more difficult to study the conditions under which they occur than it is to study their after-effects (especially when the after-effects leave such clear evidence). However, both the statistics and the studies reported in later chapters indicate that arson plays a significant role in the cause of fire.

It may now be the case that success in the control of fire is such that the majority of fire damage is the result of malicious human activity. Vreeland (in Chapter 3 of this volume) discusses the many different kinds of arson which exist. He refers to the evidence which shows that arson for profit is a significant cause of fire, but he also is careful to distinguish this from the other forms of malicious arson, typically associated with adolescents and young children. Kafry develops this argument in Chapter 4, suggesting that often young children are fascinated with fire, but they may well be unable to have as great a control over it as they could wish. As so often happens when one examines human actions and failings, the message here points to the importance of education, especially education outside the school and within the family.

It is difficult to know how far the risk of fire could be removed if arson could be eliminated entirely. It seems likely that since the beginning of human use of fire there have always, as with other technological developments, been people who have misused the technology to their own ends and the harm of others. It therefore seems likely that arson in one form or another will remain a component of human experience for many years to come. Even if it were not, there would still be the need to cope with those circumstances in which fires occur without the conscious agency of other people.

Patterns of early response activity

Once a dangerous fire has been ignited, by whatever means, the seriousness of its effects depend to a considerable extent on the reactions and patterns of activity of people associated with the fire. Chapters 8, 9, 10 and 11 of this book are concerned with providing an account of what people actually do in fires. What emerges is that, although this work is very recent, and there are still relatively few detailed accounts of what has actually happened in fires, there are nonetheless surprising consistencies. All researchers seem to be agreed upon two stages in the early response to a fire, and a third distinct stage of escape which is dealt with in the next section. The first stage of the early responses is 'recognition'. The second is the stage of 'action'.

Unless a building is very fully equipped with automatic sensing devices and fire brigades can get to it in an exceptionally fast time, it is not possible for any attempts to cope with the fire to be put into motion until someone has *identified* that the fire is present and, usually, has acted upon that identification by informing others. As can be seen throughout the detailed case studies presented by Canter, Haber, Lerup, Bickman and their colleagues in this volume, the early stage of fire recognition is typically characterized by ambiguity. Whether it be a fire in a prison such as that reported by Haber, the nursing home fires described by Edelman *et al.* or Lerup *et al.*, or those hotel or hospital fires discussed by Canter *et al.*, it is clear that early acceptance that the unusual circumstances present constitute a fire of some severity, is frequently delayed to a dangerous extent.

Once a fire has been recognized as such there is then the possibility for a range of actions. The extent of available actions depends very much on the circumstances of the fire. The case studies show that in a large organization the possibility of contacting others or becoming part of an administrative chain can have a direct influence on the patterns of behaviour which occur. In a private house the role of neighbours or access to a telephone can have their own influence. As Wood shows in Chapter 6, contacting others, fighting the fire, or leaving the scene, are all major possibilities. Many people may do any combination of these in different sequences and different stages in the development of a fire. Canter and his colleagues emphasize that an important implication of this variation in possible sequence is that no general guidance can be given about what should be the safest sequence which will be valid for all circumstances. Readers only have to compare for themselves fire warning notices in different buildings to find that where one notice may indicate that the first thing to do is to warn others, others state the first thing to do is to try and cope with the fire. Until there is a clearer picture as to the variations in sequence and their significance for fire growth and danger such confusions will continue.

It follows, then, that a central question for the student of fires is to account for the variations in the action sequences which occur. The popular account, discussed at some length by Sime in Chapter 5 is that in the face of fire, people will 'panic'. There is a growing body of comment and evidence to suggest that, at the very least, the notion of panic is an unhelpful one. Quite possibly, whatever the term means, there is very little evidence that can be found for it actually occurring in fires. Instead, what can be considered the opposite of 'panic' appears to take place. People continue to carry out their normal roles. Whenever evacuation or action in relation to a fire is necessary it takes place within the general organizational framework and the roles which people have before ignition occurs. Perhaps one of the most dramatic examples of this is provided by the Beverly Hills Supper Club fire in Kentucky (Best, 1977) in which waitresses showed people out of the building through smoke. However, it appears that they only showed out those people who were at the tables for which they were normally responsible. People at adjacent tables were shown out by their own, different waiters or waitresses. If a person had been unlucky enough to be at a table with a less-than-effective waitress, or indeed one who had been trapped by smoke, they would not have been so readily shown out of the building, although people sitting at an adjacent table could have been.

Canter and his colleagues, and Wood both indicate the maintenance of husband and wife roles in domestic fires and their consequently different actions. In the accounts of hospital and nursing home fires the more subtle differences in status and position in the hierarchy can be seen as clear influences on the patterns of fire behaviour. Coping with these situations thus becomes a matter of recognizing the potential of the role structure which already exists and of its inherent weaknesses. People, for example, who are in important positions in the communication system yet who do not appreciate the significance of their role in a crisis, such as the announcer in the control room in the Summerland fire on the Isle of Man in 1973, or the announcer in the building evacuation referred to by Pauls in Chapter 13, can have just as dangerous or lifesaving an effect during the course of a fire as the selection of the right materials, or a careful check on the electrical wiring.

Processes of escape and evacuation

Once the decision to leave the building has been made the fire situation moves into a third and different phase. The activity now can be considered as having a single focus: escape. The great majority of world fire regulations relate to making possible a means of escape in the case of a fire without hindrance or exposure to further hazards, or the need for assistance. Yet there are many conditions in which the provision of a safe escape *route* is insufficient to guarantee hazard-free egress. One clear example is provided by very large occupancy buildings, such as office blocks, department stores, or stadia. In these buildings the large number of people leaving it may mean that some people may block the evacuation of others. The effects of constrictions in the escape route and the potential of stair accidents can be disastrous when large numbers of people are involved. These issues and their related design implications are examined in detail by Pauls and Jones in Chapter 13 and recommendations from empirical studies are presented by Pauls in Chapter 14. He shows that it is possible to develop quite accurate predictions for the speed of movement and the amount of total time necessary to evacuate large office buildings. In doing this, he demonstrates that our current wisdom as enshrined in existing legislation is not only based on false premises but frequently leads to false expectations, which may well be dangerous.

A second type of buildings in which safe evacuation does not depend solely upon safe escape routes is that in which many of the occupants are unable to leave, in normal circumstances, by their own unaided efforts. In hospitals, nursing home and many other therapeutic settings the patients depend on assistance from staff in order to leave the building. In Chapter 12 Janet Hall shows the difficulties inherent in the provision of staff assistance and the dangers of many current practices.

A number of other chapters deal with the problem of fires in hospitals and nursing homes and the particular difficulty in evacuating those types of buildings. It is clear that in many circumstances there is no possibility of total evacuation. Furthermore, when total evacuation is essential many significant details need to be considered. These include both the strengths and skills of those available to help as well as the opportunities provided by the building. One point that emerges is that earlier stages in coping with the fire, those of recognition and early action, may well provide more opportunity for effectively reducing the hazard of fire in these occupancies than relying upon costly and difficult escape procedures.

Frequency, impact effect of fire

Although many individuals never actually experience a serious fire, nonetheless the preparations and precautions which are taken in case of a fire, as well as the far-reaching effects of major fires, are experienced by everyone. It is therefore particularly important that the reactions in any given incident are understood against an effective background of systematic statistical information. It is all too easy for one large fire which reaches the newspapers and is reported in dramatic terms to have an overwhelming influence on the way fires are considered. Unfortunately, few countries have a really thorough overview of the frequency that different types of fire occur. Even countries such as Britain, which has a legal requirement for the filing of reports of every fire attended by the fire brigades, have incomplete and inaccurate information in their frequency counts, both because of the number of fires which are not reported to the brigades, and because of human errors in recording those fires which are reported. Various informal checks and one or two studies referred to later in this volume, indicate that at least four out of every five fires which occur are not reported to the brigades and probably as many as nine out of ten go unnoticed in the official statistics. Furthermore, recent work discussed by Appleton (Chapter 2) indicates that the frequency of reporting fires may often be a direct reflection of public concern. It can be greatly influenced by public fears, enhanced by highly publicized major incidents. Indeed in one small study carried out by Canter (1978) it was found that the fires reported to the authorities dropped when the British fire brigade was on strike. Against this backdrop of weaknesses in the official statistics it is important that information is collected from as wide a range of sources as possible.

Even with all their weaknesses the fire statistics provide a very important lesson. As mentioned earlier the majority of fatalities in fire, well over half of them in fact, occur in small domestic incidents. The major fires which catch the newspaper headlines do not account for anywhere near as many deaths. It is argued by Marchant in Chapter 16 that the main thrust of legislation is to prevent the occurrence of major disasters. A number of deaths of single old people, each in their own private houses, away from the obvious responsiblity of any public authority, will never have as great an impact on law-makers or public officials as will the fewer number of fatalities which occurs in one incident in one public institution. This is probably one reason why the concerns of many contributors to this volume relate much more closely to fires in institutions and public buildings than they do to an understanding of *domestic* fire hazards. By contrast the study by Wood (summarized in Chapter 6), which dealt with a very large sample of fires as they occurred, emphasizes fires in houses. However, Wood's study and the others which look at domestic fires (Chapters 7 and 8), show that there is still much to be learnt about those domestic fire events, which contribute so heavily to fire fatalities. This is not to denigrate the need to do everything we can to improve the safety of our old people's homes, hospitals, hotels, schools and offices, factories and other places of work and recreation. It is to point out that the broad picture derived from the national statistics themselves indicates a gap in our understanding which we still have a long way to go towards filling.

One of the consequences of the weaknesses in official statistics is that other sources of information must be tapped. This book has many examples of innovative approaches to finding out about human response to fire. The case of domestic fat fires is a good illustration. In Chapter 7 Whittington and Wilson show how, by focusing on the frequency of fat fires (chip-pan fires, or grease-pan fires as they would be known in North America), a number of surprising results can emerge. The much greater frequency of fires on electrical cookers than on gas cookers that they reveal is a point for serious consideration. The quite different approach to domestic fires taken in Chapter 8 in which a few detailed examples are looked at, also serves to illustrate that patterns of activities of some significance can be identified from complex events when examined closely.

The study of chip-pan fires also serves to highlight another important aspect. It is essential to obtain comparisons between different countries as these will reveal the way in which hazards are increased in one country or region when compared to another. The predilection of the English for chips with their meals and the tradition of preparing these through deep frying provides a clear example of a nationally and possibly subculturally located hazard; it might even be possible to take this example further and examine fire incidents in relation to days of the week. Often the implications for fire hazard of variations in patterns of activity is known and understood by local public figures or fire officers. The difficulty comes in aggregating such knowledge, putting it together in a coherent form and devising ways of acting upon it.

The comparison of frequencies of fire occurrence between countries as discussed in Chapter 2, is fraught with difficulties. But even when these are taken into account dramatic differences in the frequency of fire incidents and their effects can be seen. The United States emerges from this comparison surprisingly badly. Thus whether it is at a local or at an international level of comparison there are strong indications that the probability of experiencing a fire is not at all evenly distributed throughout the population. The tendency, in the past, has been to assign this variation to differences in the risk presented by buildings (the 'hardware'), but the possibility must now be embraced that differences in lifestyle, or standard of living may also be at the core of these large variations. It is important to establish the basis of these variations because they do have such significant implications for the planning of fire precautions and the education of people in fire prevention. However, the descriptive statistics alone do not provide the whole picture. They need to be elaborated by the details of what happens when the fire occurs.

THE USE OF RESEARCH KNOWLEDGE

The research described in this volume, and the future research which it is hoped this publication will encourage has relevance for three distinct practical issues.

(1) The assessment of risk. Whilst this area does itself have a psychological component as discussed by Green in Chapter 15 there is a need to consider the costs and effects of providing for a reduction in fire hazard in buildings. Marchant shows in Chapter 16 that the complex process of deciding on costs and risks can be modelled in a way which facilitates decision-making.

(2) A contribution to training and preparation for the possible occurrence of fire, as discussed by Strother and Buchbinder in Chapter 17.

(3) The design of buildings, both to minimize the risk of fire and the scale of the hazard should a fire occur.

These three areas of contribution interrelate closely. An understanding of the relative risks involved has relevance for the design decisions. The behaviour which occurs in fire has relevance, together with the understanding of the risks, for both training and design.

It is possibly because of the complexity of the practical implications that decision-makers may find it difficult to respond directly to the results reported in this book. Thus, as mentioned already, fire precautions have been dealt with as a problem of providing the appropriate 'hardware' (fire escapes, smoke doors, alarms, etc.) in order to prevent some notional, 'irrational' behaviour, frequently labelled 'panic'. The exploration of the possible advantages of improving training, education and other aspects of the human component, instead of ever-more sophisticated 'hardware', has only come to be seriously considered with the great escalation of the costs of that hardware and the reduction in the funds available.

However, the possibilities for relying upon such a 'fallible' component as the human one depends upon a thorough understanding of human behaviour in fires. Such an understanding is only beginning to emerge from systematic research such as that reported in this book. This is not to imply that no information has been available before. There have been and continue to exist a number of sources of information which exert a strong influence upon thoughts and regulations: (a) 'common-sense', (b) personal experience from previous fires, (c) newspaper reports, and (d) commissions of enquiry and other forms of official investigation (Canter and Matthews, 1976). Unfortunately all of these, by their very nature, can only point to major deficiencies. They also are very dependent upon the particular experience and skills of the individuals involved. But their major disadvantage is that they do not provide the basis for a clearly articulated, cumulative framework. They react to particular instances. In general, they only work because they 'overprovide'. The lack of understanding and precision is counteracted by very expensive requirements.

There are two aspects of building today which leads to a questioning of even the strategy of overprovision. One is the speed of change, not only in the design of buildings, but also in what they are used for and how they are used. The other is the novelty and variety of buildings and usages which are emerging. The Summerland fire in 1973 and the Beverly Hills Supper Club fire in 1977 both occurred in types of building form, structure and layout, being used for a variety of purposes, which would have been unknown 50 years ago, when many present-day regulations were being formulated. As ever-more novel forms and uses of buildings emerge so the direct, past experience of the experts becomes of less value.

A caution is necessary. The argument in favour of research does not rule out the contribution of experienced people such as fire officers. This book contains a contribution from a distinguished senior fire officer. In Chapter 18 he eloquently argues for the involvement of fire officers in decision-making with regard to fire precautions. He shows the intensity of experience on which such involvements is frequently based.

Nonetheless, as is shown by the contribution from Strother and Buchbinder (Chapter 17), there is a growing need for people whose role is clearly preventive and educational. These people require to advise and teach beyond their individual experience and so require a more systematic knowledge of behaviour in fires than has been available in the past.

FUTURE REQUIREMENTS

This book can be regarded only as an initial incursion into a complex area of human activity. Yet the recent research reported in subsequent chapters has been productive enough to indicate the directions in which future research is likely to prove most fruitful.

Social consequences

Most thinking about fire precautions is influenced by the notions of balancing costs and benefits. It is possibly stronger in this area of decision-making than in many others because of the long history of fire insurance, in which risks are offset against expenditure. The consequent reduction of expenditure on insurance, through the purchase of hardware which will reduce risks, has thus been an economic calculation implicit in much thought about fires. One advantage of this approach is that its clear statement of the tradeoff which is being made makes it possible to see how limited have been the number of considerations in the equation. The time is therefore appropriate for broadening the basis of the calculations. At least three further considerations require exploration if cost-effective fire precautions are to be introduced.

(1) The first is the broad social consequence of fires. Lawrence and Melinek (1979) have started to consider the costs to the health services of fire injuries, but the costs to the broad range of social services which are drawn upon in a fire should not be left out of these studies. Beyond this, consequences through post-fire trauma and bereavement are all too rarely considered. In a recent, unpublished study by the Fire Research Unit at Surrey University it was found that a substantial proportion of people who had experienced a fire in their home had not returned to that address until some months after the fire. Whether the reasons for this were functional convenience or more profoundly psychological it is still the case that extra pressures are exerted on individuals and society generally when people are made homeless by fires.

(2) The second factor to be considered in any form of 'cost–benefit' calculation is the inconvenience and possible unwanted effects of any fire *precaution* procedures

which are introduced. It is not known how many people are injured by fire doors, nor how much it really costs to introduce good fire training procedures. Most of the other less obvious effects of fire precautions, such as compartmentalization of buildings, or the provision of protected escape routes, are unlikely to be free from disadvantages to the people who live amongst them. Are there any occasions in which these disadvantages outweigh the potential reduction in fire hazard?

(3) The answer to the question of the disadvantages of current fire prevention practices can only be given in comparison with the other alternatives which are becoming apparent. The improvement of training instead of hardware, the utilization of early warning devices instead of 'first-aid' fire-fighting equipment, the concentrating of scarce resources in high risk areas, are all possibilities which will improve the overall effectiveness of fire precautions. But they all rely upon a valid understanding of the human aspects of a fire.

Differentiating building occupancies

Part of the understanding of people in fires, will derive from a more detailed consideration of the differences between the contexts in which fires occur. Reference has already been made to the differences between domestic and institutional fires. The importance of a person's role at the time of a fire in structuring his activities has also been noted. Together, role and occupancy type provide a useful basis for developing more elaborated models which will help to predict actions in a fire incident. Such predictive models must be developed if the impetus of the research in this book is to have any significant impact.

Although these considerations may seem to be rather academic, a number of examples exist which show their utility, for instance the light which the emerging models throw upon the function of alarm systems. It is common experience that the conventional alarm bell is frequently ignored. The interpretation that 'people don't take fire alarms seriously' is of little help. Certainly making alarms louder, shriller, or testing them more often does not seem likely to make people treat them any more 'seriously'. However, understanding the way in which people use alarms and the way they contribute to behaviour in a fire, in other words defining the role which alarms play in a model of human behaviour in fires, can lead to the development of more effective early warning systems.

Redirecting official guidance

The most direct impact which research can have is on the guidance, legislation and codes of practice which official bodies provide to influence practitioners. There is still little knowledge of how research can or does have its impact. Nor has there been much study of the various forms which official guidance can take. Certainly the move towards the introduction of officially backed training procedures, as described in Chapter 17, is to be encouraged. The possibility of it replacing aspects of legislation is also well worth considering.

If this book can make some contribution to moving towards a society in which control and legislation can be replaced by education and encouragement; if this book encourages guidance for fire prevention and fire precautions to be more firmly based upon systematic understanding than idiosyncratic experience; if this book can stimulate thinking about fire, and this is encouraged to encompass all those diverse aspects of human, personal involvement in fires, then it will have achieved its purpose.

REFERENCES

Best, R. L. (1977). *Reconstruction of a Tragedy: The Beverly Hills Supper Club Fire, Southgate, Kentucky, May 28th,* Boston Nat. Fire. Prot. Assoc., NFPA No. LS-2.

Canter, D. and Matthews, R. (1976). *The Behaviour of People in Fire Situations: Possibilities for Research,* Building Research Establishment Current Paper.

Canter, D. V. (1978). *Prospects for Research,* Paper presented to the 2nd International Conference on Human Behaviour and Fires, NBS (Washington).

Langdon-Thomas, G. J. (1972). *Fire safety in Buildings* (London: Black).

Lawrence, J. C. and Melinek, S. J. (1979). ,Hospital Records Analysed to Find Medical Costs of Fire Injuries', *Fire,* 72 (889), 94–96.

Lie, T. T. (1972). *Fire and Buildings* (London: Applied Science).

Marchant, E. W. (1972). *Fire and Buildings,* (Aylesbury: Medical and Technical Publishing Co. Ltd.).

Summerland Fire Commission (1974). Report, Douglas Isle of Man, Government Office.

Fires and Human Behaviour
Edited by D. Canter
© 1980 John Wiley & Sons Ltd.

CHAPTER 2

The Requirements of Research into the Behaviour of People in Fires

IAN APPLETON*
Building Research Establishment
Fire Research Station, Borehamwood

INTRODUCTION

The author of this chapter is leading the Operational Research Section of the Fire Research Station (FRS) Borehamwood, England who, with the Home Office (HO) Scientific Advisory Branch, are the main sponsors of research into the behavior of people in fires in the United Kingdom. The research commissioned by FRS, past and present, is described in detail in chapters 8 and 15 of this book. This chapter gives background to the magnitude of the problem in terms of the losses due to fires and cost and continues with a brief summary of the UK Fire Statistics and the trends in those statistics since 1965. A comparison with other countries is given. It continues by developing how government and society attempt to cope with these losses and shows that the procedures adopted may be lacking in certain respects, which demonstrates a need for research into the behaviour of people in fires. There is also a brief summary of the research projects commissioned by the FRS to meet these deficiencies, which discusses the importance of paying particular attention to the difficulties of applying the research. The author has tried to encompass the needs of all those who use the results of the research.

COST OF FIRE IN THE UK

The total cost of fire in the United Kingdom in round numbers amounts to £1000m per annum, and this is divided amongst six component parts (Table 2.1). The direct losses are the value of the stock, machinery and buildings destroyed by fire,

*The views and ideas expressed in this paper are not only those of the author, they have been collated in discussion with a number of people, principally Roger Baldwin, Tony Howard, Bill Coggan, R. W. Bray, W. A. Allen, the late G. C. Ackroyd, to whom many thanks.

This paper forms part of the work of the Fire Research Station, Building Research Establishment of the Department of the Environment and is Crown copyright. It is reproduced here with the permission of the Controller of Her Majesty's Stationery Office. Dr. Appleton is now in the New Zealand Government Service.

Table 2.1 Cost of fires in the UK, 1977

	£ m
Direct losses	250
Consequential losses	75
Fire brigades	300
Fire protection	200
Insurance	200
Enforcement costs	50
	1075

1000 deaths
6000 non-fatal casualties

while the consequential loss encompasses a rather more vague group of losses such as loss of production, of manpower, and of trade. A recent study (*Fire*, 1978) has indicated that in the past the figures for consequential losses may have been an over-estimate, especially if losses to the nation rather than to the individual firm are counted.

The cost of local authority fire brigades is estimated at £260m (Department of the Environment). (The figure in Table 2.1 includes the cost of industrial fire brigades). The item 'fire protection' in Table 2.1 is an approximate estimate and includes not only items of hardware such as sprinklers and detectors which can be costed fairly easily, but also a contribution from the provision of such items as fire-resistant structures and means of escape which cannot be costed easily, since the proportion of the cost assignable to fire safety is not obvious. The item in Table 2.1 labelled 'insurance' is an estimate of the insurance companies overheads. In addition to the total of £1000m, there is loss of life and limb, which accounts for approximately 1000 deaths and 6000 non-fatal casualties per annum. One could say that the ob-jective of regulations, codes and fire research is to minimize this total fire cost.

FIRE STATISTICS IN THE UNITED KINGDOM

For each fire they attend, the brigades complete a report form (FDR1)*. The fire report forms are collected by the Home Office Fire Statistics Section at the Fire Research Station where they are coded and the data stored on a computer. This central collection of information enables a comprehensive data analysis to be pub-lished. At the time of writing the most recent publication is for 1976 (*UK Fire Stat-istics*, 1976). The following data are abstracted from that publication and serve to give a picture of the magnitude and complexity of the problem.

In 1976 the fire brigades attended nearly 500 000 fires, of which 96 000 occurred in occupied buildings, 47 000 in chimneys, and 20 000 in derelict buildings; the re-mainder were outdoor fires. The outstanding feature of the 1976 statistics was the

*Formerly K433

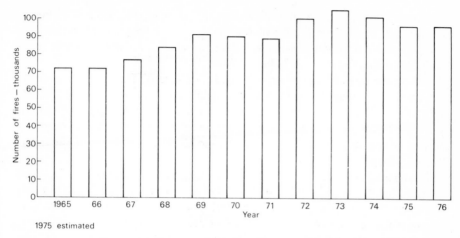

Figure 2.1 Fires in occupied buildings 1965–76 (UK)

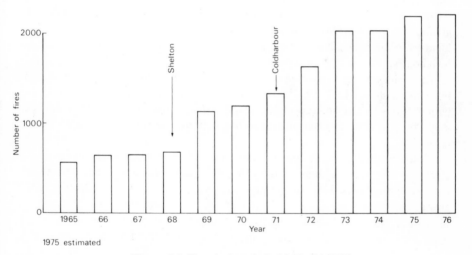

Figure 2.2 Fires in hospitals 1965--76 (UK)

exceptionally high number of outdoor fires due to the very hot and dry summer. This chapter is mostly concerned with fires in occupied buildings, and the pattern over the previous twelve years is shown in Figure 2.1. The data shown in this figure and subsequent figures for 1975 are estimates. The firemen in about half the brigades in the United Kingdom worked to rule for four months during the summer of 1975 and did not report all fires during this time. The method of estimating the number of fires in 1975 is given in *UK Fire Statistics*, 1976. From 1965 to 1976 the total number of fires in buildings increased by about 40 per cent. This increase has not been uniform across all occupancies. The number of fires reported in hospitals has increased by three-and-a-half times in this period (Figure 2.2). However, the most

Fires and Human Behaviour

pronounced increases were after major fires at Shelton in 1963 and Coldharbour in 1971, which implies that the increase in the number of fires may not be an actual one, but rather a greater awareness and willingness to call the brigades. A recent analysis supports this view (Chandler, 1978).

During the early 1960s the average number of fires reported per annum in UK schools was less than 900, with malicious ignition as a minor cause. By 1972 the number had more than doubled to 1889 and the incidence of malicious ignition had risen by a factor of 8. This increase created an entirely new situation in respect of the cost of reinstatement (£6.78m in 1974) and disruption to education (Silcock & Tucker, 1976).

The number of fires in hotels and other residential accommodation has increased one-and-a-half times since 1965. Between 1974 and 1976 many hotels and boarding houses were certificated under the Fire Precautions Act 1971. It is interesting to note that the number of deaths in these premises in 1976 was appreciably lower than in 1973 and 1974. Two large fires in recent years in old people's homes caused considerable public concern (Department of Health and Social Security, 1975; Humberside County Council, 1977) but a detailed analysis of the fire brigade data (Chandler, 1976) indicated that the increase in numbers of fires observed is directly attributable to the increase in the number of such premises. In addition, the residents of old people's homes were seen to be safer from fire than they are in their own homes.

The major sources of ignition of fires in occupied buildings, and in dwellings only, in 1976, is shown in Table 2.2. The trends in these sources of ignition over the period 1965–76 can be seen in Figure 2.3. Although some of the upward trends might be attributed to changes in reporting patterns, many of the increases are real ones.

Electricity is the main source of ignition of fires in buildings, and in Figure 2.3 electric cookers are separated from other electrical appliances. The trend in the number of fires caused by other electrical appliances may be just a reflection of the greater use of such appliances and the possible consequential overloading of wiring in pre-war premises. One cannot always distinguish between the lack of safety of appliances and their misuse. The number of fires attributed to electric cookers is

Table 2.2 Major sources of ignition 1976

Source	All occupied buildings (per cent)	Dwellings only (per cent)
All cooking appliances	21	34
Electrical appliances other than cookers	19	21
Smokers' materials and matches	9	9
Malicious or doubtful ignition	9	4
Children playing with fire	7	7

In about 13 per cent of fires in occupied buildings it was not possible to ascertain the cause.

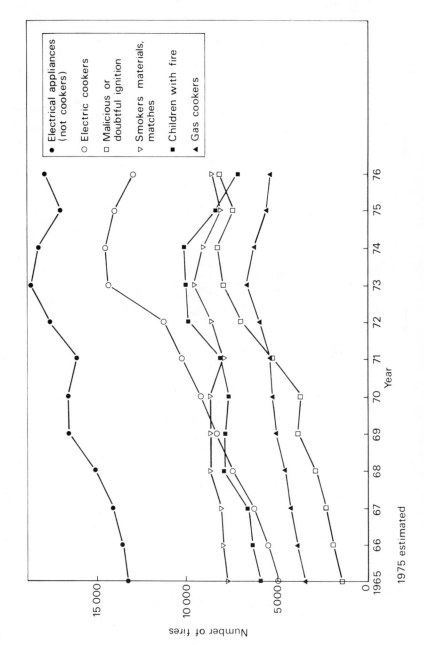

Figure 2.3 Trends in source of ignition in occupied buildings 1965–76 (UK)

rather more disturbing; these rose three fold between 1965 and 1974, with the chip-pan fire as the main culprit. (For a full discussion of this see Chapter 7). There have been several explanations put forward for this (Department of Prices and Consumer Protection, 1976), for example the present preference for using oil instead of drip-ping for cooking chips. By comparison the number of fires started in gas cookers has risen by a factor of one-and-a-half, and consequently an explanation is sought in the difference between electric and gas cookers. Electric rings have a slow response compared with gas rings—electric cooker controls are usually located at the back of the cooker making them inaccessible in case of a fire; even in normal usage it is easy to turn off the wrong ring. The Home Office have attempted to reduce the number of chip-pan fires with a monitored publicity campaign which also gave a measure of the effectiveness of the campaign (Rutstein, 1978).

The trend in the number of fires caused by gas cookers also needs comment. When natural gas was first introduced into the United Kingdom, a study of the gas explosions attended by the fire brigades in dwellings suggested that the natural gas was twice as dangerous as town gas (Fry, 1971). However, the inquiry into the spate of gas explosions which occurred in the winter of 1976–77 concluded that this is not the case (Department of Energy, 1977). Instead the increase in the number of fires caused by gas cookers may be associated with the rise in domestic gas consumption which almost trebled between 1965 and 1976 (Central Statistical Office, 1977). The number of fires attributed to malicious and doubtful ignition has increased six-fold. Such statistics may be regarded as a symptom of a wider social

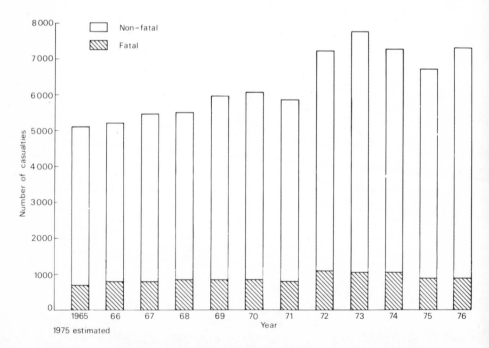

Figure 2.4 Fatal and non-fatal casualties in occupied buildings 1965–76 (UK)

problem, of poverty, inner city crowding, poor housing conditions, and inadequate facilities which are usually associated with arson and vandalism.

In 1976, 895 fatal casualties were reported as a result of fires; the figure for non-fatal casualties was 6394. Since most deaths occur in occupied buildings, the number of casualties follows a similar pattern to the number of fires in those premises, the general level for 1972–74 being about 25 per cent above that for previous years. The number of fatal and non-fatal casualties is shown in Figure 2.4. By looking at overall numbers and trends, the circumstances are often masked. Data is published (*UK Fire Statistics,* 1976) on the death rates per thousand fires and this can indicate the high risk areas. Those occupancies in 1976 which suffered a death rate higher than the average (8.1) were dwellings (13.5) and hotels (9.5) but since the majority (89 per cent) of deaths in fires in occupied buildings occur in dwellings, the data for other occupancies could be influenced by isolated incidents. For comparison the deaths per thousand fires in hospitals is 3.2, this same statistic when related to the source of ignition indicates three sources that are particularly more hazardous than average—appliances and installations fuelled by solid fuel (21.7), smokers' materials and matches (22.3), and unknown sources (14.8). The first two sources are usually related to clearly defined circumstances—the open coal fire causing clothing to be ignited, the discarded or dropped cigarette or match causing furniture or bedding to smoulder and ignite.

FIRE LOSSES

The British Insurance Association publishes fire loss data monthly. These figures are 'first estimates' of damage and are often modified at a later date. The data show fluctuations from month to month due to single incidents, but when aggregated into yearly totals, a smoother upward trend can be seen (Figure 2.5). Since 1972 figures for Great Britain and Northern Ireland have been published separately and these are shown separately in Figure 2.5. The data for 1974 indicates one very large loss—the explosion at Flixborough. Although the data demonstrates a disturbing rise in the annual losses due to fire, a number of factors should be taken into account. Firstly, the figures are at current prices, and in recent years the United Kingdom has experienced general economic inflation and rising prices. Thus some increase in fire losses would be inevitable. If the data is corrected for the decreasing value of money then a more slowly rising trend would be seen. A more useful presentation would be to express the loss as a fraction of the value at risk; losses can be related to the Gross National Product (GNP), although this is not a particularly satisfactory method as GNP does not indicate the total value at risk, nor the increase in value at risk. GNP is an estimate of the total output of production and services plus the income from abroad, and indicates the economic strength of the nation. There is currently no reliable estimate of the value at risk, but work is in hand at FRS and the Home Office to collect such information.

INTERNATIONAL STATISTICS

For a number of years the US National Fire Protection Association Fire Analysis Department collected and published (see Harlow, 1975) data involving fires, fire

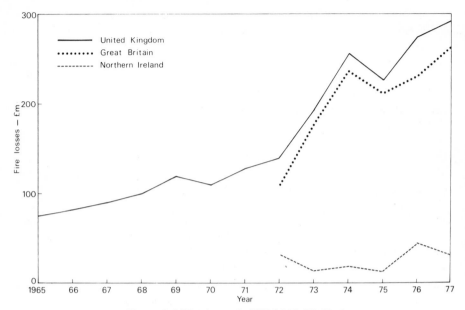

Figure 2.5 Fire losses in UK 1965–77 (£m)

deaths, injuries and losses from many countries around the world. The figures for
the different countries are not strictly comparable due to differences in the methods
of collecting and classifying the fire losses and differences in calculating the GNP.
For instance, some countries record only those fires attended by the public fire departments, while others include all those on which insurance is paid; some exclude
chimney, brush, rubbish and forest fires while others include them. Some countries
report everything except loss to government property. There are also wide differences in the values of the property subject to loss. Methods of estimating losses are
generally not known and are likely to vary from country to country. However, the
figures give at least some indication of the losses sustained and enable interesting
comparisons to be made. While comparisons between nations ought to be made
with caution, trends from year to year can usefully be studied. Fluctuations in exchange rates do not help matters.

The population of a country is a useful base to which fire losses can be related.
Estimates of per capita loss for 1965–74 are given in Figure 2.6. Data for 1972 is
uncertain due to the instability of the US dollar in that year. All countries presented
exhibited a rising trend, though comments about inflation and rising prices in the
last section are relevant. One of the factors contributing to the variation between
countries is the difference in standards of living. It is therefore less surprising that
countries like the USA, Australia and Canada suffer a greater loss per capita than
the United Kingdom. But *within* a country, the lower the standard of living the
greater the incidence of fire (Rardin and Mitzner, 1977).

The number of fire deaths per million of population for 1965–74 is given in
Figure 2.7; but again, comparisons are suspect because of reporting differences.

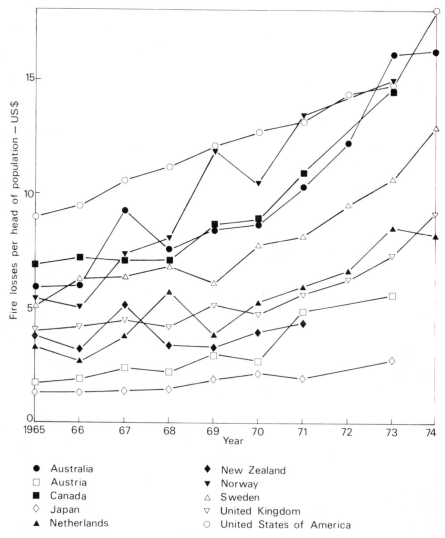

Figure 2.6 Fire losses per head of population 1965–74

Over this period of time most countries show little change, the exceptions being the USA and Australia whose death rates are declining and Japan whose death rate is gradually rising. Far more noticeable is that the USA and Canada have a very poor record compared with other countries. Bearing in mind the reservations mentioned in the previous section, data relating fire loss to GNP 1965–74 is given in Figure 2.8. The fluctuations shown are perhaps a greater reflection on the economic state of the nation than the fire losses. Increasing industrial activity would normally be expected to increase the chances of fire occurrence. However, this does not appear to be the case in Japan where fire losses are increasing at a slower rate than the GNP.

Fires and Human Behaviour

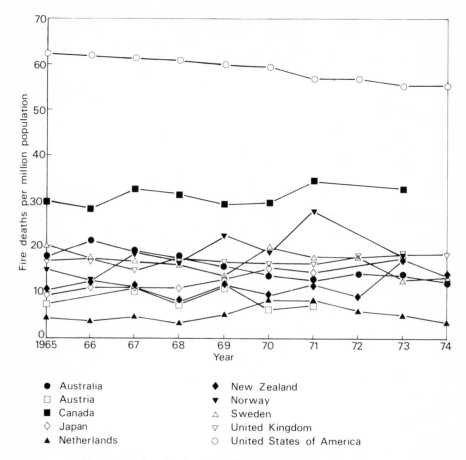

Figure 2.7 Fire deaths per million population 1965–74

Many hypotheses have been proposed to explain the difference in the fire experience. These have been studied in detail by Rardin and Mitzner (1977). The disparity between the procedures used to collect the data in various countries has already been mentioned. A second hypothesis is the difference in human activities between nations, in their social and cultural norms, and their economic and technological development. Associations have been found between fire losses and disposable income, energy consumption, crime rates, population mobility, social stability, alcohol and tobacco consumption. A third explanation lies in the weather experienced by different countries, and hence the different building construction methods employed. Hot, dry climates or cold climates are conducive to high fire losses. Warm wet climates are not. A fourth argument is the way in which fire is combated in the different countries: the differences in firefighting organization and capability, the different role of insurance companies, and the different building codes.

There is a Conseil International du Bâtiment (CIB) Commission W14 (Fire) initiative to attempt to get a common base for statistics with a view to testing the

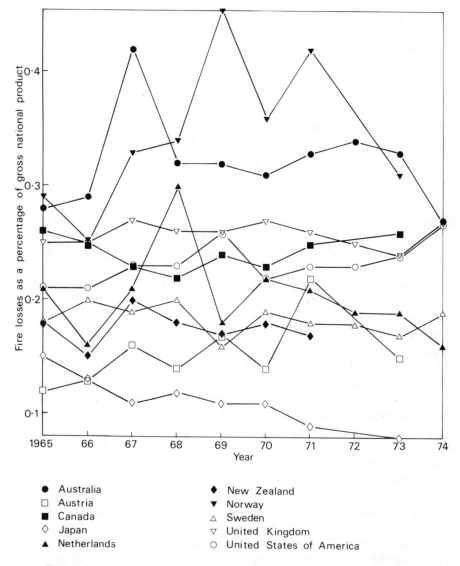

Figure 2.8 Fire losses as a percentage of gross national product 1965–74

effectiveness of the building codes of different countries. Because of the many contributing factors outlined above, it is unlikely that one element, the building codes, could be isolated.

PROCEDURE FOR CONTROL THROUGH REGULATIONS

The previous sections have described the magnitude of the losses due to fire. Government attempts to deal with this problem through statutory methods and through influencing manufacturers and the public by propaganda of various sorts. However, this chapter is not concerned with the various methods of control, laws, licences etc. in themselves, but rather with the substance of these controls and how they achieve their objectives. For example, the 1974 Health and Safety at Work etc. Act gives powers 'to make regulations with respect to the design and construction of buildings and the provision of services, fittings and equipment in, or in connection with buildings'. The main purpose is to 'secure the health, safety, welfare and convenience of persons in or about buildings and matters connected with buildings'. Thus the objectives of these regulations can be seen to be to decrease the adverse effects of fire on health, safety, welfare and convenience.

Such objectives define how the building is required to function; however, the objectives do not naturally lead to any standard of performance for buildings. Many of a building's functions can easily be envisaged; for example, the occupants wish the air temperature within a building to be comfortable—the desirable temperature ranges can be determined experimentally. It is then possible to define the standards of performance for the control of the heat supply, the ventilation and the insulation properties of the building. There are already suitable units available which measure function and compare performance. Safety, however, is much harder to define and quantify, standards of performance are elusive. Consequently the regulations for fire safety are definitions of requirements for components of buildings which are judged to achieve an acceptable standard, instead of defining what the standards are. These regulations form a set of constraints which are not expressed as a measurable standard for the whole building, but as hardware requirements for individual elements of the building.

It is worth examining the process through which regulations are derived. The word regulation (here used with a small 'r') is used to denote all forms of influence, be it Building Regulations, Codes of Practice, Standards or advice. The starting point in Figure 2.9, is a fund of knowledge and experience of fires, buildings, their contents, and materials together with people who inhabit them. The quality and relevance of this knowledge may be challenged, but from it is abstracted a set of simple assumptions about the behaviour of fires in buildings. The set of assumptions is then formulated into the regulations. Without entering into detailed examples there are a number of questions that can be asked of the process. Does the process work? Has it got the right balance between active and passive fire protection? Will it always work? Can it cope with new building types, with large buildings, with new occupancies, with changing styles? Are the concepts correct? Are the numbers correct?

In traditional buildings this system seems to work reasonably well. There are debates as to whether the regulations are too complex and are an unnecessary restriction on the designer. However, the approach is being questioned in the present non-traditional buildings which have become larger and more complex, as public concern following disasters demands a response, and as new building design, materials and technologies are introduced. The existing system is based on knowledge of traditional materials and methods; using current standards for new designs may not produce buildings which meet acceptable standards of safety.

Also relevant are recent developments in techniques of socioeconomics. The

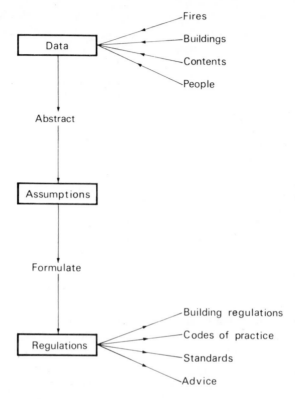

Figure 2.9 Process for writing regulations

introduction and development of cost-benefit analysis and decision theory into the policy-making machinery has resulted in the idea that there should be a net social benefit from regulations. All the costs and benefits must be included. When correctly applied these techniques demonstrate where improvements can be made and where redundancies exist.

Safety may be defined as the state of being free from danger. In practice such a state does not exist, since there is always a chance of being killed, no matter how small. Safety must be regarded as a relative term. There are many ways of defining and measuring safety or the level of risk such as the probability of being killed per year or the number of multiple fatality accidents per year. The problem is to choose an appropriate set of parameters which will describe the level of safety of a building in such a way that any building satisfying these conditions will be accepted, and after a fatal fire, will still be judged by society as having been acceptably safe.

A fundamental approach to regulations should start with a statement of the objectives of the regulations and these should not be related to the standards for the component parts of a building, like the fire resistance of a floor or column. The regulations should describe how the building is to perform in terms of the way it functions in a fire, and the functions should be expressed in numerical terms. As an example one might postulate that the building safety function should be the probability that one individual is killed by fire per annum. The performance requirement or standard is a value of this probability that is deemed to be an acceptable risk. Such a performance requirement may vary between buildings, if for example society deems that hospitals should be safer than hotels. Indeed it is probable that the performance requirements would change with the economic wealth of the nation.

In order to design buildings to this type of standard, there has to be a method of linking such probability-based standards to the existing hardware system for the building components. Marchant shows the complexities of modelling this in Chapter 16 of this volume. However, both the designers and the enforcers of regulations must agree that a particular design meets the performance requirements. To achieve this a valid procedure is needed for predicting the performance of buildings from its component parts at the design stage.

The Fire Research Station has been instrumental in initiating research programmes to meet some of the demands made by such a scheme. There are two avenues relevant to this book. The first relates to the measurement of safety and the determination of acceptable levels of risk and how much it is worth spending to achieve them. This requires research into attitudes to and perception of risk, disasters, safety, health and welfare. The FRS has let a contract to Colin Green of the School of Architecture at the Jordanstone College of Art, Dundee; Chapter 15 deals with this subject in greater detail (see also Green and Brown, 1978).

The second area relates to the behaviour of people in fires. As mentioned earlier, there is a need to develop a predictive model of fire in buildings. Some work has already been done in this area. The FRS, for example, is developing models of fire resistance, smoke spread and fire spread. A means of predicting the hazards to occupants of buildings based on their personal characteristics and the building form is required. Buildings will contain a variety of people, some who will be able to escape in most circumstances, others who would have extreme difficulty, and again those who will not attempt to escape. These facets of behaviour are influenced by psychological, physiological and circumstantial factors, as well as by the nature of the threat posed by the fire, and they can be influenced by the building form and fire protection provided. The FRS has commissioned Dr Canter of the Department of Psychology at Surrey University to research into the behaviour of people in fires. The contributions of his team appear in Chapters 5 and 8.

HISTORY OF BEHAVIOURAL RESEARCH SPONSORED BY THE FRS

A realization of the research needs outlined above led the FRS to sponsor a series of studies of the behaviour of people in fires. This section is a very brief summary of these studies since they are covered in greater detail elsewhere.

The first attempt was the study by Wood (1972, and Chapter 6 of this volume) who used a questionnaire approach. Having gained experience of this pragmatic method, the need for a structured and fundamental approach to the problem was seen. The FRS commissioned a study at Surrey University to lay the foundation for further research. The objectives were to review current knowledge about behaviour

of people in fires, review behavioural assumptions and data in drafting existing regulations, to structure the problem in the light of modern psychological research, to
review available methodology for research and suggest a strategy for future research
on human behaviour.

The report of this study (Canter and Matthews, 1976) set out a research strategy
which formed the basis of the current pilot study described in Chapter 8. The overall objective is to develop a predictive model of behaviour of people in fires based
on their psychological characteristics, how this is modified by the building form
and fire protection provided, and how they respond to various fire stimuli. From
the experience gained in the pilot study, a questionnaire or other methods can be
designed to gather more extensive, representative data to test the model and establish
a predictive capability.

The FRS has made an in-house contribution by concentrating on methods of calculating the evacuation times of buildings. Melinek and Booth (1975) made a survey
of data on the movement of crowds in corridors and stairs and developed a simple
model to calculate the evacuation times of tall buildings.

The FRS collaborated with the Department of Health and Social Security in a
study of the evacuation of hospital wards. The exercise is described by Janet Hall in
Chapter 12. The FRS contribution (*B.R.E. News,* 1976) was to analyse the evacuation process, develop a model of the evacuation of wards and relate the predictions
of the model to the time-scales of potential fires.

FUTURE NEEDS AND APPLICATION

The rather long-term needs for research have been described earier in relation to the
development and rationalizing of regulations of one kind or another. It was seen
that the regulations could not work for new building forms where there was no data
base and could not cope with the current speed of change which allowed no time
for feedback. Consequently there is a need to develop predictive models of fires in
buildings.

However, these are long-term and perhaps rather esoteric needs. It is essential
also to ensure that the research is applicable to the more short-term needs. The views
of three practitioners, an employer, an architect and a hospital administrator, have
been sought. The issues they raise are given below as a way of illustrating the types
of questions that research has to answer.

The employer

The employer needs to know whether his employees can be relied upon to use 'first
aid' fire-extinguishing appliances, or whether they will merely attempt to escape.
To what extent can people in a fire situation be relied upon to take the necessary
action to alert the fire brigade? Some employers will totally discount help from
those exposed to a fire and rely on automatic equipment and automatic means of
calling the brigade. Where large numbers of people are involved, there is a feeling
that their normal behaviour as crowds will hamper firefighting. More specifically,

when individuals safely emerge from buildings involved in fire they automatically become interested bystanders and so hamper the work of the brigades trying to control the fire.

The architect

The architect has a different perspective. What he needs is a behavioural picture of what goes on in the minds of those who find themselves suddenly in a fire. Architects usually try to put themselves into their own buildings in order to anticipate how they might be used in many circumstances. For example, will a theatre be able to cope with the audience leaving at the end of a performance? The architect, when thinking of fire, has to evaluate what hazards precede the visible evidence of a fire. People will not perhaps react with the same appreciation as they would at the appearance of smoke and flames. As he works out a design, the architect attempts to provide for environmental factors, such as warmth and light and circulation, as well as the hazard from fires. The research results must be in a format which enables the architect to include the fire hazard. Since the architect's own thought processes in the design stage are often visual, in the form of shapes and spaces, then a portrayal of the research material in this format will be of greatest benefit. If such a portrayal is not made then the benefits of research may be meagre. A number of provisions have already been made along these lines. The FRS computer model of smoke movement in buildings (Appleton, 1975) has a graphic output which demonstrates how smoke from a fire spreads. As part of the research commissioned by the FRS, Canter *et al.* have produced a Fire Scenario Display (*Building Research Establishment Information Sheet*, 1977) or 'game' which demonstrates the interaction of a fire's development and people's ability to cope with them. Lerup and his associates at Berkeley have used a method of mapping the behaviour of people; this consists of a series of cartoons alongside plans of the building showing the extent of the fire and location of the people at particular times (Lerup, 1977, and Chapter 10 of this volume).

Architects are not able to use research results as presented in the scientific literature only because the volume is too large. Logical and self-consistent design rules must be derived which guide design decisions, permitting flexibility.

The hospital administrator

There is currently a great deal of activity in fire safety for hospitals and nursing homes. The hospital administrator needs to know why the staff are reluctant to attend fire precaution instructions, why fire drills are often treated jocularly by the participants and why many people are reluctant to respond to a fire alarm, when a more appropriate response is promptly to evacuate the building. If the staff are liable to lose the ability to organize and control the patients, what is the chance this will happen? What are the fire conditions in which it will happen? Will the staff be physically able to handle the patients? Will the patients be able to respond to instructions? How will different types of patient—general, geriatric, psychiatric—differ in their response? Are notices and signposts adequate under fire conditions?

In the Health Service there are a considerable number of examples where the fact that the staff had previously received training has really 'saved the day' when a fire occurred. Indeed, the value of training staff and the methods used are currently being examined by the health service. Perhaps better use can be made of the media, especially television, to give training and guidance on 'what to do' in various circumstances.

While it is important to foster the more fundamental needs for rationalizing the way in which regulations are formulated, as outlined earlier, it is imperative to keep the day-by-day practitioners in mind. The examples cited above, demonstrate that the emphasis of research must be on application and that the format of the research results must be in a style which those practitioners can readily handle. It is the responsibility of both sponsors and researchers to ensure that the benefits of the application of research are reaped.

REFERENCES

Appleton, I. C. 'A model of smoke movement in buildings'. CIB symposium on the control of smoke movement in building fires. Garston 4–5 November 1975.

Building Research Establishment Information Sheet IS 7/77. Fire Scenario Display.

B.R.E. News (1976). 'Fire evacuation exercise at Hackney Hospital', **35**, 2.

Canter, D. and Matthews, R. (1976). *The Behaviour of People in Fire Situations: Possibilities for Research.* Building Research Establishment Current Paper 11/76.

Central Statistical Office (1977). *Annual Abstracts of Statistics.* (London: Her Majesty's Stationery Office).

Chandler, S. E. (1976). *Fires in residential personal social service building.* Building Research Establishment Current Paper 62/76.

Chandler, S. E. (1978). *Some Trends in Hospital Fire Statistics.* Building Research Establishment Current Paper 67/78.

Department of Energy (1977). *Report of the Inquiry into Serious Gas Explosions.* (London: Her Majesty's Stationery Office).

Department of Health and Social Security (1975) Cmnd 6149. *Report of the Committee of Inquiry into the fire at Fairfield Home Edwalton.* (London: Her Majesty's Stationery Office).

Department of Prices and Consumer Protection (1976). *A Commentary on Electrical Fatalities in the Home 1974–75* (London).

Fire (1978). 'Consequential losses have been overestimated says report', **71** (879), 168.

Fry, J. F. (1971). Gas explosions attended by the fire brigade in dwellings. *Journal Institute Fuel,* **Aug**, 470–471.

Green, C. H. and Brown R. (1978). *Life Safety, What Is It and How Much Is It Worth?* Building Research Establishment Current Paper 52/78.

Harlow, D. W. (1975). 'International fire losses 1974'. *Fire Journal,* Nov, 43.

Humberside County Council (1977). *Report of the Committee of Inquiry into the Fire at Wensley Lodge on 5 January 1977.*

Lerup, L. (1977). 'Mapping fire, a technique for depicting dynamic aspects of environment and behaviour'. *Fire Research* **1**, 87–95.

Melinek, S. J. and Booth, S. (1975). *An Analysis of Evacuation Times and the Movement of Crowds in Buildings.'* Building Research Establishment Current Paper 96/75.

Rardin, R. L. and Mitzner, M. (1977). 'Determinants of international differences in reported fire loss. A preliminary investigation'. Georgia Institute of Technology. **June.** Sponsored by NFPCA.

Rutstein, R. (1978). 'Measuring the effects of fire prevention publicity'. *Fire,* **71 (879)**, 22–25.

Silcock, A. and Tucker, D. M. (1976). *Fires in Schools: An Investigation of Actual Fire Development and Building Performance.* Building Research Establishment Current Paper 4/76.

UK Fire Statistics (1976). Home Office S3 Division, 50 Queen Anne's Gate, London, SW1H 9AT.

Wood, P. G. (1972). *The Behaviour of People in Fires.* Fire Research Station Fire Research Note 953.

Fires and Human Behaviour
Edited by D. Canter
© 1980 John Wiley & Sons Ltd.

CHAPTER 3

Psychological Aspects of Firesetting

ROBERT G. VREELAND*
University of North Carolina, and

BERNARD M. LEVIN
National Bureau of Standards

Arson and firesetting have caused increasingly serious economic, social, and health care problems. Data provided by Teague (1978) provide some indication of the rapid increase in arson in the United States. In 1964 there were 30 900 fires of confirmed or suspected incendiary origin, causing 60 million dollars in direct losses. By 1974 the figures had grown to 114 000 fires and 550 million dollars in direct losses. In 1975 there were 144 000 fires of confirmed or suspected incendiary origin (an increase of 26 per cent over 1974 figures) and 633 million dollars in direct losses (an increase of 15 per cent over 1974).

These figures do not include the large proportion of fires (36 per cent in 1975) classified as due to 'unknown causes' (National Fire Protection Association, 1976). Assuming that some fires in the 'unknown causes' category were, in fact, due to arson, the direct costs of arson are even higher. The *total economic costs* (direct and indirect) from fires due to arson, while difficult to estimate, are still higher, and would include unemployment losses, tax losses, fire-fighting systems costs, insurance systems costs, court costs, and health care costs. The social and health effects of incendiary fires are best estimated in terms of human disruption and misery. Clearly, firesetting represents a problem of major proportions, and an examination of the kind of person who sets fires is warranted.

The purpose of this chapter is to examine some of the psychological factors involved in firesetting. A wide variety of medical, criminological and fire publications have dealt with the psychopathology of firesetting over a long period of time. There were, for example, at least 130 articles on the subject prior to 1890 (Lewis and Yarnell, 1951). It is perhaps not a little surprising that, with such an early start in research, we still know very little about the aetiology of firesetting and even less about its treatment.

*The preparation of this chapter was supported in part by National Bureau of Standards Grant G7-9021 to the University of North Carolina, Marcus B. Waller, principal investigator. This is a contribution of the National Bureau of Standards and is not subject to copyright.

The literature on firesetting contains a paucity of systematic, well-controlled studies which present reliable data. Instead, most studies are rather conjectural, based on the authors' experiences with firesetters over a number of years, and presenting case examples with incomplete data. There are probably many reasons for this sad state of affairs, and many factors may be out of the researcher's control. Often cases come to the attention of researchers through the courts, and the data available may serve legal rather than scientific purposes. For example, lawyers for the prosecution and defence both have an interest in securing psychiatric and psychological evaluations favourable to their purposes, and there are legitimate differences of opinion among professionals.

Factors such as age, social status, psychiatric diagnosis and the nature of the offence may all influence the disposition of criminal cases. In addition, many cases do not pass through the legal system at all, but come to the atttention of therapists and researchers working in clinics. Thus, data on firesetting come from many sources. There are studies on firesetting in children (Kanner, 1957; Nurcombe, 1964; Vandersall and Wiener, 1970; and Yarnell, 1940), adolescents (Lewis and Yarnell, 1951; Macht and Mack, 1968; and Yarnell, 1940), adults (Lewis and Yarnell, 1951), females (Lewis and Yarnell, 1951; Tennent *et al.*, 1971), prisoners (Wolford, 1972) and hospitalized patients (McKerracher and Dacre, 1966).

With studies often dealing with a narrow sample and incomplete data, it is not surprising that conclusions have often been contradictory and comparisons between groups have been extremely difficult. The only study covering a large sample (over 1300 cases) and a wide range of types of firesetters has been the classic study by Lewis and Yarnell (1951), and this study will serve as the anchor-point for many of the conclusions drawn in this review.

TYPES OF FIRESETTERS

It should not be surprising that, because of the legal implications for firesetting, motives have played an important part of the study of firesetters. Levin (1976) identified three major groups of firesetters: (a) arson-for-profit firesetters, (b) solitary firesetters, and (c) group firesetters. The latter two categories exclude cases of arson-for-profit. These three categories may be broken down further into specific incendiary motives.

Arson-for-profit

This category includes those who set fires for material gain, and probably represents the most rapidly increasing group of arsonists. People who set these fires include the following examples: the businessman who sets fire to his own business to collect insurance, or who hires a paid arsonist; the paid arsonist who sets fire for a fee; the affluent housewife who sets a smokey fire to collect money for redecorating: the welfare recipient who sets fire to his apartment to collect relocation expenses; and the criminal who sets a fire to destroy evidence (Levin, 1976).

There is a scarcity of reliable data on the relative incidence of arson-for-profit or the types of structures or businesses which are at high risk. It is generally agreed that the risk of fraud arson is especially high where changing economic conditions have created overinsurance. For example, changing market conditions may reduce the value of business inventories, so that the owner may be able to collect more from the insurance than could be obtained if the items were sold on the open market. The owner of a business may not have money to make needed repairs, and a fire represents a ready source of income. The possibilities for motives are as varied as are the economic conditions which make it possible to realize gains or minimize losses from a fire, and arson investigators are, in fact, very aware of the relationship between economic variables and arson-for-profit. Improved data on types of fires would serve to confirm this relationship and predictions from changing market conditions would help alert investigators to changes in risk conditions for various types of fires.

It has been suggested that for certain types of entrepreneurs, arson may be treated as a part of normal business activities, with no greater perceived risks associated with it than with many other business activities. There is little direct evidence to support or refute this view. However, the conviction rate for all types of confirmed or suspected arson is quite low (cf. Moll, 1974). The risk of detection and conviction for well-planned professional arson jobs is likely to be extremely low. Within the business community, the low risks involved may result in many people actually treating arson as though it were somehow legitimate. Rice (1956), for example, described five different criminal business conspiracies, incuding arson. Each was operated by a *bona fide* businessman functioning within the spirit of free enterprise and supported by the complicity of so-called 'honest' men. If arson indeed occupies a place on the fringe of 'legitimate' business, then one response of law enforcement agencies and insurance companies should be to take prompt action to clearly change that image. Emphasis on arson investigation and improved technology in arson detection should result in greater risk to the potential arsonist. An upstanding member of the community may be reluctant to engage in an activity which may result in his being branded as a criminal. Community educational programmes may also be beneficial in emphasizing the seriousness of arson as a costly criminal activity.

There is little in the psychological and psychiatric literature about arson-for-profit, presumably because it is considered a rational act, and thus not of great interest from a psychological standpoint. However, if we can ask what are the high-risk economic factors for arson-for-profit, we may also wish to consider what are the high-risk psychological factors. Are there types of *people* who, because of the types of activities in which they engage, represent a high risk for arson? What makes a person turn to criminal acts to solve financial problems, and why do some people become 'torches' or arsonists-for-hire rather than engage in other criminal acts. While the economic explanations are obvious, there may also be other contributing factors, and these factors are likely to be different for the businessman than for the 'torch'. For example, an entrepreneur, in order to realize rapid economic gains or to support an expensive lifestyle, may engage in poor business practices and make very risky investments. Because of these activities, his business may do poorly, and as a result he sets fire to his business to collect insurance. The factors which influence his poor business practices and expensive lifestyle may be *more* significant than the immediate economic incentives in explaining why he set fire to his place of business. At this point we simply do not know much about the psychological aspects of arson-for-profit, and this is an area which needs a great deal more research.

Solitary firesetters

Firesetters who set fires in secret (excluding those with a profit motive) are the group most widely studied by researchers. In their classic study of over 1300 firesetters, Lewis and Yarnell (1951) identified a number of different types of firesetters and their assocoated motives. Included were jealousy-motivated fires, revenge-motivated fires, fires centring against the mother and other members of the family, and suicidal firesetters; 688 cases were classified as 'pyromaniacs': people who repeatedly set fires for no practical reason other than an 'irresistible impulse' to set fires, and perhaps some sort of sensual sastisfaction obtained from the fire and associated events. Other types of firesetters include 'would-be-hero' firesetters who start fires and then help firefighters put them out, volunteer firemen who set fires, psychotic firesetters, and firesetters who associate sexual gratification with fires.

Group firesetters

Moll (1974) described three types of fires which are set by groups: political fires, vandalism fires, and riot fires. Political fires are premeditated and set to dramatize an issue, embarass authorities or political opponents, or intimidate or extort for political reasons. Vandalism fires are often set by pairs or groups of boys in the presence of others from a peer group. Many such fires are set by adolescent groups, and Moll believes that they are set for excitement rather than destruction. Lewis and Yarnell (1951) made a similar observation, but they also noted groups of adolescent firesetters who seemed to set fires for reasons similar to those of the solitary firesetters. Their study found about twice as many adolescent firesetters worked in groups as set solitary fires. Many adolescents seem to set fires in pairs, one boy in a dominant role and the other submissive, a relationship similar to that observed in homosexual pairs (Shaw, 1966).

Riot fires are set during outbursts of group violence when there is tension and social unrest: a mob action. However, individual members of the group may show different actions: some loot, some attack police, and some set fires (Conant, 1970). There is little information as to why one particular activity is selected over others.

A fourth type of group firesetting, described by Lewis and Yarnell (1951), is firesetting by groups of volunteer firemen. These fires may be set because of the community recognition, as well as excitement, that they generate for the firemen. Lewis and Yarnell described cases in which fires were set as a result of intense rivalry between fire companies.

Children and adolescent firesetters

Although serious firesetting in children is considered rare, a number of studies have investigated firesetting in this group. In general, motives fit into the classification scheme described earlier. Yarnell (1940) found that children usually set fires in and around the home with associated fantasies to burn some member of the family who had withheld love or was a serious rival for parental attention. However, the fires appeared to be chiefly symbolic, caused little damage and the children usually put the fires out themselves. Associated with firesetting were terrible dreams involving the devil and ghosts, and, not too surprisingly, a rich fantasy life involving aggression and magical powers over adults. These children also showed other types of problem behaviour such as running away from home, truancy, stealing, general hyperkinesis, and aggression.[1] Kanner (1957), Vandersall and Wiener (1970), and Nurcombe (1964) further noted that children who set fires virtually never have what would be considered a 'normal' relationship with their father. Often the father is absent, frequently away from home, indifferent, unresponsive, or rejecting.

While malicious firesetting in children may be relatively rare, the incidence increases after the age of twelve. With increasing age also comes a shift from predominantly home-centred fires to those set away from the home. In Lewis and Yarnell's (1951) sample, frequent targets of males between twelve and sixteen years of age included homes of strangers, churches and schools, factories, and rubbish or brush. Of children who set fires to schools, the greatest incidence occurred in the 12-14 year age group, and only a few instances of this type of fire were set by children under ten years of age. According to Lewis and Yarnell, the stated motives for firesetting by these children were usually revenge connected with school-related problems. Frequently, firesetting was preceded by stealing, vandalizing, and harassment of the teacher.

In contrast to children, adolescent firesetters usually set fires away from home, and chiefly for excitement (Yarnell, 1940). Also, adolescent firesetters often work in pairs or groups. Macht and Mack (1968) studied four adolescent firesetters extensively and, as in the children's group, found in all four cases that the father was absent from the family. They further attached great significance to the fact that in each of the cases the father had some significant involvement with fire which the children knew about and in some cases had actually been involved themselves. Macht and Mack suggested that firesetting had served to 're-establish a lost relationship with the father'.

Female firesetters

Females made up only 14.8 per cent of Lewis and Yarnell's (1951) sample, and very few females have appeared in other case studies of firesetters. Arson, therefore, appears to be a male-dominated crime. This was not always the case, however. Lewis and Yarnell report that in the earliest literature, during the eighteenth and nineteenth centuries, arsonists were often servant girls who set fire to their master's house. It was generally assumed they were suffering anxiety over menstruation, depression of puberty, separation from family and harshness of their lives.

Tennent *et al.* (1971) studied 56 female arsonists in the English Special Hospitals along with a control group matched for age (over 14) and length of hospitalization. Most of the firesetters were adults, with a mean age of 25.8 and a mean length of hospitalization of 20.1 months. The motives for firesetting were most often identified as revenge and 'response to conflict with authority'. As in the children and adolescent groups described earlier, a high incidence of early parental separation was noted, although separation from both or either the father and the mother were equally common. As in Lewis and Yarnell's study, 'the majority of acts were against their own property, or at least the area of property symbolically owned by them within an institution'. The presence of behaviour problems other than firesetting was also common, a finding typical of male arsonists as well. By various measurements, female arsonists appeared to be more depressed than controls. Suicide and self-mutilation had been attempted by more than half, although the incidence did not differ from controls.

Lewis and Yarnell (1951) found motives for female firesetters to be similar to those of males. However, there were some notable differences. In contrast to males, females rarely turned in alarms or helped to fight the fires. Instead, the 'heroism' motive often involved a dramatic fire in which they were the 'heroic victims'. Lewis and Yarnell report that some women 'staged a bizarre fire (usually set on their own property, it should be noted) where they appeared, sometimes fantastically dressed for the occasion, as part of the drama'. Occasionally, fires appeared to result from anxieties over pregnancy. Anxiety over menstruation was found only in the older female firesetters, apparently because 'the removal of the secrecy and superstition about physiological phenomena seems to have reduced the adolescent anxiety about menstruation'.

Some general comments on motives and types of firesetters

Firesetting appears to be a behaviour primarily of the adolescent and young adult. While the distribution of firesetters in Lewis and Yarnell's (1951) sample covered a broad age range, the modal age was 16–18 years. However, it is difficult to assess how representative their sample was of the population of firesetters, both in terms of age distribution and the distribution of types of firesetters. The sample covered an extended period of time, and the choice of cases may have been made more on the basis of accessibility than on the use of appropriate sampling methods. Moreover, just as the distribution of age and types of firesetters had apparently changed over the decades prior to the Lewis and Yarnell study, it is likely that further changes have taken place since that time. An impediment to current data characterizing the population of firesetters is the lack of agreement on a consistent and meaningful classification system.

In the studies of specific groups of firesetters already mentioned as well as in the comprehensive study by Lewis and Yarnell (1951), differences have been found in the type of fire set, the location of the fire, etc. Fires set by those classified as revenge motivated were generally well-planned and carefully timed. Fires set by 'pyromaniacs', on the other hand, were generally set on impulse with whatever materials were available. 'Pyromaniacs' apparently experienced mounting anxiety until they finally went on rampages of firesetting, terrorizing entire neighbourhoods, until the 'irresistible impulse' subsided and they went home and slept. In 'would-be-hero' type firesetters, the major motive seems to be the consequences of the act: the general confusion, the fire engines, and often praise from the authorities for reporting the fire or helping to put it out.

Yet there is nothing exclusive about these differences. Elements of the revenge motive appeared in the 'pyromaniacs' as well. Inordinate fascination with fire, the 'irresistible impulse', and the importance of the consequences of the act were all elements which appeared throughout the sample, perhaps with differences in emphasis rather than differences in substance. Lewis and Yarnell originally set out to study the pure 'pyromaniac'. However, they were so impressed by the similarities across motives that they finally excluded only arson-for-profit from their sample. At the risk of overgeneralizing, the remainder of this chapter will concentrate on some of the common factors among firesetters.

INTELLECTUAL AND ACADEMIC PERFORMANCE OF FIRESETTERS

Lewis and Yarnell (1951) reported that 48 per cent of their sample could be classified by IQ, as 'mentally defective'. Only 17 per cent could be classified as of average to superior intelligence. The rest ranged from borderline to low average intelligence. Lowest intellectual performances were found among the true 'pyromaniacs'—those who set fires with little planning and for no apparent reason. Wolford (1972) found the educational level and IQ scores of arsonists to be significantly lower than those of non-arsonists. In neither of these studies were the testing instruments specified. More than likely the researchers relied on whatever data were available from case histories.

Nurcombe (1964), in his study of 21 children, found two of above-average intelligence; ten were of average intelligence, five were classified as 'dull' and four were considered 'borderline'. Again, specific tests were not specified. Kaufman, Heims, and Reiser(1961) found over 70 per cent of their children and adolescent firesetters to be retarded by one or more grade levels. Yarnell (1940) reported that approximately half of her children firesetters had specific learning disabilities.

Intellectual tests are not precise measurement instruments. Different tests may yield different results. Moreover, age of testing, circumstances surrounding testing, manner of test administration, and time intervening since testing may all affect the confidence we should place in a test as a measure of intellectual ability. Even under ideal conditions, there are many problems associated with evaluating intellectual ability. In the foregoing studies, testing instruments and other measures of intellectual development were not specified. While it seems clear that firesetters as a group tend to be intellectually and academically below average, future research should be more precise in exploring the significance of this finding.

PSYCHOPATHOLOGY AND FIRESETTING

While fascination with fire may be an almost universal phenomenon, the setting of destructive fires is considered abnormal behaviour. Most explanations of firesetting have relied heavily on a psychoanalytical view, and the sexual roots of firesetting have been particularly emphasized. Freud (1932) conjectured that 'in order to possess himself of fire, it was necessary for man to renounce the homosexually tinged desire to extinguish it by a stream of urine'. Urinating, especially to extinguish fires, is seen as erotic, and the creation of fires is seen as a symbol for sexual activity. Stekel (1924a) pointed to the relatively young age of most firesetters as indicative of 'how strongly the awakening and ungratified sexuality impels the individual to seek a symbolic solution of his conflict between instinct and reality'. Simmel (1949), in an analysis of a 21-year-old firesetter, saw firesetting as the result of repressed masturbatory impulses. Gold (1962) emphasized firesetting as both a resolution of repressed sexual impulses and as a primary sexual excitant. However, Lewis and Yarnell (1951) found relatively few cases in which fire appeared to be a stimulus for a physical orgasm; most firesetters were not aware of any sexual motivation for their acts. Thus, as Lewis and Yarnell point out, the commonly held view that firesetters are sexually stimulated by fires does not appear to be supported by available data.

The association between enuresis and firesetting has often been cited as supportive of the urethral–erotic interpretation of firesetting. However, statistical data on the incidence of enuresis in firesetters come from only a few studies, and the results are inconclusive. Kaufman, Heims, and Reiser (1961) and Nurcombe (1964) found about half their subjects to be enuretic, while studies by Yarnell (1940) and Vandersall and Wiener (1970) showed the incidence of enuresis to be 15 per cent and 20 per cent, respectively. The latter two figures do not exceed the probable population incidence of enuresis (Bakwin and Bakwin, 1960), casting doubt on the correlation between enuresis and firesetting.

Analysis of firesetting as a result of repressed sexual impulses has generally assumed a fixation at the phallic–urethral level, and places firesetting in the general class of neuroses. Kaufman, Heims, and Reiser (1961), in a study of 30 firesetters ranging in age from 6 to 15 years, concluded that firesetters were more likely to be fixated at the oral (psychotic) level rather than the more advanced phallic–urethral (neurotic) level. With respect to psychiatric diagnosis, Lewis and Yarnell (1951) found that most firesetters were given a diagnosis of psychopathic personality. However, they also noted that the list of diagnostic possibilities was almost endless. Neurotic, psychotic and psychopathic traits were present to one degree or another in nearly all cases. Actual diagnoses appeared to meet legal and statistical requirements and were therefore not very useful for classification purposes.

While they have often been used in interpreting the cases of individual firesetters, psychoanalytical approaches are of questionable value in explaining firesetting. Of particular importance is the failure of psychoanalytical explanations to find support in empirical evidence. For example, not only is there no convincing evidence of the relationship between enuresis and firesetting, but enuresis has been linked to more generalized disturbance of character (Michaels, 1955). Furthermore, a triad of childhood enuresis, firesetting, and cruelty to animals has been suggested as a predictor of adult violent crime (Hellman and Blackman, 1966; and MacDonald, 1968). While there is some controversy surrounding these findings, the important point is that they tend to disconfirm the existence of a special relationship between enuresis and firesetting, and they do not lend support to the notion that firesetting and sex are linked via urination. It will be useful to examine some characteristics which distinguish firesetters from individuals with other abnormal behaviour patterns.

Wolford (1972) compared incarcerated arsonists and non-arsonists from three southeastern states on a number of demographic, criminological and psychological variables, including Minnesota Multiphasic Personality Inventory (MMPI) performance. Among other things, Wolford concluded that arsonists 'exhibited personality characteristics more closely associated with persons undergoing psychic stress possibly due to an external control situation' than did non-arsonists. This conclusion was based on observed differences in mean profiles between the arsonist and non-arsonist groups. However, since significant group differences for most of the subscales were not found, Wolford concluded that psychopathology did not play a significant role in the aetiology of arson for the sample that he studied. What Wolford failed to point out was that, although he found few *differences* between groups, *both* groups showed highly deviant profiles, compared to what would be expected in the normal population. Thus, it could be argued that psychopathological factors were present in both groups.

There is also a serious difficulty in Wolford's analysis of mean MMPI profiles, since important differences in individual profiles might be obscured. A further consideration is the extent to which more immediate factors influenced MMPI performance, the prison environment being the most obvious such factor. Despite these issues, Wolford's study represents an important attempt at measurement of personality variables associated with firesetting. Furthermore, there is much evidence to support Wolford's conclusion that firesetters, as a group, are in many ways similar to other types of criminals.

For example, a number of studies have shown that firesetting is hardly an isolated antisocial behaviour. Lewis and Yarnell (1951) reported that half of their subjects had been in trouble with authorities for other types of antisocial activities. Yarnell (1940) and Vandersall and Wiener (1970) both reported that most of the children they studied were referred for problems other than firesetting, and firesetting 'most often emerged as only another symptom among other indications of poor impulse control and a more generalized behaviour disturbance'. Among adult firesetters, poor marital and occupational adjustment, and alcoholism have been identified as common problems (Inciardi, 1970; Lewis and Yarnell, 1951).

Of particular importance here is the *type* of antisocial behaviour shown by firesetters. A number of authors found that arsonists had committed a significantly higher number of crimes against property and a lower number of crimes against persons than had non-arsonist controls (Hurley and Monahan, 1969; McKerracher and Dacre, 1966; Tennent *et al.*, 1971; Wolford, 1972). Tennent *et al.* also found their arsonist group to be more depressed than controls and suggested that arsonists have difficulty externalizing aggression, following the lead of McKerracher and Dacre (1966) who suggested that firesetting might represent displaced aggression. In a similar vein, Lewis and Yarnell (1951) recognized the possibility that firesetting may result from fear of aggression towards others, being attacked by others, or the inability to effectively attack or fight back.

The importance of fire as a symbol, and of firesetting as a displacement of sexual and aggressive drives cannot be overlooked, since human behaviour is rich in symbolism. However, another aspect of firesetting deserves some emphasis. Firesetting, associated antisocial behaviours, sexual, marital and occupational maladjustment, and alcoholism can all be considered parallel indicators of a general lack of self-control, self-confidence, and the skills, particularly social skills, necessary to obtain rewards from the environment in an appropriate manner. Certainly the available data appear to support this picture of the firesetter. Most firesetters grow up in environments which are not likely to support normal social development; they typically do not fare well in school; and by the time they are adolescents or young adults they have usually experienced a great deal of failure in activities which society deems important. Lewis and Yarnell (1951) also noticed a relatively high incidence of physical abnormalities among these people.

Since a great many people undoubtedly experience a similar development, it is appropriate to ask: Why does the firesetter choose fire? One answer, of course, is that firesetting is only one behaviour among many maladaptive behaviours seen in these individuals. In fact, it might be said that the major difference between firesetters and other criminals is that firesetters set fires. But beyond that trite statement, two major reasons for firesetting may be stated.

First, firesetting offers immediate consequences which may be rewarding to the individual. These include the sensory stimulation of the fire itself, the commotion of crowds, sirens and bells, and praise and recognition derived from the community if the firesetter turns in the alarm or helps put out the fire.

Second, an avoidance mechanism is likely to be involved. If the individual is lacking in self-confidence and has been unsuccessful at interactions with other individuals in the past, then such social interactions are likely to be aversive. When problems with other people arise, he is not likely to solve them in a socially acceptable manner. If he chooses to engage in aggressive behaviour, he responds in a way which will avoid confrontation with the victim, firesetting being one such way. This analysis is similar in many ways to the displaced aggression analysis.

The analysis presented here is more or less consistent with a social learning theory of behaviour (cf. Bandura, 1976). According to social learning theory, behaviour develops through the selective action of its consequences in the natural environment. Some responses are rewarded, while others are not or are even punished. The consequences of behaviour need not be experienced directly for learning to take place; they may also be experienced vicariously through observing others perform a behaviour, and through pictorial or verbal experiences. These vicarious learning processes are generally referred to as *modelling* processes. A person may be likely to perform a response which he observes someone else perform; he is even more likely to perform the response if he observes the model being rewarded for that behaviour. Several studies provide examples of how experiences with fire may have influenced the behaviour of firesetters; Wolford (1972) found that arsonists in his sample of prisoners tended to come from more rural settings than did non-arsonists. It is likely that persons growing up in rural settings were exposed to large fires when land was cleared and trash was burned. Macht and Mack (9168) found that fathers of adolescent firesetters in their sample all had some significant involvement with fire in their employment (fireman, furnace stoker, gas burner repairman, scrap car burner at a junkyard). In one case, a son had watched his father set fires to vehicles for one year, and the following year was allowed to help set the fires himself. With this group, both vicarious and actual experiences with fire may have contributed to firesetting. Yarnell (1940) considered the possibility of prior experience with fire as a contributing factor to firesetting, but she did not believe that earlier experiences with fire in her sample of child firesetters were much different than those which might be found in normal children. On the other hand, such factors as modelling may have a great influence only when other conditions are favourable.

We have already discussed what some of these conditions favourable to firesetting are likely to be. In general they revolve around an environment which has not been very rewarding. Yet beyond the quality of the immediate external environment, the perceived level of reward may depend largely upon cognitive factors which involve the individual's perceptions and expectations (Bandura, 1976). A thorough analysis of these factors would be beyond the scope of this paper, but some basic principles can be outlined here. The firesetter's general ineffectiveness in obtaining rewards may affect his perception of the relationship between his behaviour and its consequences, his evaluation of his performance relative to that of others, and his evaluation of his performance against his own standards. A history of failure may result in an expectancy of failure in new situations. Even if he is successful in producing rewards, the firesetter may be inclined to attribute success to factors beyond his control. In a similar vein, he may see his performance as poor, relative to others and relative to whatever personal standards he may have set for himself, even though such a low self-evaluation may have little basis in reality. This kind of pattern is largely self-perpetuating, and it may persist far beyond the boundaries of the environment which produced it. The act of firesetting often appears to be far out of proportion to its motive. From the firesetter's point of view, however, that may not be the case. Firesetting may offer a kind of control over the environment which the firesetter has been unable to obtain in other ways.

At this point, we offer this intrepretation of firesetting as an adjunct to, rather than a replacement for, other accounts of firesetting which have been presented. We have not done justice to any of them in this short space. As Macht and Mack (1968) have suggested, firesetting is multiply determined, and biological, symbolic, and adaptive process may all have to be taken into account.

TREATMENT OF FIRESETTING

Very little has been written about the treatment of firesetting. Stekel (1924b) presented an extensive account of the psychoanalysis of a firesetter. The approach relied heavily on the analysis of dreams in bringing to light the patient's unconscious motives. Stekel reports the analysis to be incomplete, and its effectiveness is not clear. None of the previously reviewed studies of firesetters in clinics or institutional settings have outlined what, if any, treatment procedures were employed. Schmideberg (1953) insisted that it was possible to effectively treat firesetters, but stated that an account of treatment was too detailed to give in her paper. It seems safe to assume, given the heavy emphasis on psychoanalytical accounts of firesetting, that psychoanalytical methods have been heavily employed. On the other hand, recent published accounts of successful treatment have used behaviour modification approaches with children.

Welsh (1971) used 'stimulus satiation' to eliminate firesetting behaviour in two seven-year-old boys. Each child was given the opportunity to strike as many matches as he wished, one by one, for one hour each day. The procedure was carried out in the therapist's playroom under the supervision of the therapist. In one case, the child was required to hold each match until the heat was felt on the fingertips; in the other case, each match was required to be held over the ashtray at arm's length until most of the match had burned, and the extended arm could not be supported on the table or by the other arm. When the child asked to stop, he was requested to light a few more matches and then allowed to engage in play activities. Firesetting was eliminated in both the playroom and the home in just a few sessions. Follow-ups over the next six months or so revealed no recurrence of firesetting behaviour in the home. It should be pointed out that the results cannot be considered as due merely to satiation of firesetting; in both cases there was also an aversive component in the procedure.

Holland (1969) employed the parents as therapists in the treatment of firesetting in a seven-year-old boy. The boy was asked to bring the father any package of matches that he found around the house. The father rewarded the child with money for any package brought. To ensure that the desired behaviour initially occurred and was rewarded, the father left one package with no matches, so that when the child found it there would be no reason to keep the package rather than return it to the father. When bringing matches to the father was firmly established, a second procedure was designed to strengthen non-striking behaviour. The child was given the opportunity to strike twenty matches, but was rewarded with one penny for each match he did not strike. By the third such opportunity the child did not strike any matches. Firesetting was eliminated and follow-up over the next eight months revealed no recurrence.

Using parents as therapists was an important part of Holland's procedure and was designed to increase the number of rewards the child received from the parents, thus improving his relationship with the parents. By working through family members in a *triadic* model of therapy (Tharpe and Wetzel, 1969) instead of directly with the patient (a *dyadic* model), the therapist attempted to bring about improvements in the client's natural environment and eliminate those factors which might have been maintaining the maladaptive behaviour. The effectiveness of this approach lies in the fact that it teaches people around the client skills in dealing with the client. If these skills are continually employed in the natural environment, positive changes are more likely to be maintained.

Recently a movement has evolved within the fire-fighting community to provide counselling services for youthful firesetters and their parents.[2] A counselling programme will have two major components. First of all, the firefighting/counsellor must screen the case to determine whether or not a referral for professional therapy is warranted. Where the youthful firesetter and/or his home environment are sufficiently disturbed, the firefighter/counsellor has neither the time nor the expertise to handle the case, and failure to make proper referrals could have serious legal consequences. However, in many cases youthful firesetting is more or less an outgrowth of a natural fascination with fire, sometimes combined with transient or less serious disturbances in the youth or his home environment. In these cases, the counsellor may elect to carry out the second component of the programme. Here, the counsellor will act as an educator in providing the client with information on the proper use of fire and on the possible consequences of firesetting. With the parents, the counsellor will often work out a programme to reward the youth for proper use of fire and for not setting inappropriate fires. The counsellor is also instrumental in setting up restitution programmes in which the youth makes amends for damage caused by his firesetting, and in helping the parents set up home fire safety programmes for which the youth may be given primary responsibility. The opportunity to exhibit and be rewarded for responsible behaviour can be helpful where the client seems to be lacking in self-confidence and may perceive himself as not being accepted by family and friends.

If the counselling programme is to be successful, cooperation of the parents or other responsible individuals is usually essential. The optimal programme will also depend on the age of the client and individual circumstances surrounding each case. Training and experience are important in the counsellor's ability to deal with these individual aspects. Counsellors should participate in workshops and training seminars, and should regularly share information with one another. It is also desirable that professional consultation be readily available. Although still in its infancy, the counselling model presented here shows a great deal of promise in dealing directly with less serious cases of youthful firesetting, and in seeking help in treatment of the more serious cases.

With adult firesetters, the triadic model may be more difficult to employ, since it appears that for many adult firesetters, the constellation of surrounding people may be continually changing. In this case, the direct approach may be more appropriate. Assertion training (cf. Cotter and Guerra, 1976) to teach skills in social competence, and job skills training, where appropriate, may help the individual to obtain rewards from the environment in a more appropriate manner. There are many variants of the assertion-training approach, all of which stress the learning, rehearsal, and use of specific skills in dealing with other people. As the client uses these skills and is successful, he becomes more self-confident and the social skills are maintained by the success they produce. If, as we have suggested, social ineffectiveness is at least one component of firesetting behaviour, then such an approach to treatment seems reasonable. Mathie and Schmidt (1977) recently reported the successful employment of a similar approach to the treatment of two incarcerated male arsonists, although precise details of therapeutic procedures were not given. It should also be noted that there is no *a priori* reason why assertion training could not be employed along with other approaches which the therapist may deem important.

CONCLUDING REMARKS

In this paper we have glossed over many important details about individual firesetters and various groupings of firesetters in favour of a more general picture of the 'typical' firesetter. The picture turns out to be one of an individual with several maladaptive behaviour patterns, of which firesetting is one. We have identified social ineffectiveness as a common factor in the general tendency of firesetters to have drinking problems, marital, occupational and sexual problems, and to exhibit a variety of other criminal and antisocial behaviours. Levin (1976) has suggested that there are no clearcut differences between criminals 'who set malicious fires and those who do not; the similarities are more striking than the differences'. One difference, if it holds up in future studies, which does seem to be important is the finding that arsonists appear to commit more crimes, firesetting excluded, against property than do other criminals. This finding supports the social ineffectiveness view and also supports the view that firesetters avoid activities which involve confrontation with another person. More data exploring this hypothesis are needed.

Much of the information in this paper comes from studies completed over 25 years ago. Most studies have been case studies very rich in relevant clinical material, but very scarce in reliable data. This situation has to change if our knowledge about firesetting is to increase significantly. We earlier described some of the problems likely to be encountered by the researcher. Nevertheless, in many cases reliable data could have been gathered but were not; meaningful hypotheses could have been tested but were not; relevant control groups could have been included but, with the exception of a few studies, were not. There is also a glaring lack of information on treatment procedures for firesetters, even in cases where the client was seen in a clinical setting. These deficiencies in the current literature present extensive opportunities for future research.

The clinical literature on firesetting contains many descriptions which provide rich clinical material on individual firesetters. Much of the material provides insights into the behaviour of the individual firesetter which cannot be easily characterized in terms of measurement and statistical information. We are not suggesting that researchers should ignore this type of material in favour of systematic data collection; rather, systematic, well-controlled studies should become an integral part of the research process, and should be an addition to other important clinical material.

NOTES

1. More recently, Patterson (1978) has suggested that defiance, lying, 'running around', stealing, and firesetting form a progression of delinquent behaviour categories, such that a child exhibiting a given category of behaviour will also be likely to exhibit categories of behaviour falling lower on the progresssion. Thus, a child who exhibits lying is likely also to be defiant; a child who 'runs around' is also likely to lie and to be defiant, and so on. Since firesetting is the highest category in this progression, children who set fires are likely to show the greatest number of delinquent behaviour symptoms.

2. The ideas in the next two paragraphs are largely taken from a recent workshop on youthful firesetters sponsored by the National Fire Prevention and Control Administration and the Los Angeles County Fire Department Protective Services Panel. The sponsors hope to publish a training manual which presents in detail the counselling model which is only briefly presented in this paper.

REFERENCES

Bakwin, H. and Bakwin, R. M. (1960). *Clinical Management of Behavior Disorders in children.* (Philadelphia: W. B. Saunders Co.).

Bandura, A. (1976) 'Social learning theory'. In J. T. Spence, R. C. Carson, and J. W. Thibaut (eds.), *Behavioral Approaches to Therapy.* (Morristown, New Jersey: General Learning Press), pp. 1–46.

Conant, R. W. (1970). 'Rioting, insurrection, and civil disobedience'. In R. Hartogs (ed.), *Violence: Causes and Solutions.* (New York: Dell Publishing Co.), pp. 105–115.

Cotter, S. B. and Guerra J. J. (1976). *Assertion Training.* (Champaign, Ill.: Research Press).

Freud, S. (1932). 'The acquisition of power over fire'. In E. Jones (ed.), *Collected Papers, Vol. 5, Sigmund Freud.* (London: The Hogarth Press and the Institute of Psycho-Analysis), 1956, pp. 288–294.

Gold, L. (1962). 'Psychiatric profile of the firesetter', *Journal of Forensic Sciences,* 7, 404–417.

Hellman, D. S. and Blackman, N. (1966). 'Enuresis, fire setting, and cruelty to animals: a triad predictive of adult crime', *American Journal of Psychiatry,* 122, 1431–1435.

Holland, C. J. (1969). 'Elimination by the parents of fire-setting behavior in a 7-year-old boy', *Behavior Research and Therapy,* 7, 137.

Hurley, W. and Monahan, T. (1969). 'Arson: the criminal and the crime', *British Journal of Criminology,* 9, 4–21.

Inciardi, J. A. (1970). 'The adult firesetter: a typology', *Criminology,* 8, 145–155.

Kanner, L. (1957). *Child Psychiatry.* (Springfield, Ill.: Charles C. Thomas).

Kaufman, I. Heims, L. W., and Reiser, D. E. (1961). 'A re-evaluation of the psychodynamics of firesetting', *American Journal of Orthopsychiatry,* 31, 123–137.

Levin, B. (1976). 'Psychological characteristics of firesetters', *Fire Journal,* March, 36–41.

Lewis, N. D. C. and Yarnell, H. (1951). 'Pathological firesetting (pyromania)', *Nervous and Mental Disease Monographs,* no. 82.

MacDonald, J. (1968). *Homicidal Threats.* Springfield, Illinois: Charles C. Thomas).

Macht, L. B. and Mack, J. E. (1968). 'The firesetter syndrome', *Psychiatry,* 31, 277–288.

Mathie, J. P. and Schmidt, R. E. (1977). 'Rehabilitation and one type of arsonist', *Fire and Arson Investigator,* 28(2), 53–56.

McKerracher, D. W. and Dacre, J. I. (1966). 'A study of arsonists in a special security hospital', *British Journal of Psychiatry,* 112, 1151–1154.

Michaels, J. J. (1955). *Disorders of Character.* (Springfield, Ill.: Charles C. Thomas).

Moll, K. D. (1974). *Arson, Vandalism and Violence: Law Enforcement Problems Affecting Fire Departments.* US Department of Justice, Law Enforcement Assistance Administration, National Institute of Law Enforcement and Criminal Justice, Washington, DC.

National Fire Protection Association, Fire Analysis Department. (1976). 'Fire and fire losses classified, 1975, *Fire Journal,* November, 17.

Nurcombe, B. (1964). 'Children who set fires'. *Medical Journal of Australia,* 18 April, 579–584.

Patterson, G. R. (1978). 'The aggressive child'. In G. R. Patterson and J. B. Reid (eds.) *Systematic Common Sense.* (Eugene, Oregon: Castalia Press).

Rice, R. (1956) *Business of Crime.* (Westport, Conn.: Greenwood Press).

Schmideberg, M. (1953). 'Pathological firesetters', *Journal of Criminal Law, Criminology, and Police Science,* 44, 30–39.

Shaw, C. R. (1966). *The Psychiatric Disorders of Childhood.* (Englewood Cliffs, NJ: Prentice-Hall).

Simmel, E. (1949). 'Incendiarism'. In K. R. Eissler (ed.) *Searchlights on Delinquency.* (New York: International Universities Press), pp. 90–101.

Stekel, W. (1924a,). 'Pyromania'. In *Peculiarities of Behavior, Vol. 2.* (New York: Boni and Liveright), pp. 124–181.

Stekel, W. (1924b,). 'The analysis of a pyromaniac'. In *Peculiarities of Behavior, Vol. 2.* (New York: Boni and Liveright), pp. 182–232.

Teague, P. (1978). 'Action against arson', *Fire Journal,* **March,** 46.

Tennent, T. G., McQuaid, A., Loughane, T., and Hands, N. J. (1971). 'Female arsonists', *British Journal of Psychiatry,* **119,** 497–502.

Tharpe, R. J., and Wetzel, R. J. (1969). *Behavior Modification in the Natural Environment.* (New York: Academic Press).

Vandersall, T. A. and Wiener, J. M. (1970). 'Children who set fires', *Archives of General Psychiatry,* **22,** 63–71.

Welsh, R. S. (1971). 'The use of stimulus satiation in the elmination of juvenile firesetting behavior'. In A. M. Graziano (ed.), *Behavior Therapy with Children.* (Chicago: Aldine Publishing Co.), pp. 283–289.

Wolford, M. (1972). 'Some attitudinal, psychological and sociological characteristics of incarcerated arsonists', *Fire and Arson Investigator,* **22(4),** 1–30; **22(5),** 1–26.

Yarnell, H. (1940). 'Firesetting in children', *American Journal of Orthopsychiatry,* **10,** 282–286.

Fires and Human Behaviour
Edited by D. Canter
© 1980 John Wiley & Sons Ltd.

CHAPTER 4

Playing with Matches: Children and Fire

DITSA KAFRY*
Department of Psychology, University of California, Berkeley

Most children develop fascination and interest in fire at an early point in childhood striving to achieve control over fire and to acquire fire skills. A first attempt to engage in fire behaviour is often inititated by the children themselves; they either ask for parental permission to light matches or do it without permission or in defiance of explicit parental prohibition. Unfortunately, the lighting of matches without parental approval and/or supervision (labelled 'fire-play') often results in fires causing injury and property damage. Actuarial data show that indeed fire is a major hazard for children both as setters and as victims. Fire data concerning age and sex suggest that *young boys* represent an especially hazardous group.

This chapter summarizes research (Kafry, 1978) which investigated fire behaviour and knowledge in a sample of young normal boys, and studied these behaviours in relation to the characteristic of the children and their environment.

A random sample of 99 boys from the school district of Berkeley, California, participated in the study. The boys were enrolled in kindergarten (N = 33, mean age 5.9), second grade (N = 33, mean age 7.9) or fourth grade (N = 33, mean age 9.9). Children and parents (mostly mothers) were interviewed in their homes. In addition, both parents were asked to complete a research questionnaire. The interviews and questionnaire provided information about fire behaviour, knowledge and attitudes, family background, child health and accident history, child rearing practices, child cognitive and personality characteristics. The interview also included a fire experiment in which the mother taught the child to boil water on a camping stove.

PREVALENCE OF FIRE-PLAY

Fire interest was found to be almost universal and fire-play was performed by 45 per cent of the boys studied. The high prevalence of fire-play is alarming; it supports Lewis and Yarnell's (1951, p. 285) observation that 'undoubtedly the

*Acknowledgement is made to the Pacific Southwest Forest and Range Experiment Station, Forest Service, United States Department of Agriculture, who financed the research involved in this work.

incidence of children who play with fire is far greater than any statistics show', and indicates a severe problem with fire prevention efforts of parents and social institutions. The existence of the problem is further exemplified by the direct connection between the number of incidents of fire-play and the probability that they will result in actual fires. In this study it was found that single instances of fire-play resulted in fires for 33 per cent of the cases while 81 per cent of the repeater fire-players caused fires. The fires, set by 21 per cent of the boys studied were mostly small fires which were easily extinguished. However, the severity of the fire is often affected not only by the child's behaviour but by chance factors such as availability of combustible materials and weather conditions. (For instance, the one fire in the study which resulted in severe damage to a house was caused by the first fire-play of a child.) These results suggest that education efforts should be addressed not only toward the prevention of repeated incidents of firesetting but toward the prevention of fire-play in general. Gross underestimation of our statistical information about children's firesetting is reflected by the percentage of fires reported to the fire department. Only 9 per cent of the fires set by the boys were reported, and 33 per cent of the fires set by other members of the family, so that on average 24 per cent of the total fires were reported. The latter percentage is closely similar to the one found by Globerson and Grossman (1971); in their survey of 500 staff members of the University of California at Berkeley sampled randomly, 21 per cent of the total number of fire incidents were reported to the fire department. Human hazardous behaviour obviously accounted for all fires set by the children who played with matches, and was the direct cause of 31 per cent of the other fires reported by the families.

Children's interest in fire was found to start at an early age; 18 per cent of the fires set by them were set before the age of three. This result supports Nurcombe's (1964) finding that interest in fire starts between the ages of two and three, and Block, Block, and Folkman's (1976) generalization that fire assumes increasing salience between the ages of three and five. The present study showed a decreasing tendency in the number of fires beyond the age of seven. Very few fires were set after this age (none after age eight), even when the number of children who were older than seven at the time of the interview is considered. These results indicate that although a relatively small percentage of children play with matches at a very early age, fire-play is essentially dangerous and often results in actual fires. However, the decrease in the number of fires by no means indicates a decrease in fire interest. Comparisons of fire interest among the three age groups as reflected by parental reports showed the same percentage of boys expressing interest in all groups (Figure 4.1). The same results are reflected by the reported frequencies of different fire behaviours which showed that only fire competence increased with age while other attraction and avoidance behaviours were relatively stable within the studied age groups, with the peak of the attraction in the second grade. Furthermore, attraction, avoidance and competence were found to be three independent components of fire behaviour: children who play with fire appeared to be more attracted to it, but also more competent and less cautious. These results show a need to deal with attraction to fire in positive and constructive ways which encourage competence rather than attempt to eliminate fire attraction.

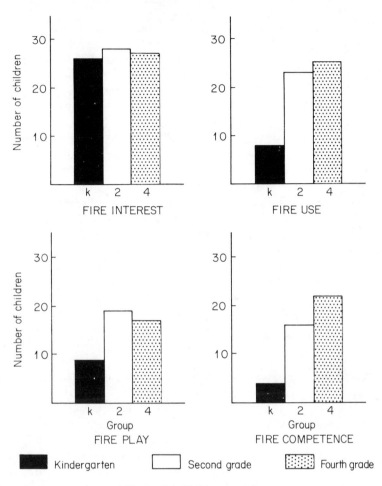

Figure 4.1 Children and fire

Analysis of the fires set by the boys revealed that they had, in most cases, a high level of understanding about the possible consequences of their violation of behavioural codes involved in it. This seems evident from the fact that they set the fires in places which could not be easily detected by adults. These places were either outside the home, such as yards, fields or forests; or inside the home, for instance bedrooms, bathrooms or basement. The same trend was found for boys who played with matches without setting fires. In several cases, the play was never detected by the parents and the child reported them to the interviewer confidentially. Over half of the fires were set in groups as part of chidren's adventurous play and sometimes involved reactions of excitement to the fire. Most of the boys admitted their participation in fire-play and were aware of the risk involved in it. Some of them made comments such as 'I am through with matches', 'I don't play with matches any more, I learned my lesson', 'I played with matches only when I didn't realize they could start fires', or 'most of the boys I hang around with play with matches', 'my friend tried to light a pack of matches. I tried to stop him but he wanted to know how matches work', 'I don't get in trouble if they [parents] never find out [about it]'. These comments indicate both the awareness of danger and the fascination with fire and reflect the development of ambivalence toward fire. Other comments made by the children reflected the punishment attached to playing with matches. Examples of such comments were: 'Grandmother found me and beat me', 'I got a lot of spankings and I just can't do it anymore', 'Mom made us stay in for a few weeks'. In fact, 65 per cent of the boys interviewed had some comments when asked 'have you ever played with matches?'. These comments reflect their own involvement or the involvement of other children in playing with matches. Only 35 per cent of the boys said a blunt 'no' when asked this question. Furthermore, fire-play was not at all related to fire knowledge. There were no significant differences between the children who played with matches and those who did not on any of the indices measuring knowledge about fire situations. These findings indicate lack of correspondence between the cognitive and behavioural aspects of fire and suggest that increase in knowledge and understanding about fire is not a sure route to decrease the prevalence of fire-play and firesetting.

FIRE KNOWLEDGE AND SKILLS

In addition to the investigation of fire interest, hazardous play and firesetting, the other side of the coin, namely fire knowledge and skills of parents and children, were also investigated. Parents showed an acceptable level of fire knowledge and information, though many deficiencies were found especially in the area of prevention, preparation and skills (for example conducting home fire-escape drills, instructing babysitters about fire, leaving small children alone, knowledge of fire department telephone number, owning fire extinguisher and smoke detectors). Only 43 per cent of the mothers reported that they had had some form of fire education, usually in primary school. Parents' reports also showed a high level of awareness of fire hazards coupled with problems in handling them; 68 per cent of the mothers indicated that the danger of fire is the main thing they want to teach their children about, and 44 per cent expressed the fear that fire is their major worry when they leave children unattended. When asked, however, about the way they handled the fire interest of their children, 38 per cent of the mothers of boys who expressed some interest in fire dealt with it by either ignoring or by warning or forbidding the child to use any kind of fire. The existence of parents' problems in dealing with the fire behaviours of their children is also indicated by their reactions to the fires set by their children. Many mothers reacted either by expressing fear and anxiety ('I was horrified', 'I was paralysed and freaked out', 'I was very upset and called "Parental Stress" unit'), or by rage and severe punishment ('I gave the child a whipping', 'I was very angry and apprehensive', 'I was more concerned about the child than putting out the fire'). The above data indicate that though there is a high level of awareness about the risks of fires, a large percentage of parents do not deal with it directly by giving constructive instructions and providing their children with fire skills and knowledge, a result which replicates previous findings (Block, Block, and Folkman, 1976).

Lack of appropriate instructions is obviously reflected in the children's knowledge, and indeed many gaps were found in this area; about half of the boys could not assign plastic, rubber, chalk and aluminium items correctly to burnable or non-burnable materials. In addition, many children did not know the correct responses to various fire situations (such as clothes catching fire, or smoke in the room). As for the actual fire skills observed during the fire experiment, the results showed that only 42 per cent of the children were reasonably competent in dealing with matches (competence was defined using a lenient criterion as the ability to strike matches under supervision and extinguish them immediately), though only a few were competent in conducting the experiment as a whole (including a safe use of the camping stove). Competence increased with age and was found to be high mainly for two groups of children: those who were allowed to light matches under supervision and those who were involved in fire-play without firesetting. Most boys in these two groups were allowed to light matches under supervision at the time of the study. Most of the mothers who forbade their children to use matches were mothers either of children who had never used matches previously or of children who set fires. All the firesetters who did not light matches during the fire experiment were forbidden to do so by their mothers. The latter findings supports the results of Siegelman and Folkman (1971) which indicated that mothers of repeater firesetters tried to prohibit the use of fire materials under any conditions. These findings indicate that the overprotective mothers who would not allow their sons to light matches were either mothers of children who did not have any fire experience or who had firesetting experience. Most boys in these two groups were not normally allowed to light matches at all. It should also be noted that several boys had exceptionally high awareness of fire safety; two boys who were allowed to light matches under supervision actually had insisted that their parents purchase a smoke detector and a fire extinguisher.

These results point to the need to develop and implement fire skills and prevention programmes which are readily available to both teachers and parents. Educators who are involved in parent counselling and who write child-rearing guides should also be encouraged to include fire information as an integral part of their work. An indication that parents will be receptive to such guidance is provided by the responses of parents during the interview. Many parents expressed fear of fire as a major concern when leaving children alone, several parents asked the interviewers for advice in this area, one parent called the 'Parental Stress' unit and one the fire department for help after their children set fires. Ray Walters (personal communications) reported that most of the children who participated in the city of Hayward, Wisconsin, fire-players programme did so because their parents called the fire department and asked for advice on handling their children's fire-play. A more rigorous research support for parents' awareness of the fire hazard was provided by Winget and Whitman (1973) who asked 300 adults sampled randomly: 'If you had a child who repeatedly set fires, what would you do about it?' Their results showed that 54.3 per cent of the responses indicated some kind of professional referral and 13.7 per cent indicated reference to police, fire department or control at home.

CHARACTERISTICS OF CHILDREN WHO PLAY WITH MATCHES

The comparison between children who played with matches and those who were never involved in fire-play showed several interesting and consistent results in terms of the characteristics of the children and their family background. The analysis of the child's characteristics revealed that playing with matches is not just one stage in the development of normal children, but also a symptom which is highly related to other problem areas in the children's life. The personalities of the children who played with matches showed a surprisingly consistent picture. This picture can be reduced to one 'non-scientific' term by describing a child who plays with matches as a 'rascal'. This child is more mischievous, energetic, adventurous, exhibitionist, aggressive and impulsive than his peers. He is more often involved in adventurous plays leading to accidents and has more conduct problems reflected by his interpersonal relationships with parents and friends. This picture of the child fire-player was drawn from all evaluations collected during the interview, evaluations performed by parents and interviewers, reported orally or in writing, evaluations which were responses to structured questionnaires and which were spontaneous descriptions of the child, all showing the same pattern. Hence, playing with matches was not found to be an isolated problem but rather a part of the child's mischievous behaviour.

The above picture of the fire-players is similar to the picture of firesetters which emerges from the literature (Hellman and Blackman, 1966; Kaufman, Heims, and Reisner, 1961; Lewis and Yarnell, 1951; Nurcombe, 1964; Wilmore and Pruitt, 1972; Scott, 1974; Siegelman and Folkman, 1971; Yarnell, 1940). This similarity is surprising especially in the light of the fact that the present study surveyed the behaviour of a random sample of normal children, whereas other studies of firesetting investigated children who had a variety of other problems as well. The similarity suggests a behavioural continuum, which has on one extreme the repeater firesetters, close to it are the non-repeater firesetters, then those who play with matches and then, at the other extreme, those who are devoid of behavioural problems and risk-taking behaviours.

This pattern of personality of the match-players and firesetters was also found to be very similar to the personality of the accident-repeater children. This picture was presented by Manheimer and Mellinger (1967), and supported by Block and Block (1975), that children who are frequently involved in accidents are mischievous, aggressive and impulsive, traits which lead them to high accident exposure. The 'rascal' is not only prone to hazardous fire behaviours and accidents but also has similar personality characteristics to those of the hyperactive child as well as the delinquent child. The former has recently attracted the attention of educators and child psychologists. Hyperactivity, which is synonymous with hyperkinetic behavioural patterns was defined as a long-term childhood pattern characterized by excessive restlessness and inattentiveness (Safer and Allen, 1976). Hyperactivity has been described as a developmental disorder which begins in early childhood and fades during adolescence; it is sometimes also described as a predelinquent pattern. The major features of hyperactivity are inattentiveness, learning impediment and behavioural problems, and the minor features are impulsivity, peer difficulties and low self-esteem. Coleman (1972) provided clinical descriptions of antisocial personality.

One of these descriptions stressed irresponsible and impulsive behaviour with low frustration tolerance, proneness to thrill-seeking, deviant and unconventional behaviour and seeking immediate pleasure and gratification. This clinical picture also describes what he called psychopathic delinquents who are again defiant, resentful, devoid of guilt or remorse, unable to profit from experience, lack conscience and reality control, are at the mercy of inner impulses, seek stimulation and excitement, live in the present and are incapable of establishing and maintaining interpersonal ties. This picture of boys was presented in a number of classical studies in the area of child delinquency. Glueck and Glueck (1952) compared 500 delinquents to a matched group of non-delinquents. Among their findings they reported that delinquents are extremely restless, energetic, aggressive, stubborn and adventurous, they express uninhibited, untamed and unreflective characteristics, and inability to cope with the general process of socialization and adjustment and with the realistic demands of life. They also have interpersonal problems with peers, teachers and parents. Bandura and Walters (1959) compared 26 delinquents to 26 non-delinquents. The delinquent–aggressive boys were openly antagonistic to authority and less positive in their feelings towards their peers. Conger and Miller (1966) who compared 271 delinquents to non-delinquents in Denver public schools found that delinquents were overly aggressive, less well adapted, resented authority and manifested problems in social, academic and emotional areas. All studies found that delinquents came from more deprived backgrounds relative to non-delinquents, especially in terms of relationships with their fathers.

The discussion of firesetters, accident-prone children, hyperactive children, and delinquent children showed, therefore, that they are characterized by similar behavioural and personality patterns, the 'rascality' pattern (or what Siegelman (1969) called the 'impulsive' syndrome). It is customary to view this pattern as a highly negative one since it is related to so many negative consequences. However, embedded in some of the above studies is also a list of positive characteristics expressed by the 'rascals'. The first general positive trait is their level of energy. A high level of activity should not be viewed exclusively as negative. Is the opposite trait, dullness, exclusively positive? In studies of delinquency, some delinquents were described as more mature, independent, assertive, vivacious, curious and adventurous, and less neurotic, submissive, insecure and conformist relative to their non-delinquent peers. Undoubtedly, these characteristics, expressed by problem children are highly desirable in our society. However, they are sometimes 'hidden' in research reports and receive no explicit discussion. Exceptions to this rule do not exist, psychologists pointed to the differences between assertiveness and aggression and discussed the philosophical issues involved in over-control and risk-avoidance at the price of spontaneity and creativity in our value system (Manheimer and Mellinger, 1967). Others claimed that efforts can be directed to reduce exposure to hazard while retaining 'the positive personality characteristics that are correlates of risk-taking, curiosity, openness to experience, vitality' (Block and Block, 1975, p. 116). Hence, it is suggested that the 'rascals' are not exclusively negative in their behavioural pattern but also possess several personality assets. The 'rascality' pattern can be defined on several planes each including several features: the physical plane includes high arousal level (vivacity, alertness, energy), adventurous behaviours (physical risk-taking) and physical sensation seeking; the cognitive plane may involve alacrity (psychological zest and need to exert mental effort), knowledge or information-seeking, and the more negative feature of distractability and short attention-span. The motivational plane may include such features as curiosity or exploration, non-physical risk-taking and mischief or destruction (defying of social norms); the personality–social plane includes features such as impulsivity, social aggression and feelings of omnipotence or over-control over the environment. These four planes may serve as a rudimentary framework for the definition and study of 'rascality', its validity is obviously unknown, and the relations among the features are still unclear. The focal point of this argument is the need to consider the positive components of what is customarily viewed as 'mischievous' behaviour.

The 'rascality' syndrome may help in understanding the apparent contradiction in the personality descriptions of repeated firesetters (Siegelman, 1969). While they are perceived as more mischievous, boisterous and temperamental relative to the non-repeaters they are also perceived as more curious, enthusiastic, imaginative, masculine and original. These latter traits reflect the positive aspects of 'rascals', the traits which make them likeable and attractive, just as their other traits make them a nuisance to their parents. The combination of the negative and positive components of 'rascality' may explain the emotionally ambivalent reactions of the mothers to the repeated firesetters in Siegelman's study.

The present study did not find any difference on activity level for children who play with matches and those who do not. Results suggest, however, that lack of impulse control is the major problem of fire–players. Acceptance of this suggestion implies the need for the development of constructive ways for impulse expression, ways that may replace fire-play and other hazardous behaviour patterns of 'rascals'. It may be speculated that the conduct problems shown by 'rascals' do not stem exclusively from their personality make-up as suggested by psychoanalytical theory but from their personality as it interacts with their environment. The high-energy children are difficult to cope with and pose problems to their environment, including parents and teachers. Hence, their background should be considered as interacting with their personality traits if a psychological rather than a sociological explanation is sought. Traditionally, conduct problems were viewed as dependent variables with family background as an independent variable. The point made here is that, as much as parents affect the behaviours of their children, children affect the behaviours of their parents. High-energy children pose problems as much as challenge to parents and other educational agents, in that they require more channelling of energy, more direction and guidance for impulse expression in a productive manner. The problem is not in their high activity level *per se* but in the mechanisms for channelling it to positive directions and in the mechanisms for impulse control.

FAMILY CONDITIONS

Several studies have dealt with the familial conditions necessary for the development of impulse control. Though it is not attempted to cover this literature completely, part of it will be used as the basis for explaining the interaction between the characteristics of the child and the features of his family background as affecting 'rascality'. Block (1976) presented a comprehensive review of the literature in the area of familial factors associated with affective disorders. In discussing undercontrolled, impulsive and aggressive children, she concluded that 'they appear to have grown up in homes where the parents are hostile and rejecting and set few limits' (p. 28). This conclusion has two very important implications for the present study; it presents the importance of warmth and support on the one hand, and firm and structured constraints on impulse expression on the other. Though the present study was not systematically designed to test these conjectures it includes several indirectly relevant findings. Mothers of children who did not play with matches reported better family relationships, better relationships between themselves and their sons, better relationships between their spouses and their sons, and better relationships between their sons and other children when compared to mothers of those who played with matches. Parents of the 'play' group also reported more frequent use of harsh punitive methods such as shaking the boy and taking away many privileges than parents of the 'non-play' group, and some also reported severe physical punishment for match-play. (Punitive methods are also relevant to the topic of setting limits.) Good family relationships and appropriate disciplinary procedures are imperative for feelings of warmth, acceptance and support. In addition to the reports of the parents, the interviewers also observed the parents' behaviours during the experimental tasks. The results showed that the mothers of the 'non-play' group provided more support to their children during the experimental interaction. Content analyses of the experiment's transcripts also showed that mothers of the 'non-play' group provided their children with less criticism relative to the 'play' group. Several matchplayers were also under situational stress, which may have interfered with the feeling of warmth and support in the home, examples being recently divorced parents and hospitalization of the mother. A focal point in the family interaction is the son–father relationship. This relationship was perceived as of crucial importance in firesetting and delinquency studies. A partial indication for the importance of this variable in the present study is the finding, that fathers of the 'play' group perceived their sons to be more mischievous and negative on many more behaviours than the mothers did, and mothers perceived their husbands to have worse relationships with their sons relative to the 'non-play' group. These findings, though incomplete, indicate the existence of problem relationships between sons and fathers for the boys who play with matches.

The above combination of findings indicates that children who had conduct problems as expressed in fire-play suffered from more rejection and less support and warmth relative to their non-problematic peers, supporting the first part of Block's statement, this part dealt with the quality of interpersonal relationships in the family, while the second part dealt with 'setting limits' (understood as the provision of constraints in terms of the conditions under which certain behaviours should not occur). This conjecture has also been promoted by other researchers and has been summarized by Block and Block (1975, p. 109) by stating that 'children develop self-confidence and impulse control in a family setting in which the parents feel themselves neither helpless not intimidated and are able to manage their own lives and to deal with the child authoritatively' (see also Baumrind, 1973). Though the present study was not designed to test this conjecture there is some indirect evidence to support it. Fathers of the 'play' group indicated the more frequent use of ignoring the problem as a punitive method. Ignoring mischievous behaviour is not always a constructive way to cope with the child's problems. Fathers of the 'play' group also reported more often shaking the boy and telling him not to move than fathers of the 'non-play' group. The latter punitive method was also used more often by mothers of the 'play' group, together with making the boy do extra chores and taking away things he wanted. These methods are not necessarily effective as a constraint on the child's mischievous behaviour. In addition, during the fire experiment, the mothers of the 'non-play' group provided clearer explanations for the task, explanations that better presented the appropriate uses and misuses under supervision indicated that they spent time in training their children about fire skills and also allowed their children to participate in the use of fire in the house. These parents took the time to clarify to their children what is an appropriate or inappropriate behaviour. These findings support Block's conclusion (1971, p. 263) that 'in order for the child to become appropriately controlled someone has to invest time and trouble'. In analysing data from the Berkeley longitudinal studies Block (1971) surveyed the development of ego-control through life and concluded that the adult undercontrolling male was neglected by both parents as a child. His parents 'did not invest the time, exhibit the constraints needed to deliver the precepts of self-regulation to their child' (p. 264). His mother was found to be neurotic and self-indulging rather than selflessly maternal, and his father was a detracted, indifferent, self-absorbing man. Block (1971 p. 262) also noted that because of the essential selfishness of the parents in pursuing their own lives, little effort was expended on socializing the boy via verbal, rational means, or on showing him skills or on encouraging his interest, or on simply being present. Though all the above components were not studied here, the present findings are in the same spirit as Block's conclusions and suggest future research questions concerning parent–child interaction for both children who are and who are not involved in fire-play.

Parental warmth and their quality of intervention efforts in 'setting limits' seem to be highly crucial for 'rascal' boys. These boys who express a high level of vigour, physical activity and risk-taking behaviours often exceed the parents' tolerance for impulse expression. Hence, the parents of 'rascals' need to exert more effort in their interpersonal relationships with their sons, and attempt to channel the energy of their children into constructive impulse expression. This is not an easy task, and many parents fail to perform it due to the high level of demand placed on them. The parents have to accept the task on a moment-to-moment basis, to provide their children with skill training and development of their ability through a supportive and structured environment both inside and outside the home, so that their energy will be directed toward constructive channnels. These channels should provide satiation for the children's activity and exploratory behaviour in a socially functional way without overinhibition of creativity and alertness.

The preceding discussion, scanty and incomplete as it is, provides several suggestions concerning the effects of parent–child interaction on the child's behaviour and mainly on his ego-control which affects his exposure to hazard situations. The features of the child's environment and especially his family background should be considered in the light of the above. It was found that match-players came from more deprived families relative to their non-peers. This relative deprivation was reflected by parents' marital status, education, occupation and residential area as well as in terms of family interpersonal relationships. Though it is impossible to relate all these sociological variables to the psychological mechanisms presented above, several general speculations may be offered. The relative unavailability of the two biological parents found for the 'play' group may account for some of the problems in interpersonal relations and lack of stability in the house and absence of tolerance for the child's impulse expression. Several mothers of the 'play' group were young single women who struggled for financial survival and for the enhancement of their own lives. They may lack child-rearing skills and are often helpless when they attempt to cope with the burdens of parenthood. As to the significant effect of education of the parents, there are some indications in the data that suggest that the more educated mothers perceive their sons as more competent, responsible, and intelligent implying that they may have a more rational and problem-solving approach to child-rearing. Family stability and high level of parental occupation and education may also provide the child with positive identification figures who give him support and reinforcement. It is obvious that an intact family with highly educated, professionally successful parents are not necessarily better performers of parental roles but the data indicate that their children have a higher chance of getting support and direction. A note of caution must be repeated here that these conclusions are speculative in nature and need further research before they are clarified and fully accepted. Future research effort should deal with the relations between sociological variables, and actual socialization procedures adopted by parents should clarify such general concepts as 'setting limits', 'restrictiveness', 'warmth', and tie them to actual parent–child interaction (for systematic implications for research in this area see Block, 1976).

IMPLICATIONS FOR FIRE PREVENTION

For many children fire is an object of fascination and attraction at a very early age, and the interest in it increases some time prior to the development of caution and competence. This finding is corroborated by results of previous research and suggests that intensive fire prevention efforts should be addressed to preschool children. Since these children are not exposed to organized school programmes, part of the prevention efforts should be undertaken by parents, babysitters, day-care workers, family counsellors and paediatricians. Educators and psychologists who deal with parents and their youngsters should be aware of fire prevention and should include it as an integral part of their work. Authors of child-rearing guides should be encouraged to include information about fire skills, attraction and preventive measures. Such efforts are especially imperative since fire interest and fire-play were found to be more prevalent among young boys than any statistics show.

Fire information gathered in this study indicated that almost all fires set by normal children in the age groups studied were the result of simply playing with matches, which sometimes results in a fire, mostly depending on chance conditions. Lack of consistent differences between the children who set fires and those who played with matches without setting fires also indicates the importance of playing with matches rather than firesetting as the focal variable for future research and prevention programmes.

Fire knowledge and preparation for hazard situations of children and parents indicated that many parents were not exposed to any systematic prevention programmes. Existing programmes should be available to a larger section of the population than they are at present. With the tremendous effort and resources that are being channelled in this direction, future development and application of fire prevention programmes hold the promise of reaching most members of the society, adults and future adults alike.

Fire attraction was found to be independent of fire competence and fire avoidance. This finding suggests the practical impossibility of solving the child fire problem by focusing on preventive measures while ignoring the curiosity and fascination attached to fire. As previously suggested, 'concentration on the prohibitions may well heighten interest and stimulate experimental play' (Block and Block, 1975, p. 118). Hence, prevention efforts should pay more attention to the satiation of fire interest by allowing the children to take part in a variety of fire situations which normally occur in the home, and by discussing with the children the details of potential hazards, procedures for prevention and actions needed to cope with hazards once they occur.

Children who played with matches were not found to be inferior to their non-playing peers in any of the cognitive or fire-knowledge variables. Most of the match-players were aware of the potential hazard involved in their behaviour and of the fact that this behaviour is forbidden. These results imply that prevention programmes should include training of actual behaviours in addition to transmission of cognitive verbal information. The behaviours should focus on appropriate handling of fire as well as on appropriate reactions to hazardous fire situations if they occur.

The lack of relations between the cognitive and behavioural fire variables also suggest that existing fire prevention programmes should incorporate behavioural measures in their assessment procedure. Such measures will provide information as to the effectiveness of educational programmes on actual decrease of the prevalence of firesetting.

Boys who were involved in fire-play were also found to be involved in accidents resulting from hazardous play. They were found to be more mischievous, disobedient, aggressive and impulsive. These boys were not only highly active and energetic but lacked impulse control in line with the 'rascality' behavioural pattern. Bearing in mind that most of these children *seek* activity and engage repeatedly in exploring their environment, prohibitive measures are not the surest route to fire prevention. The 'Don't play with matches' theme does not seem to be a plausible solution for these children, and may be replaced with the 'Use matches safely' theme. Children who develop interest in fire may be provided with controlled experiment-ation and direction about the safe use of fire within established constraints, and with information about appropriate reactions in times of hazard. This suggestion is not as revolutionary as it may seem to many fire-prevention specialists, parents and teachers. It has already been promoted by both researchers and fire-prevention personnel, it has been adopted by Montessori Educare Inc. in Boston, Massachusetts, and by 25 per cent of the teachers involved in the fire-prevention programme in the Riverside County Headstart Project (Folkman and Taylor, 1972). Recently the city of Hayward Fire Department won the grand award of the National Fire Protection Association for its Juvenile Fireplayers Programme which was judged to be the best of the nation's 'Learn not to burn' programmes. The programme was initiated with the purpose of channelling the fire-players' curiosity into interest in fire department and fire safety. Later, the programme changed its approach by adopting the 'Use matches safely' theme. According to this programme children are taught by the parents who are trained by fire specialists where, when and how to use matches safely. None of the 75 match-players who were trained in the programme were reported to have set fires later (Fire Inspector Ray Walters, personal communications). The present study also showed that children who had received supervised fire training were not prone to be involved in hazardous play. In spite of the anecdotal support for 'Use matches safely' as replacing the 'Don't play with matches' theme, it is not yet recommended as a common educational practice; the doubt stems from the lack of cause and effect data on the influence of fire training on fire behaviours and about the limitations that it may involve. An example for potential limitation was presented by Block and Block (1975, p. 108): 'Children may construe parental permission to light a fire in the presence of a parent as a blanket permission to use matches outside the context of parental supervision'. The ongoing survival skills acquisition study tests the effects of the two themes more rigorously in an experimental-control group design with the hope that its results will shed more light on the intricacies of the fire prevention problem and will enhance improvement of prevention programmes. Researchers should be encouraged to explore the conditions under which different training methods are effective in fire prevention.

Match players were found to be frequently embedded in deprived family backgrounds in terms of parental education, occupation, residential area and family relationships. It is hence suggested that special effort should be directed to these children. This finding and the derived implication were suggested by previous studies of firesetters; they are not, however, compatible with the study of normal children reported by Block, Block, and Folkman (1976). It is noted that their sample was skewed toward the upper-middle class, a fact which could account for the discrepancy in the findings.

Playing with matches was not found to be an isolated problem, but rather one which is related to other problem areas in the child's life. Hence it is recommended that future fire prevention efforts be part of a more general context which is designed to channel the children's energy into constructive directions in socially functional ways. This could be done by structuring the children's environment and by providing them with opportunities for constructive impulse expression via physical activities, exploratory plays, and development of hobbies. This kind of impulse expression combined with training the children in other survival skills will hope-fully enhance their development into becoming active and creative members of society.

CONCLUSIONS

The above implications can be briefly summarized in concrete terms.

(1) Increase the availability of fire prevention training to more adults and future adults in our society.

(2) Develop guidance for parents and specialists who deal with young children.

(3) Train children in actual preventive behaviours in addition to teaching know-ledge.

(4) Allow children constructive channels for expressing their interest in fire.

(5) Study the effects of existing preventive programmes on actual fire behaviours including fire-skills and fire-play.

(6) Compare the effects of the 'Don't play with matches' theme with the 'Use matches safely' theme in terms of fire knowledge, skills and play.

(7) Focus on fire-play rather than on firesetting as the major variable for future research and prevention efforts.

(8) Channel the energy of the 'rascals' to other constructive directions not neces-sarily involving fire, but involving physical activity, exploratory plays, and de-velopment of hobbies.

(9) Direct intensive prevention efforts to children who are embedded in deprived family backgrounds.

(10) Study fire variables and the effectiveness of prevention approaches in dif-ferent populations and social settings and relate them to the general context of socialization research.

(11) Study the effects of prevention programmes on behavioural as well as cog-nitive indices.

It is realised that the study opened more questions than it has answered. However, it is a small step in the investigation of the complex issue of fire-related behaviour in normal children. It is hoped that future cooperation among fire specialists, edu-cators, therapists, teachers, parents and researchers will increase our understanding of the problem, will decrease the hazardous and agonizing use of fire and will en-hance its proper use for warmth and pleasure.

REFERENCES

Bandura, A. and Walters, R. H. (1959). *Adolescent Aggression: A Study of the In-fluence of Child-training Practices and Family Interrelationships.* (New York: The Ronald Press Co.).

Baumrind, D. (1973). 'The development of instrumental competence through social-ization'. In Anne D. Pick (ed.), *Minnesota Symposia on Child Psychology, Vol. 7.* (Minneapolis: University of Minnesota Press).

Block, J. (1971) *Lives Through Time.* (Berkeley: Bancroft Books).

Block, J. H. (1976). *Familial and Environmental Factors Associated with the Development of Affective Disorders in Young Children.* Paper presented at the National Institute of Mental Health Conference on Affective States in Early Childhood, Washington, DC, November.

Block, J. H. and Block, J. (1975) *Fire and Young Children: A Study of Attitudes, Behaviours, and Maternal Teaching Strategies.* Technical Report for Pacific Southwest Forest and Range.

Block, J. H., Block, J., and Folkman, W. S. (1976) *Fire and Children: Learning Survival Skills.* USDA Forest Service Research Paper PSW-119,.

Coleman, J. C. (1972). *Abnormal Psychology and Modern Life.* (Glenview: Scott, Foresman and Co.).

Conger, J. J. and Miller, W. C. (1966) *Personality, Social Class, and Delinquency.* (New York: John Wiley and Sons).

Folkman, W. S. and Taylor J. (1972). *Fire Prevention in California's Riverside County Headstart Project. . . an evaluation.* USDA Forest Service Research Paper PSW-79.

Globerson, S. and Grossman, E. R. F. W. (1971). *Berkeley Fire Incident Survey—Initial Results.* Fire Research Group, University of California,Berkeley. Report No. UCB FRG 74-2, August.

Glueck, S. and Glueck, E. (1952). *Delinquents in the Making: Paths to Prevention.* (New York: Harper and Brothers).

Hellman, D. S. and Blackman, N. (1966). 'Enuresis, firesetting and cruelty to animals: a triad predictive of adult crime', *American Journal of Psychiatry,* **122,** 1431–1435.

Kafry, D. (1978). *Fire Survival Skills: Who Plays with Matches?.* Technical Report for Pacific Southwest Forest and Range Forest Service, Experiment Station, United States Department of Agriculture.

Kaufman, I., Heims, L. W., and Reisner, D. E. (1961). 'A re-evaluation of the psychodynamics of fire-setting', *American Journal of Orthopsychiatry,* **31,** 123–137.

Lewis, D. C. and Yarnell, H. (1951). 'Pathological firesetting (pyromania)', New York: *Nervous and Mental Disease Monographs, No. 82.*

Manheimer, D. I. and Mellinger, G. D. (1967). 'Personality characteristics of the child accident repeater', *Child Development,* **38,** 491–513.

Nurcombe, B. (1964). 'Children who set fires', *Medical Journal of Australia,* **1,** 579–584.

Safer, D. J. and Allen, R. P. (1976). *Hyperactive Children: Diagnosis and Management.* (Baltimore: University Park Press).

Scott, D. (1979). *The Psychology of Fire.* (New York: Charles Scribner's Sons).

Siegelman, E. Y. (1969). *Children Who Set Fires: An Exploratory Study.* Conducted for the Resources Agency of California Department of Conservation, Division of Forestry.

Siegelman, E. Y. and Folkman, W. S. (1971). *Youthful Firesetters: An Exploratory Study in Personality and Background.* USDA Forest Service Research Note PSW-230.

Wilmore, D. W. and Pruitt, B. A. (1972). 'Toward taking the fat out of the fire', *Medical World News,* **October,** 76.

Winget, C. N., and Whitman, R. M. (1973). 'Coping with problems: Attitudes toward children who set fires', *American Journal of Psychiatry,* **130,** 442–445.

Yarnell, H. (1940). 'Firesetting in children', *American Journal of Orthopsychiatry,* **10,** 272–286.

Fires and Human Behaviour
Edited by D. Canter
© 1980 John Wiley & Sons Ltd.

CHAPTER 5

The Concept of 'Panic'

JONATHAN D. SIME
Department of Psychology, University of Surrey

The concept of panic is a powerful one. It figures both explicitly and implicitly in fire regulations, the organization of building evacuations, the conversations of fire officials and press reports of fires. Deaths in major fire disasters are often attributed to panic. For example, British newspaper reports on the Beverly Hills Supper Club fire in Kentucky, USA. on 28 May 1977, included the following headlines: 'Panic Kills 300' (*The Sun*); 'Panic and 300 Stampede to Death' (*Daily Mail*); 'A Killer called Panic' (*Daily Express*). The existing literature on the subject reflects an international concern with panic in fires (see for example, Anon, 1971; Barlay, 1972; Chandessais, 1969; Coscarino, 1971; Haure, 1969; and Luciani, 1974). The present chapter is concerned with the way in which the concept of 'panic in fires' is used in the newspapers, building regulations and academic literature.

The concept of panic is used in a variety of contexts. Its *retrospective* use is reflected in newspapers and accounts people give of a fire which has occurred. During an actual event panic may be used to refer to someone else's behaviour as it happens. An *anticipatory* use of the concept is reflected in an individual's anticipation of his own anxiety and behaviour, expectations about what other people might do in a fire, and fire safety precautions or building regulations in general which are directed towards preventing panic. Standard fire instructions often have 'DO NOT PANIC' in block letters. Although the concept of panic figures so prominently in people's commentaries on fires, there has been no systematic appraisal of the way the concept is used. It will be argued here that the use of the concept has actually delayed systematic research of people's behaviour in fires. Recent research indicates that people's responses in a fire can best be summarized by a detailed description of the sequence of behaviour of individuals and their understanding of the fire situation at different stages (see Canter *et al.*, Chapter 8 of this volume). Preconceptions about 'irrational' behaviour in fires have in the past precluded any proper consideration of the understanding of a fire by the people to whom panic has been attributed.

The use of the concept is not of course limited to fires. It has a long history. The word 'panic' itself has its origins in the mythical figure of Pan in Ancient Greece. Panic is ascribed in legend to the 'contagious emotion' instilled by Pan in the Persian army who outnumbered their Greek enemies in the Battle of Marathon. The use of the concept serves as a precursor to an extensive literature on panic in a military context (see for example, Strauss, 1944). Social scientists have more recently been unable to find evidence for the widespread panic associated with disasters such as earthquakes, explosions and bombings (Baker and Chapman, 1962; Janis, 1954, Janis *et al.*, 1955; and Quarantelli, 1954, 1957, 1960). Indeed, Quarantelli (1973)has suggested that research should be carried out into the reasons why planners and designers continue to believe 'myths' about panic. Before one can consider his suggestion in relation to fires, it is necessary to examine the beliefs about people's behaviour in fires which underlie the concept, and if there are weaknesses why this is so. Young's (1946) definition of a social myth as 'an imaginary interpretation of a past, present or future event' is worth bearing in mind in the following discussion. For it is argued that it is the interpretation placed on people's behaviour in a fire which underlies many misconceptions about 'panic'.

USE OF THE CONCEPT IN THE NEWSPAPERS AND RECENT STUDIES

It is assumed that newspapers both reflect and influence current attitudes towards fires and are therefore an important source of information on the way the concept is used. In this section recent newspaper reports and case studies of fires are examined in terms of logical weaknesses in their use of the concept.

Reports of panic and fire tragedies

The use of the concept is best illustrated by some specific examples from newspapers and recent case studies of fires. Figure 5.1 (taken from a German newspaper) illustrates an incident which occurred on 12 April 1973 at the Löwenbräukeller in Munich. This was not a fire, but the report demonstrates how the outcome of pressure at an exit is often taken as primary evidence for panic. In this case approximately 3000 teenagers were attending an evening's entertainment in a hall filled far beyond the recommended capacity. As they made their way out two girls were crushed to death and a number of others injured in pressure created at the main exit. Under such headlines as *'Panik unter Teenagern'* (Panic Amongst Teenagers) and *'Mädchen zu Tode getrampelt'* (Girls Trampled to Death) (*Abendzeitung*, 13 April 1973), the newspapers attributed the behaviour to panic. The principal reason cited was the fact that the other eight exits were hardly used. No objective threat to the safety of people besides crowd pressure was apparent. The use of the term panic in this context may have seemed particularly appropriate. The lack of use of the exits available appeared to be 'irrational'. No account was taken in these reports, however, of the functional consequences of people leaving at the same time, of their unfamiliarity with the other exits and the fact that those at the back of the crowd leaving may have been unaware of the pressure at the front. In evaluating behaviour after the event one cannot assume without further evidence that the crowd acted according to a corporate irrationality of its own. 'Panic' in this sense is unlikely to have been the sole determinant cause of the deaths and injuries.

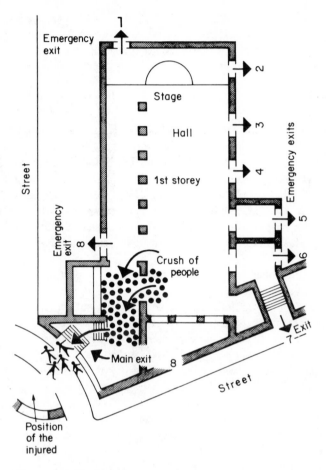

Figure 5.1 Plan of the Löwenbräukeller, Munich (from the *Süddeutsche Zeitung,* 14/15 April 1973)

A circular emphasis on the *outcome* of a fire as evidence for panic behaviour is evident in many newspaper reports. Any research which begins with an *a priori* assumption that panic was a cause of people being killed and injured in a particular instance is also invalid. Stevens (1965) attributes the mass panic cited as a primary cause of death in the Church Oyster Roast Fire, Arundel Park Hall in 1956, to the smoke, the fire itself, the failure of the lighting system, congestion and alcohol. No attempt has been made, however, to examine the patterns of behaviour which made it clearly distinguishable from flight behaviour in highly constrained circumstances. In this fire roughly 1200 people were present when fire broke out at the corner of the building where there was an outdoor roasting fire; 11 people died and 250 were injured. Bryan (1956) has examined patterns of escape behaviour (that is, exits by which people left) in this fire on the basis of police interviews of 61 self-selected respondents involved. He cites the pattern of use of exits and windows by these respondents as confirmation of the panic. Since there is no systematic analysis of the accounts of the fire by individuals assumed to have panicked and of the reasons why they used certain exits, or indeed of the full sequences of individual acts, the study reflects the prior assumption that panic led to the behaviour without examining this directly.

A further example of the way in which the concept of panic is attributed to people, on the basis of the outcome of a large scale fire tragedy, is provided by a comparison of the extensive reports on the Beverly Hills Supper Club fire, Kentucky, USA (28 May 1977) and the corresponding newspaper coverage of the fire in Britain. The reports by Kentucky State (1977) and The National Fire Protection Association (1977) are based on a data source of 630 interviews and questionnaires filled out by 1117 other people. An estimated 2400 to 2800 people were involved in the fire in which 164 died. The British newspapers, some of which were cited at the beginning of this chapter, both exaggerated the number of deaths and attributed them to panic. This is in direct contrast to the conclusions of the NFPA report (Best, 1977).

Besides the functional weaknesses of the building complex itself contributing to the fire spread, the deaths are attributed in the report to other human factors. The main ones were the delay in notifying people of the fire and the fact that the cabaret room (Figure 5.2) in which all except two of the people who eventually died were trapped, was filled far beyond the recommended capacity. An estimated 1350 people were present here as opposed to the recommended maximum occupancy of 536. It is considered that there was as much as twenty minutes' delay from the time heavy smoke was discovered in the room of fire origin to the time a public announcement was made in the cabaret room for people to leave. Directions to the exits were eventually given in an announcement from the stage. Only a few minutes later thick smoke entered the room. The cause of death was smoke and carbonmonoxide inhalation. The lack of panic, at least prior to the smoke, is attributed in the report (Best, 1977) to insufficient appreciation of the seriousness of the emergency and acceptance by the staff of their responsible role in directing people to the exits. Evacuation prior to the appearance of smoke was orderly. It is concluded (p. 173) that 'panic is not considered a major contributing factor to the large loss of life, but such behaviour probably did occur when people knew they could not escape'. While the evidence for panic occuring after people knew they could not escape is inconclusive, the fact is clear that there was *no* panic while there was reasonable access to the exits.

Fires in which panic is attributed to people by the newspapers are characterized by a rapid fire spread following a delay in people becoming aware of the fire's existence. Another example of a fire of this kind is provided by newspaper reports of the hotel fire in Boras, Sweden (June 1978) in which twenty people died (see *Dagens Nyheter* 1978). It is estimated in the newspaper coverage that 150 people were present in a building of five storeys; many of them belonged to a party of young people aged eighteen to twenty years who were celebrating the end of their exams. At 2.30 am when most people appear to have become aware of the fire, there was a large number dancing in a ballroom on the second floor. The routes downstairs, via a

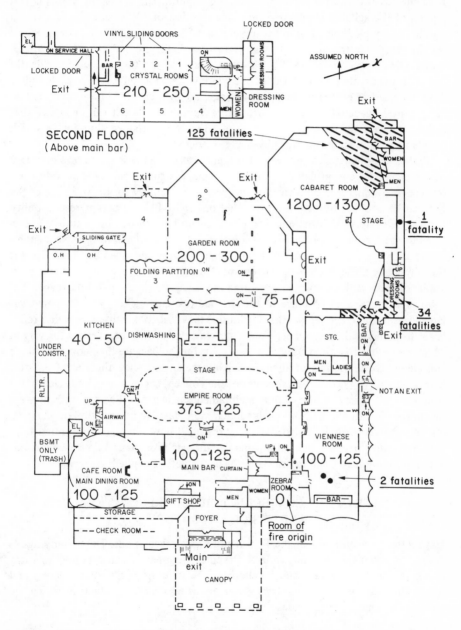

Figure 5.2 Plan of the Beverly Hills Supper Club, Kentucky: estimated numbers of people in each area at time of fire, and location of fatalities ▨▨ (from Best, 1977)

main staircase and escape exits, were blocked by the encroaching fire. The main criteria cited as evidence for panic, precipitating inappropriate behaviour, were indirect reports of the agitated states of people, the position of bodies and the fact that a number of people jumped from the upper floors. 'In their panic several young people sought refuge in the toilets. It was there that eight charred corpses were later found' (*Die Welt Zeitung*, 1978). 'However, many of the trapped people evidently lost their heads: they jumped out of the windows immediately. It is possible that if they had tried to escape to the upper floors, some of them could have been rescued by the fire brigade' (*Frankfurter Allgemeine Zeitung*, 1978). It can be argued that the deaths were an inevitable outcome of a highly constrained situation in which people's access to an escape was physically restricted.

Rather than being clear evidence for panic causing a fire tragedy, or a summary of the behaviour encapsulated in the term panic, newspaper accounts provide a retrospective commentary on a fire. In this way responsibility for the outcome of the fire can be avoided by people who would otherwise have to accept responsibility for people's safety. Veltfort and Lee (1943), who have carried out a study of reactions to the Cocoanut Grove Night Club fire in 1942 in which 488 people died, indicate that the building was filled far beyond its recommended capacity. Thus it was physically difficult for people to escape. The two researchers suggest that certain public officials and the owners were used as scapegoats (in other words, blamed) in subsequent newspaper coverage of the fire and correspondence, whereas panic was largely responsible for the deaths. Curiously, they fail to see that the earlier references to panic in the newspapers and their own use of the term reflect the way in which panic itself is often used as a scapegoat. Indeed because the people killed are considered to be victims of their own panic behaviour they themselves are also scapegoats. At worst this *evaluative* use of the concept serves as a 'let-out clause' for situations in which, although it is maintained sufficient safety precautions were taken, people still die. Without having much more detailed information on the pattern of behaviour in relation to the fire spread and how survivors interpreted their own and other people's behaviour, one cannot establish that panic was the major determinant of deaths and injuries in a fire as is so often implied.

USE OF THE CONCEPT IN BUILDING REGULATIONS AND DESIGN LITERATURE

This section concentrates on the British and French regulations. The underlying assumptions about people's behaviour in fires are often made clearest in the literature also considered here which seeks to clarify the regulations. While the regulatory and design literature in other countries is not discussed here the manner in which the concept of panic is used is widespread.

Emphasis on panic and limited research

The emphasis in newspapers on panic being the major psychological factor contributing to large-scale fire disasters is paralleled in building legislation. In Britain, the

Home Office (1934) Manual of Safety Regulations provides numerous examples of fire disasters in theatres in which panic is assumed to have been a primary cause of deaths. In France, the dangers of panic occurring in public buildings such as department stores, multistorey hospitals and cinemas are stressed. Recommendations are made to try and minimize the likelihood of its occurrence and its effects (Haure, 1968, 1969; Haure and Dussy, 1968; and Leloup 1972). Before outlining the assumptions about panic which exist, it should be made clear that references to panic in the regulations and associated literature are not based on systematic research of people's behaviour in fires. A tendency to associate panic with escape behaviour has generally ruled out attempts to examine directly people's experiences of coping in a fire situation. Phillips (1951) in his authoritative guide to the British set of regulations, *The Fire Grading of Buildings: Means of Escape, part 3*: Personal Safety (Ministry of Works, 1952) suggests the impossibility of research of this kind. He writes (p. 23); 'Mental vagaries are not capable of precise measurement and we must use much imagination in making provision for this abstract consideration'. He also adds (Phillips, 1950, p. 639): 'It will always remain essential to couple an extensive knowledge of building design and construction with a lively imagination of the effect of fire'.

Phillips is not simply suggesting the unreliability of considering or trying to measure people's experiences of a fire. He uses the term panic synonymously with the psychological reactions of people to a fire threat (hence the 'mental vagaries' referred to). Unfortunately, he fails to see that the imagination he applies through the use of the concept is a poor replacement for psychological research. The assumption that it is self-evident that panic occurs in fires in certain situations means that 'panic' is used to justify aspects of the design of a building, when there is no reliable evidence for panic consistently occuring in the way suggested.

For example, it is claimed in the Ministry of Works (1952) *Fire Grading of Buildings, part 3*, that in theatres, 'if the auditorium is cleared in 2½ minutes there will be no serious risk of panic in the event of a fire' (paragraph 243). Canter and Matthews (1976) point out that this figure of two-and-a-half minutes is an inadequate basis for exit-width requirements. The assumption that 'panic' necessarily occurs after a prescribed period of time, or when a crowd reaches a certain density (2.5 sq. ft per person) followed by immobility according to Haessler (1977), is questionable without further evidence. While there are numerous references to the fact that a cry of fire should be avoided since it would automatically provoke a panic (Home Office, *Manual of Safety Requirements*, 1934, p. 83; Leloup, 1972), there is no reliable evidence for this. If there is an immediate response it is undoubtedly linked to other constraints and factors in the situation. Despite assumptions that a fire alarm can provoke a panic (Ministry of Works, 1952: *Fire Grading of Buildings, part 3* paragraph 165), research shows a marked reluctance to respond immediately to an alarm without additional information (Hudson, McDavid, and Roco, 1954; Mack and Baker, 1961).

Panic is assumed in the *Fire Grading of Buildings* (Ministry of Works, 1952), to occur when there are excessive distances to be travelled, limited access to exits, number and size of exits, widely spaced and inconveniently spaced stairs and unfamiliarity with a building. It is also associated with situations in which large numbers of people are in a confined space, or when particular types of people such as children are present (Haure and Dussy, 1968; Haure, 1969, Home Office, *Manual of Safety Requirements,* 1934). Actual changes in the immediate social or physical environment assumed to 'precipitate' panic include a scuffle, smell of smoke, smoke itself, absence of light, a cry of alarm or 'fire', a sudden fire alarm particularly when people are asleep or in a crowded area of a building (see Haure, 1968; Home Office, 1934; Leloup, 1972; Ministry of Works, 1952). There is considerable overlap here in the British and French regulatory literature.

However, not only is there insufficient evidence to support the assumption that 'panic' occurs as an automatic response to these factors. Recommendations are directed towards dealing with 'panic' rather than more important psychological problems involved in a fire situation. Before considering the actual problems individuals face in a fire, some of the recommendations made in the regulations are briefly outlined.

Solutions to panic suggested in the design and regulatory literature

Certain forms of fire safety facilities have been recommended in the building regulations and design guidance because it is assumed they will help to avoid or minimize the effects of people 'panicking'. Physical design solutions include: adequate number of exits, anti-panic bolts, proper and conspicuous lighting and signing systems (Ministry of Works, 1952). In commentaries on the French fire regulations, it is suggested that adequate ventilation should be provided in department stores to clear smoke 'which might cause a panic' (Haure, 1968). For the same reason adequate safety lighting in public buildings is also recommended (Haure, 1968). Britton (1972) writes that the physical design 'will exert a strong psychological influence against panic'. A guidance system of phosphorescent lights has been suggested as the key to reduction of the possibility of panic disaster by Storey (1973), as have exits of a suitable width (Peschl, 1971).

The characteristic strategies adopted for dealing with 'panic' are ones in which the amount of information available to people (for example in a public building) is minimized and an attempt is made to control people's movements (that is, an organized evacuation) (Chandessais, 1971b; Taylor, 1965). This is reflected in recommendations that normal activities should be continued as long as possible in areas not affected by panic (Haure and Dussy, 1968). In the Home Office (1934) *Manual of Safety Measurements*; it is recommended that in places of entertainment 'the orchestra will if possible continue to play' (p. 83), that telephones should be situated so that the public cannot overhear a call to the fire brigade and be alarmed: 'this might cause panic' (p. 82). The dual-alarm system operating in department stores in Britain, in which an alarm first sounds in the staff area, is another example of a provision based on similar assumptions that information about a fire should be restricted if panic is to be avoided (Strother-Smith, 1976).

Underlying assumptions about panic: irrational and uncontrollable behaviour

The notion of panic which underlies the use of the concept in the regulations is that of 'irrational' behaviour. Since the regulations in Britain are considered to be a 'rational solution' (Ministry of Works, 1952 paragraph 248 of the *Fire Grading of Buildings*), lack of use of facilities tends to be interpreted as 'irrational' panic behaviour. In the French literature Haure (1968) suggests that 'the human element makes it difficult to forecast the outcome of a fire with certainty and to ensure total protection from the provisions'.

Not only is it implied that panic is impossible to control (*'le facteur humain incontrôlable'* according to Haure and Dussy, 1968), but that it occurs even when the objective threat from the fire is limited: Ministry of Works (1952) *Fire Grading of Buildings*, paragraph 100: 'a smoke filled atmosphere can cause panic when there is comparatively little danger from fire spread'.

> Panic in an assembly audience results in a crowd jamming the exits and causing injuries quite apart from injury by fire. In the type of building now being considered, individuals as well as groups may become panic-stricken. Lives may be lost, for example through fear of using staircases in which there is some smoke but which would actually give safe passage out of the building (paragraph 164).

The notion that panic can be even more of a problem than the fire itself is extended by Phillips (1951) in his guide to these regulations. He writes (p. 23): 'Panic has been the cause of more loss of life than burning by fire or suffocation by smoke . . . . Panic may be aroused when there is not the least danger from fire and in an undisciplined rush to escape many may suffer serious injury or death.'

A similar account of people's behaviour appears in Barlay (1972): 'Fire experience all too clearly shows that the fear of being burned to death or suffocated, the rush of smoke and hot gases and the sight of spreading flames completely alter the pattern of logical human behaviour. Under such conditions people do not behave like thinking human beings.'

Numerous references have and are still being made in authoritative sources to a primitive 'involuntary' instinct for survival or flight which is assumed to occur in fires when 'panic' overcomes 'man's capacity for rational thinking' (Haessler, 1977, p.36; Marchant, 1972; Phillips, 1951, p. 23; Strother-Smith, 1976, p. 52). The remainder of the chapter examines the validity of this common assumption about irrational behaviour in fires, for it underlies the use of the concept.

PROBLEM IN THE USE OF THE CONCEPT: THEORETICAL CONSIDERATIONS

The use of non-human metaphors

References to 'blind', 'illogical' or 'irrational' behaviour are usually considered justified because panic is assumed to be a highly emotional automatic response to noxious stimuli such as smoke and flames. This is an over-simplification of human responses and the actual constraints of a fire situation. The tendency to imbue people with non-human qualities and emphasise the emotional aspects of their behaviour means that the limited knowledge people often have of the fire spread in its early stages, the layout of a public building, and the potential outcome of pursuing particular options, are all ignored.

Limiting the concept to crowd situations reflects the assumption that emotional or irrational behaviour is most likely in a crowd. It is a viewpoint which was adopted by crowd behaviour theorists in the earlier part of this century beginning with Le Bon (1895). It is less accepted nowadays in the psychological and sociological literature as an explanation of the range of behaviour occurring in crowds (Brown 1954, 1965; Milgram and Toch, 1969; Turner and Killian, 1957).

Unfortunately, the emphasis on the outcome of 'crowd' behaviour and the neglect of the individual's perspective in a fire has been extended in a whole range of animalistic and physical metaphors. People are often referred to in animalistic terms, for example, Britton (1972): 'A pack of animals obsessed by a frenzied desire to escape they know not what.' Or more graphically by Hirschfield (1962) 'Many of them had heel prints ground into their faces and the flesh was trodden from their bones as if a herd of cattle had trampled over them.'

The use of disease metaphors appears in references to panic as a contagion (Le Bon, for example), to the therapeutic measures required to deal with people who are described as 'panic germs' by Chandessais (1971b), and the notion that 'adequate exits' are an *antidote* for panic' (Britton, 1972). In the design literature, people who 'are panicking' are often equated with mechanized movement, for example, 'a granular mass' (Peschl, 1973) or the 'flow of corks in a channel of water' (Phillips, 1951). It is assumed that this is justified since people are 'not in full possession of their mental and physical capacity' (Peschl, 1971). However, references to instinctive self-destructive actions, and by analogy the non-human components of behaviour, mean that panic is a concept which can often only be 'inferred' from behaviour and the outcome of a fire.

Panic and flight behaviour

An important area of confusion in the use of the concept is the way in which panic is mistaken for any form of flight behaviour. As we have seen, panic is commonly associated with a fear response resulting in flight (Foreman, 1953). This behaviour is assumed to be no longer under cognitive direction and control (Vernon, 1969). The notion that a flight response is disorganized, ruled solely by the emotional state of people and animalistic in nature is a doctrine that has been rooted in popular thought for a long time. Theories of human emotion in recent years have placed much less of an emphasis on a strict division between emotion and cognition (Lazarus, 1966; Miller, Galanter, and Pribram, 1960; Schachter and Singer, 1962). Leeper (1948) points out that 'emotion as a determinant of flight behaviour is organizing'. It is only disorganizing in the sense that it interrupts the normal activity which preceded the fire. Flight is not the normal way of leaving a building. Because of this, it looks much more disorganized to independent commentators on the fire or even individuals in the situation than in fact it is.

Quarantelli (1957) has made a distinction between 'adaptive flight' (withdrawal behaviour) and panic or 'non-rational flight'. Non-rational flight is distinguished from rational flight by the fact that 'alternative courses of action are not weighed up'. Yet, as Quarantelli points out, even when flight is non-rational this is not the same as irrational behaviour. 'Panic' flight, he argues, is not necessarily non-functional or maladaptive, for flight appears appropriate to the individual as he perceives and defines the situation. It may be that one should not equate the concept at all with a particular form of flight behaviour unless it can be readily distinguished as such.

Mintz (1951) in a psychological experiment often cited in the academic literature on panic has demonstrated how behaviour which is rational in the individual's own terms may be non-adaptive for a group of people. He devised an experiment considered analagous to a theatre fire situation in which cooperation rather than competitive behaviour is necessary if people are to escape through an exit limited in size. As a measure of 'non-adaptive behaviour' he used the average time it took for a group of people to remove cones from a glass bottle before they got wet. Water was fed into the bottom of the bottle. He found that in a cooperative group (no reward or fine to the individual) as opposed to a competitive group, 'no serious jam occurred at the neck of the bottle'.

Although his experiment can be criticized for its oversimplification of a fire, it does demonstrate weaknesses in attributing panic in the sense of 'irrational' behaviour to everyone involved in a crowd situation. Other experiments have attempted to extend Mintz's experiment by incorporating a clearer element of danger or threat using electric shocks (Guten and Vernon, 1972; Kelly *et al.*, 1965; Schultz, 1966, 1967). Unfortunately, they fail to recognize a more fundamental implication of the original experiment—the way in which the concept of panic is imposed as an interpretation of a pattern of behaviour. In the subsequent experiments it is ironically the experimenters themselves who use the concept of panic in this way since there is limited consideration of the perspective of the individuals taking part.

In cases where people in flight to an exit appear to be acting in a way which under normal circumstances would be considered anti-social, (pushing, shoving or trampling, for instance), it is often assumed that this is because they are behaving irrationally'. The behaviour may appear to be both self-destructive and destructive towards other people. However, it is misleading to assume this was irrational behaviour as opposed to a pattern of flight behaviour under highly constrained circumstances.

If the behaviour of other people in a flight situation conflicts with the individual (that is, reduces his chances of escape) he too, may consider them to be panicking. Yet the individuals to whom he attributes panic may be as unlikely to consider that they, themselves, were behaving irrationally as he would himself. 'Optimum' behaviour in a crowd, in the sense of normal movement out of a building, is only possible under optimum circumstances, where it is clear that there is sufficient time for everyone to leave. This characterizes a building in normal use, not a situation in which there is a serious threat to people's safety. It is the disparity between the perspective of the person using the concept of panic to judge the appropriateness of behaviour, and the individuals to whom panic is attributed, which underlies the attribution of 'irrationality' to people. It is the fundamental problem in the way the concept is used. This disparity in perspectives is now considered further together with the criteria by which panic is commonly judged to have occurred.

External and internal criteria

While a person evaluating behaviour could be correct in his assessment, external and internal criteria for assessing the rationality of an act are often confused when the concept of panic is used. According to Turner and Killian (1957, p.17) behaviour can be called rational, based on external criteria, when it is an efficient way of achieving some goal. Using internal criteria behaviour is irrational when the individual does not weigh all possible alternatives of which he can be aware in deciding his course of action (similar to Quarantelli's notion of non-rational behaviour). It is essential when assessing people's behaviour in fires to distinguish between the options actually available *vis-a-vis* those perceived to be available at different points in time, the objective versus subjective degrees of freedom in a fire situation referred to by Breaux (1977).

The lack of use of exits which are available and competition for a single exit, are often cited as evidence for panic. Yet these are external criteria. Examination of the behaviour against internal criteria would make it less likley that the concept of panic would be used to summarize or explain this kind of crowd behaviour. As Turner and Killian (1957, p.10) write:

> When people, attempting to escape from a burning building pile up at a single exit, their behaviour appears highly irrational to someone who learns after the panic that other exits were available. To the actor in the situation who does not recognize the existence of these alternatives, attempting to fight his way to the only exit available may seem a very logical choice as opposed to burning to death.'

One reason panic may be attributed inaccurately to crowd flight is that attention is directed away from considering the individuals's perspective. The lack of attention to internal criteria is not limited to crowds, however. Considering any pattern of actions in a fire in isolation such as dressing, opening a door to a fire, returning into a building, jumping or flight to an exit, as evidence of 'panic' would only be justifiable if one could demonstrate that particular actions are reliable indicators of an extreme emotional state seriously limiting an individual's ability to cope. There is a tendency to associate with panic actions which appear in retrospect to have been inappropriate. Yet none of these actions, even jumping, are clear or consistently reliable evidence for panic. For example, if an individual exposed to a fire as it spreads into a room is faced with a choice between perishing in the flames or jumping from a window, the latter would be a rational choice. The individual's knowledge of what was happening has to be considered, as well as the actions he engaged in. Both must be related to the pattern of fire spread, other people's actions at different stages, and the individual's knowledge of the building layout itself.

Panic has been associated with perceptual theories of sensory deprivation or overload of environmental inputs (Abe, 1976; Levitt, 1968; Schultz, 1964). Most of this research has been carried out in the laboratory. While there is some evidence that perceptual narrowing or tunnel vision occurs in anxiety-provoking situations (Panse, 1971; Ross, 1974) it is difficult to distinguish 'panic' from other patterns of behaviour which might occur in such situations. A certain amount of anxiety in a fire is likely to precipitate action by people who could increase their chances of escape. While there have been a few attempts to measure 'non-adaptive' behaviour in the laboratory (Hamilton, 1911, 1916; Patrick, 1934, a and b) most laboratory research places too great an emphasis on the behaviour rather than the individual's experience of a situation. The use of animals in the kind of extreme stress situations in which panic is assumed to occur tends to perpetuate animalistic metaphors when one extrapolates to a real life fire situation. In the laboratory, as in the assessment of real fires, there has been a failure to recognize the way the concept is used. The validity of the individual's perspective is consistently ignored (see, for example, Cantril, 1940).

ALTERNATIVE APPROACH: CONSIDERATION OF THE SUBJECTIVE EXPERIENCE

It is the subjective response of an individual which should be the ultimate criteria for validating any statements about or measures of panic in fires. In recent research of fires (Canter *et al.*, Chapter 8 of this volume) the starting-point for both practical and theoretical reasons has been one in which it is argued that if people's behaviour is to be properly explained, it is necessary to obtain an 'account' by the individual of his experience, (see Brown and Sime, 1977; Harré and Secord, 1972). These accounts are used as the basis for a systematic analysis of the sequences of actions. Panic is often associated with situations in which people have died. Examination of their experience is automatically precluded. Unfortunately, accounts of survivors are rarely examined properly. There is little attention to the way misinterpretations can be made of other people's behaviour. The use of the concept of panic has replaced a more systematic psychological appraisal of the behaviour of people in fires.

In clinical psychology a pathological panic response is considered to have three components: subjective, physiological and behavioural (see Rachman, 1968). A pathological panic (for example, in the case of a person who suffers from claustophobia) implies an excessive fear reaction which is persistent and unrealistic in terms of the situation. The concept of panic in this context is used to denote the *subjective* component: 'an alarming feeling of intense fear, tension or full panic'. The motor response 'is usually one of flight but some patients become inert or frozen and feel too weak to move' (Rachman, 1968). Accounts of people's 'panic experiences' are often considered essential in a clinical context. Several points are worth considering in relation to the use of the concept.

While there is limited opportunity to examine physiological reactions in a fire, attention can be directed towards the subjective aspects of an individual's experience of a fire as in clinical psychology. Accounts of fires that have been collected by the author (see Chapter 8 of this volume) suggest that the concept of panic is used by individuals in relation to subjective anxieties occurring in a fire, as well as other people's behaviour. It is possible that the concept is more likely to be used synonymously with anxiety or fear in fires than it would in other situations. A greater understanding of the ways in which the concept is used by people with different degrees and kinds of involvement in fires, is essential if one is to distinguish between the internal and external criteria people use to evaluate certain behaviour and/or experience as 'panic'.

Comparison of panic in a fire and pathological panic remains at the moment a confused area of debate (Farber, 1976; Heiser and De Francisco, 1976). Unlike normal situations in which a pathological panic might occur, examination of panic in a fire is complicated by the fact that the external environment to which people are reacting normally constitutes an objective threat. This may be one reason why the amount of 'panic' occurring in fires seems to be greatly exaggerated. It is all the more reason to examine the retrospective use of the concept both by people referring to their own experience and commenting on other people's behaviour in a fire. The distortion between different people's perspectives is more likely to remain in the assessment of behaviour if the experience of the individual is ignored. Perhaps the most informative study would be one in which the validity of people's assumptions could be systematically examined by cross-referencing between accounts of the same fire (Brown and Sime, 1977). In this way, for example, the use of the concept by particular role groups, such as firemen, could be compared with the accounts of the experience by the people to whom panic was attributed in a particular instance.

CONCLUDING REMARKS

There is evidence that 'panic' in fires is much more intangible than is popularly assumed. Yet it is still confused with flight behaviour, even by researchers of fires who have suggested that panic in fires is 'rare' (Bryan, 1956, 1958, 1970; Wood, 1972). Contradictory arguments and invalid assumptions exist in references to panic in the newspapers, building regulations and academic literature. It has been argued that an important reason for this is the lack of consideration of the way in which the concept of panic in fires is used.

The use of the concept can now be characterised as follows: 'Panic' in a fire is a concept attributed in a retrospective, contemporary or anticipatory fashion by and to different role groups (such as firemen, staff, public, journalists) with differing degrees and kinds of involvement in a fire (as participant, observer or commentator). It is used as a description, explanation or evaluation of a state of anxiety, pattern of behaviour in a fire on the part of an individual, group or crowd. The anxiety or behaviour is assumed to be 'irrational' since its outcome is likely to be unfortunate (for example, jumping, blockage at exit, escape without family members).

The present problem is that 'panic' is frequently equated with *any* 'apparently' ineffective behaviour in a fire; whereas it is more important to consider under what circumstances behaviour in fires is likely to be ineffective than to immediately attribute the behaviour to panic (Sime, 1978). The use of the concept as a description, explanation or evaluation of behaviour reflects its 'utility' as an all-embracing term. But, as has been stressed, behaviour in a fire can only be properly judged against the individual's awareness of the options available at different stages of the fire, his familiarity with the building and general constraints of the fire situation. Unfortunately, the primary criterion adopted for panic or irrationality is the inappropriateness of behaviour as judged frome someone else's perspective.

Reference was made at the beginning of the chapter to a question posed by Quarantelli (1973), namely, why planners and designers continue to believe myths about panic. While it seemed premature to consider this directly without establishing what are the beliefs about panic, it is clear that many of the existing assumptions underlying the use of the concept are questionable. Perhaps one reason for the strong belief in panic apart from its historical origins, is its utility in justifying decisions and enabling people to come to terms with the unfortunate outcome of a fire. The concept is evidently useful in simplifying the human problems faced in a fire. It is also one way in which *responsibility* for deaths and injuries in a large-scale tragedy is minimized.

Unfortunately, the current emphasis on panic leads to strategies which can exacerbate the objective dangers of a fire and increase the constraints of the situation in which people are eventually required to act. As Williams (1964) notes, 'the sterotype image of panic as a reaction to disaster leads to delays in warnings and ambiguous messages'. The strategy of limiting the information available to people in a fire in the early stages is currently advocated in the building regulations and design literature. Yet this strategy can be self-defeating. The kind of flight behaviour described retrospectively as panic, may eventually be necessary if people are to have any chance of escape. People may be forced into a position where they are unable to act immediately or on their own initiative, because their awareness of a potential danger is delayed and the options available are thereby reduced until it is too late. This is what happened in the Beverly Hill Supper Club fire. A similar pattern of delay in notifying people of a fire characterizes the Summerland fire, Isle of Man, in 1973, in which 50 people died (Sime, 1979; Summerland Fire Commission 1974). In general, a retrospective allusion to the lack or misuse of safety facilities in a fire 'because people panicked', reflects the way in which the concept is used. It is less likely to provide an adequate explanation of people's behaviour in a fire.

REFERENCES

Abe, Kitao (1976). 'The behaviour of survivors and victims in a Japanese nightclub fire:a descriptive research note', *Mass Emergencies*, **1**, 119–124.

Abendzeitung (München) (1973). *'Panik unter Teenagern'* (transl. Panic Amongst Teenagers), *'Mädchen zu Tode getrampelt'* (transl. Girls Trampled to Death), **13 April**.

Anon (1971). 'Panic—det kan undvikas genom information' (in Swedish transl. Panic Can Be Avoided Through Proper Information), *Brandforsvar*, **8 (1)**, 27.

Baker, G. W. and Chapman, D. W. (1962). *Man and Society in Disaster*, (New York: Basic Books Inc.).

Barlay, S. (1972). *Fire, An International Report*. (London: Hamish Hamilton Ltd.).

Best, R. L. (1977). *Investigation Report: The Beverly Hills Supper Club Fire*,

Southgate, Kentucky, May 28th 1977, NFPA, (National Fire Protection Association) Fire Investigations Department (in cooperation with Nat. Fire Prevention and Control Admin. and NBS, National Bureau of Standards), draft report.

Breaux, J. J. (1977). *Analysing Complex Data: The Description and Analysis of Dynamic Behaviour in Fire Situations,* paper presented at the British Psychological Society's Annual Conference, Exeter, 31 March–4 April.

Britton, J. W. (1972). 'Adequate exits—antidote for panic', *Fire Engineering,* **115** (5), 386–388 and 418–419.

Brown, R. W. (1954). 'Mass phenomena'. In Lindzey, G. (ed.) *Handbook of Social Psychology Vol. 2, Special Fields and Applications,* Chapter 23, pp. 833–876.

Brown, R. W. (1965). *Social Psychology.* (New York: The Free Press).

Brown, J. M. and Sime, J. D. (1977). *Accounts as a General Methodology,* paper presented at the British Psychological Society's Annual Conference, Exeter, 31 March–4 April.

Bryan, J. L. (1956). *A Study of the Survivors' Reports on the Panic in the Fire at the Arundel Park Hall in Brooklyn, Maryland, on January 29th, 1956,* unpublished manuscript, University of Maryland.

Bryan, J. L. (1958). *Psychology of Panic,* 13th Annual Fire Department Instructors Conference, Memphis, Tennesse, 18–21 February.

Bryan, J. L. (1970). *Is Panic Inevitable in Department Store Fires?,* paper prepared for the International Study Conference on 'Fire Protection in Department Stores and Supermarkets; Men and their Behaviour', 2–3 April.

Canter, D. V. and Matthews, R. (1976). *Behaviour in Fires: The Possibilities for Research,* CP11/76, BRE Fire Research Station, Borehamwood.

Cantril, H. (1940). *Invasion from Mars: A Study on the Psychology of Panic,* (Princeton: Princeton University Press).

Chandessais, C. (1969). 'La panique dans les perspectives de la psychosociologie contemporaire', *Protection Civile et Sécurité Industrielle,* **172,** 5–24.

Chandessais, C. (1971a). 'La panique? Quelle importance?, *Allo,* **18, March,** 14–20.

Chandessais, C. (1971b). 'Conditions pratiques de lutte contre la panique', *Revue tech.,* **February, 12,** (110), 24–29; also 'Practical measures against panic', in *J. Brit. Fire Serv. Assoc.,* Ind. Fire Prot. Assoc., **1972, 2(1)** 264–267.

Coscarino, A. (1971). 'La sicurezza contro i rischi di incendio et di panico nelle costruzioni ospedaliere' (in Italian transl. Safety Against the Risks of Fire and Panic in Hospital Buildings), *Anticendio Prot. Ind.,* **23 (2),** 84–86.

Dagens Nyheter (Swedish newspaper) (1978). 'Brandoffren kunde räddats' **Monday 12 June,** 1, 9.

Die Welt Zeitung (1978). 'Sie stürzten wie lebendige Fackeln heraus' (transl. They Jumped out Like Living Torches), **Monday 12 June.**

Farber, I. J. (1976). 'Is panic normal?' (in the same issue: Drs. Heiser and DeFrancisco reply), *American Journal Psychiatry,* **134 (5), May,** 588.

Foreman, P. B. (1953). 'Panic theory', *Sociological and Social Research,* **37,** 295–304.

Frankfurter Allgemeine Zeitung (1978). 'Zwanzig Todesopfer bei Feuer in Hotel' (transl. Twenty Fatalities in a Hotel Fire), **Monday 12 June,** 7–8.

Grosser, G. M., Wechsler, H. and Greenblatt, M. (eds.) (1964). *The Threat of Impending Disaster* Cambridge: MIT Press.

Guten, S. and Vernon, L. A. (1972). 'Likelihood of escape, likelihood of danger and panic behaviour', *Social Psychology,* **87,** 29–36.

Haessler, W. M. (1977). 'Many factors influence design of emergency exit requirements', *Fire Engineering,* **September,** 36–39.

Hamilton, G. V. (1911). 'A study of trial and error reactions in mammals', *Journal of Animal Behaviour,* **1,** 33–36.

Hamilton, G. V. (1916). 'A study of perseverence reactions in primates and rodents', *Behaviour Monographs,* **3,** Serial No. 13.

Harré, R. and Secord, P. F. (1972). *The Explanation of Social Behaviour,* (Oxford: Basil Blackwell).

Haure, E. (1968). *Règlement francaise relative à la prévention contre les risques d'incendie et de panique dans les grands magasins de vente,* (Ruschlikon–Zurich: Switzerland Gottlieb Duttweiler Institute for Economic and Social Studies), GD1-topics No. 29, Ruschlikon-Zurich.

Haure, E. (1969). 'Le règlement de securité contre les risques d'incendie et de panique dans les établissements recevant du public', *Revue Tech.,* **February, 10 (89),** 38–41.

Haure, E. and Dussy, M. (1968). 'La securité contre les risques d'incendie et de panique (dans les immeubles sanitaires de grande hauteur)' *Revue Tech.,* **February 9 (81),** 37–46.

Heiser, J. F. and De Francisco, D. (1976). 'The treatment of pathological panic states with propanolol', *American Journal Psychiatry,* **133 (12), December,** 1389–1394.

Hirschfield, S. (1962). 'Panic—the indiscriminate killer', *Fire Engineering,* **115 (8).**

Home Office (1934). *Manual of Safety Requirements in Theatres and Other Places of Public Entertainment,* (London:HMSO).

Hudson, B., McDavid, J., and Roco. M. (1954). 'Response to the perception of threat', in Symposium on Human Behaviour in Natural Disasters, *American Psychologist,* **9,** 503.

Janis, I. L. (1954). 'Problems of theory in the analysis of stress behaviour', *Journal of Social Issues,* **10,** 12–25.

Janis, I. L., Chapman, D. W., Gillin, J. P. and Spiegel, J. P. (1955). 'The problem of panic', (National Research Council Statement on behalf of Office of Civil Defense), repeated in Schultz, D. P. (ed.) (1964). *Panic Behaviour: Discussion and Readings* (New York: Random House), pp. 118–123.

Kelly, H. H., Cowdry, J. C., Dahlke, A. E., and Hill, A. H. (1965). 'Collective behaviour in a simulated panic situation', *Journal Experiential Social Psychology,* **1,** 20–54.

Kentucky State (1977). 'Beverly Hills Supper Club Fire, May 28th, 1977. Investigative Report to the Governor', **September.**

Lazarus, R. (1966). *Psychological Stress and the Coping Process,* (New York: McGraw-Hill).

Le Bon, G. (1960, 1st ed. 1895). *Psychologie des Foules* (transl. The Crowd), (New York: Viking).

Leeper, R. W. (1948). 'A motivational theory of emotion to replace emotion as disorganized response', *Psychology Review,* **55,** 5–21.

Leloup (1972). 'Sorties de secours, panique . . . et resquille', *Revue Tech.,* **February, 13 (115),** 17–20.

Levitt, E. E. (1968). *The Psychology of Anxiety,* (London: Staples).

Luciani, R. (1974). 'Contributo allo studio del panico' (in Italian transl. Contribution to the study of panic), *Antincendio Prot. Civ.,* **26 (1),** 17–21.

Mack, R. W. and Baker, G. W. (1961). *The Occasion Instant: The Structure of Social Responses to Unanticipated Air Raid Warnings,* (NRC publication 945), National Academy of Sciences, Washington DC.

Marchant, E. W. (ed.) (1972). *Fire and Buildings,* (Aylesbury: MTP Medical and Technical Publishing Co. Ltd.)

Milgram, S. and Toch, H. (1969). 'Collective behaviour: crowds and social movements', in G. Lindzey and E. Aronson (eds.) *Handbook of Social Psychology, 2nd Edition, Vol. 4.,* (New York: Addison-Wesley Publishing Co.), ch. 35, pp. 507–610.

Miller, G. A., Galanter, E. and Pribram, K. H. (1960). *Plans and the Structure of Behaviour,* (New York: Holt, Rinehart and Winston Inc.).

Ministry of Works (1952). 'Fire Grading of Buildings, means of Escape, part 3: Personal Safety', *Post-War Building Studies No. 29* (London: HMSO).

Mintz, A. (1951). 'Nonadaptive group behaviour', *Journal Abnormal and Social Psychology,* **46**, 150–159.

National Fire Protection Association (NFPA) (1977). *Reconstruction of a Tragedy. The Beverly Hills Supper Club Fire.* (Boston: NFPA), LS-2.

Observer Magazine (1974). 'Inferno in an office block', 5 May 36–47. (Peagram, A.: A day of terror and its lessons, 43–47).

Panse, F. (1971). 'Angst und Schreck', in *Katastrophenreaktionen* (ed. C. Zwingmann), pp. 3–17, (Frankfurt: Akad. Verlagsgesellschaft).

Patrick, J. R. (1934a). 'Studies in rational behaviour and emotional excitement. 1: Rational behaviour in human subjects', *Journal Comparative Psychology,* **18**, 1–22.

Patrick, J. R. (1934b). 'Studies in rational behaviour and emotional excitement. 2: The effect of emotional excitement on rational behaviour in human subjects, *Journal Comparative Psychology,* **18**, pp. 53–195.

Peschl, I. A. S. Z. (1971). 'Flow capacity of door openings in panic situations', *Bouw,* **26 (2)**, 62–67.

Phillips, A. W. (1975). 'The physiological and psychological effects of fire in high-rise buildings', *Factory Manual,* 8–11.

Phillips, B. G. (1950). 'Means of escape from fire: exit widths', *Journal of the Institute Mining Engineers,* **76 (10)**, 633–658.

Phillips, B. G. (1951). *Escape from Fire: Methods and Requirements.* (London: E. and F. N. Spon. Ltd.).

Quarantelli, E. L. (1954). 'The nature and conditions of panic', *American Journal of Sociology,* **60**, 267-275.

Quarantelli, E. L. (1957). 'The behaviour of panic participants', *Sociology and Social Research,* **41**, 187–194.

Quarantelli, E. L. (1960). 'Images of withdrawal behaviour in disasters: some basic misconceptions', *Social Problems,* **8**, 68–79.

Quarantelli, E. L. (1973). *Human Behaviour in Disaster,* paper delivered at Designing to Survive Disaster Conference, Chicago, 6–8 November, 53–73.

Rachman, S. (1968). *Phobias: Their nature and Control.* (Springfield, Ill.: Charles C. Thomas).

Ross, H. E. (1974). *Behaviour and Perception in Strange Environments* (London George Allen and Unwin Ltd.).

Schachter, S. and Singer J. E. (1962). 'Cognitive, social and physiological determinants of emotional State', *Psychology Review,* **69**, 379–399.

Schultz, D. P. (ed.) (1964). *Panic Behaviour: Discussion and Readings,* (New York: Random House).

Schultz, D. P. (1966). *An Experimental Approach to Panic Behaviour,* Office of Naval Research Contract, Number NONR-4808(00), AD. -637604.

Schultz, D. P. (1967). *Individual Behaviour in a Simulated Panic Situation,* Final Tech. Report, Contract NOOO14-67-C-0131 (Charlotte: North Carolina University), AD-661356.

Sime, J. D. (1978). *The Concept of Panic in Fires.* Paper presented in the panel on 'Panic' session at the Conference on Behaviour in Fires, National Bureau of Standards, Washington DC 30 October–1 November 1978.

Sime, J. D. (1979). *The Use of Building Exits in a Large-scale Fire.* Paper presented at the International Conference on Environmental Psychology, University of Surrey, 16–20 July 1979.

Stevens, R. E. (1956). 'Church oyster roast fire panic', *Quarterly of National Fire Protection Association* (NFPA), **49 (4)**, 277–285.

Storey, M. R. (1973). 'The reduction of panic possibilities', *Journal of the British Fire Service Association/Independent Fire Protection Association,* 2 **(4)**, 94–95.

Strauss, A. L. (1944). 'The literature on panic', *Journal of Abnormal and Social Psychology,* **39**, 317–328.

Strother-Smith, N. C. (1976). *Behaviour of People in Fire Situations,* 5th International Fire Protection Seminar, Karlsruhe, 22–24th September, Conference Proceedings Vol. 2, pp. 51–69.

Süddeutsche Zeitung (1973). 'Die Katastrophe im Löwenbräukeller', Samstag/Sonntag, 14–15 April, Nr. 88.

Summerland Fire Commission (1974). Report, Douglas, Isle of Man Government office.

Taylor, J. W. (1965). 'Human reactions to fire hazards', *Institution Fire Engineers Quarterly,* **25 (57),** 80–87.

Turner, R. H. and Killian, L. M. (1957). *Collective Behaviour,* (Englewood Cliffs, NJ: Prentice-Hall).

Veltfort, H. R. and Lee, G. E. (1943). 'The Cocoanut Grove fire: a study in scapegoating', *Journal of Abnormal and Social Psychology,* 138–154.

Vernon, M. D. (1969). *Human Motivation.*(Cambridge: Cambridge University Press'.

Williams, H. B. (1964). 'Human factors in warning-and-response systems'. In Grosser *et al.* (1964), pp. 79–104.

Wood, P. G.. (1972). *The Behaviour of People in Fires,* Fire Research Note 953, *British Research Establishment* Fire Research Station, Borehamwood.

Young, K. (1946). *Handbook of Social Psychology.* (London: Routledge and Kegan Paul Ltd.).

Fires and Human Behaviour
Edited by D. Canter
© 1980 John Wiley & Sons Ltd.

CHAPTER 6

A Survey of Behaviour in Fires

PETER G. WOOD
University of Surrey

INTRODUCTION

The research described in this chapter was undertaken to provide some knowledge about the sort of things which people do when they are involved in a building fire.

Prior to the commencement of the project, almost all the previous literature on people's behaviour in fires had concerned itself with the more dramatic manifestations. Not unnaturally, anecdotal accounts tended to concentrate upon the horrifying, 'panic' reactions.

In order to obtain some detailed information on the range of actions which fire-participants undertook, it was decided to conduct a large-scale survey, interviewing those involved in the fire. These interviews, which were undertaken at the scene of the incident, were conducted by fire service personnel, using a questionnaire format.

METHOD

Although the development of a fire may take many hours from the first dropped cigarette end to the full-scale conflagration, the period of time during which building occupants are aware of the fire, and take some action as a consequence of it, is usually much shorter, probably measured in minutes. Since for a given individual a building fire will be a rare occurrence, the behaviour may engender feelings of excitement or fear, within an environment which may include unusual heat, smoke and toxic fumes.

In attempting to study behaviour under such difficult conditions the selection of a suitable research technique is one of the main preoccupations of the early part of the investigation. Simulation of a fire situation was discarded as a possible method due to the ethical considerations implicit in such an experiment. The decision was therefore made to study fires which occurred during the course of the research. Both interview and questionnaire techniques were evaluated during the pilot phase of the research, a questionnaire administered by fire brigade officers at the scene of the fire being utilized for the main body of the study. By this method data was collected from nearly 1000 fire incidents, and from more than 2000 people who were involved in them. Behaviour itself was examined at two levels, a general analysis of the things which people did, and a more intensive study of two particular aspects, evacuation of the building and movement through smoke. These latter aspects were selected for study in view of their obvious importance in relation to the provision of means of escape in buildings. We thus have a general picture of overall behaviour in fires, and detailed knowledge of these two aspects. To construct a detailed model of behaviour in fires, future studies will be required to concentrate on other aspects, such as raising the alarm, contacting the fire brigade and firefighting behaviour. However, it was not, nor will be, possible to conduct a single large study on all these factors using the present method of data collection. It would be unreasonable to expect fire service personnel to conduct extensive and lengthy interviews at the scene of the fire, when the utilization of men and machines is at such a premium. This disadvantage was known and accepted before the main study was undertaken, it being considered that the advantages obtained from on-the-spot data collection far outweighed any restriction necessary on the volume of information collected.

SUMMARY OF MAIN FINDINGS

Of the 952 fire incidents visited, slightly more than 50 per cent occurred in dwelling-houses, 17 per cent in factories, 11 per cent in blocks of flats or other multi-occupancy dwellings, 7 per cent in shops and 4 per cent in institutions (schools, hospitals, etc.). The remainder occurred in a variety of occupancies, each at less than 2 per cent of the sample total. As was indicated in the introduction, behaviour was examined both at a general level and with particular reference to two specific behavioural variables, evacuation of the building and movement through smoke. It is useful to continue this distinction in our summary of the main results.

GENERAL BEHAVIOUR

Over the course of the incident there are three general types of reaction to fire: they were, in order of frequency:

(1) Concern with evacuation of the building either by oneself or with others.
(2) Concern with fire-fighting or at least containing the fire.
(3) Concern with warning or alerting others, either individuals or the fire brigade.

The majority of behaviour falls either exclusively into one of these categories or some combination of them. The most frequent course, of action appeared to be directed solely to one end, either leaving the building or fighting the fire.

In general terms the majority of people appeared to have behaved in what might be considered an appropriate fashion, although some 5 per cent of the people did something which was judged by the author to 'increase the risk'. There was little evidence of irrational, non-social behaviour (Table 6.1).

The following are the effects of other variables upon the first action taken.

(1) The more serious a person considered a fire to be, the more likely that he would immediately leave the building and the less likely that he would attempt to fight the fire.

Table 6.1 First actions in fire

Behaviour category	Percentage of participants undertaking this as their first action
1. Take some firefighting action	15
2. Contact fire brigade	13
3. Investigate fire	12
4. Warn others	11
5. Do something to minimize danger	10
6. Evacuate oneself from building	9.5
7. Evacuate others from building	7

These seven classes of action describe almost 80 per cent of the first actions taken.

(2) Familiarity with the layout of the building did not affect whether or not a person attempted to immediately leave the building. People who were less than completely familiar with the building were more likely to try and save personal effects.

(3) The more frequently people had received training or instruction on what to do in a fire, the more likely they were to raise the alarm or organize evacuation as a first action. In other respects, frequency of training did not affect first action taken.

(4) People who had been previously involved in a fire incident were no more likely to contact the fire brigade than those who had not. They were, however, more likely to fight the fire or minimize the risk in some way. They were less likely to leave the building immediately.

(5) Women were *more* likely to take the following first actions:
>   (a) warn others
>   (b) immediately leave the building
>   (c) request assistance
>   (d) evacuate their family

They were *less* likely to take the following first actions;
>   (a) fight the fire
>   (b) minimize the risk

(6) An increasing proportion of people fought the fire, from age ten years to age 59 years.

EVACUATION OF THE BUILDING

The single variables which led to increased evacuation were, in *descending* order of importance (relevant figures are shown in Table 6.2):

>   (1) extensive smoke spread as opposed to less extensive
>   (2) home environment as opposed to work environment
>   (3) lack of previous involvement in a fire as opposed to previous involvement
>   (4) women as opposed to men
>   (5) younger people as opposed to older people

Table 6.2 Statistics of people leaving

		Number of people leaving	
Smoke spread	Extensive	65%	(810 out of 1293)
	Less so	40%	(395 out of 932)
Environment	Home	67%	(665 out of 1000)
	Work	40%	(256 out of 642)
Previously involved	No	60%	(914 out of 1521)
	Yes	41%	(269 out of 646)
Sex	Female	60%	(583 out of 952)
	Male	49%	(605 out of 1239)
Age	'Young'	59%	(718 out of 1225)
	'Old'	49%	(467 out of 945)
Training	None	56%	(949 out of 1684)
	Some	49%	(211 out of 455)
Familiar with building	Completely	56%	(1023 out of 1838)
	Less so	49%	(159 out of 327)
Smoke	Present	56%	(1057 out of 1903)
	Absent	48%	(129 out of 269)

(6) untrained people as opposed to trained people

(7) people completely familiar with the building as opposed to those less than completely familiar with it

(8) incidents in which smoke was present as opposed to those in which it was absent

The single variables which led to increased re-entry into the building were in *descending* order of importance (relevant figures are shown in Table 6.3):

(1) men as opposed to women

(2) daytime incidents as opposed to night-time

(3) incidents in which smoke was present as opposed to those in which smoke was absent

(4) previous involvement as opposed to lack of previous involvement

The following variables also applied:

(1) Differences in training only affect evacuation at lower levels of smoke spread.

(2) Differences in training have no effect upon either evacuation or re-entry in night-time incidents.

(3) The presence of smoke increases the proportion of older people who leave, but does not affect the proportion who return into the building.

(4) The time of occurrence of the incident did not affect whether or not people left the building.

(5) People were more likely to leave the building if they stated they did not know a means of emergency escape.

Table 6.3 Statistics of people re-entering

		Number of people re-entering	
Sex	Men	53%	(321 out of 605)
	Women	34%	(201 out of 583)
Time	Day	51%	(232 out of 456)
	Night	38%	(276 out of 735)
Smoke	Present	49%	(133 out of 269)
	Absent	42%	(386 out of 914)
Previously involved	Yes	47%	(109 out of 230)
	No	42%	(299 out of 717)

(6) The more extensive the smoke spread the more frequently exits other than normal were used.

MOVEMENT THROUGH SMOKE

In incidents where smoke was present, 60 per cent of the people attempted to move through it. Nearly 50 per cent of these people moved 10 m (10 yards) or more.

The single variables which led to increased movement through smoke were in *descending* order of importance (relevant figures are shown in Table 6.4):

(1) men as opposed to women
(2) extensive smoke spread as opposed to less extensive
(3) home environment as opposed to work environment
(4) daytime incidents as opposed to night-time
(5) people completely familiar with the building as opposed to those less than completely familiar with it.

Table 6.4 Movement through smoke

		Movement through smoke	
Sex	Men	64%	(684 out of 1064)
	Women	54%	(448 out of 836)
Smoke spread	Extensive	64%	(738 out of 1156)
	Less so	53%	(396 out of 744)
Environment	Home	64%	(590 out of 928)
	Work	52%	(265 out of 511)
Time	Day	65%	(484 out of 739)
	Night	56%	(650 out of 1161)
Familiar with building	Completely	61%	(988 out of 1618)
	Less so	51%	(145 out of 282)

Other variables are as follows:

(1) Neither previous involvement in a fire incident nor training by themselves affect movement through smoke, however together they act to increase the proportion of people moving through smoke.

(2) Trained people are more likely to move through smoke when it spreads less extensively, and less likely to move through it when it spreads extensively.

(3) Previous involvement in a fire incident increases the proportion of people moving through smoke in a work environment.

(4) The combination of night-time and extensive smoke spread increases the proportion of people moving through smoke.

(5) Whether or not a person left the building was not related to whether or not they moved through smoke.

(6) The correlation between the distance a person was prepared to move through smoke and the distance he could see ahead was imperfect. In some conditions people were prepared to move much further through smoke than their range of visibility. The distance moved through smoke was not related to the age of the person.

DISCUSSION AND CONCLUSION

The project which has been described above attempted to examine behaviour in fires at two levels, a general overview involving descriptions of the actions people made in sequence, and a particular view of factors which affect evacuation of the building and movement through smoke. Many of the tentative hypotheses have been based upon assumptions concerning 'correct' behaviour in fires. What has not been attempted, however, is to assess explicitly the 'adequacy' of the behavioural response in relation to the hazard. This lacuna results largely from the difficulty in arbitrating in any specific instance as to what exactly the right course or sequence of actions should be. Each incident represents an almost unique set of circumstances, the number of variables being so large that control or examination of all of them would be practically impossible. From the human viewpoint a possible measure of the 'adequacy' of the response might be whether or not the incident involved injury to someone. Inevitably obtaining information from the person injured was not often possible, particularly if the person was hospitalized. It was, however, quite feasible to examine some of the building variables and the behaviour of the *other* people in the incident, and thus draw comparison with incidents which did not involve injury. Such an analysis might be thought to be even more relevant in cases where a fatality occurred, but as we have seen the numbers in this category are small, this difficulty being further compounded by the reluctance of other parties to be interviewed, largely because of the fear of incriminating themselves in any subsequent official investigation.

INCIDENTS INVOLVING CASUALTIES

Returning to incidents involving non-fatal casualties, a number of the factors act in a direction which would have been predicted. For instance, such incidents were often frequently rated by the fire participants as 'extremely serious'. As smoke spread increased, so did the proportion of people injured, similarly for smoke density. The proportion of people who had never received training is significantly greater in casualty-producing incidents, the proportion of people who knew a means of escape was significantly smaller and the the proportion of people who were completely familiar with the building was less in incidents where two or more people were injured. Other factors which might be thought to affect the incidence of casualties appeared to be unrelated to it. The time of the incident, the presence or absence of smoke and the age of the other people in the building appeared to have little effect upon whether or not casualties resulted. In our sample casualties largely occurred in the 'home' environment, with a secondary group occurring in hotels.

With regard to actions taken, apparent differences arose between casualty type and non-casualty incidents. In casualty incidents a smaller percentage of people contacted the Fire Brigade or fought the fire, whilst a larger percentage investigated, warned others, tried to save personal effects and moved towards the exit. However, of these differences only in moving towards the exit did the two groups differ significantly under statistical analysis. The points of difference do perhaps indicate that the other people in the building are rather less 'socially orientated' in the casualty incidents, although it would be quite wrong to draw firm conclusions from such an analysis.

It should be emphasized that the above comments refer of course to the reactions of the *other* people involved in the casualty-producing incident, and therefore might be considered a somewhat artificial way of distinguishing between appropriate and non-appropriate behaviour. In addition, the small numbers involved would indicate caution in drawing any general conclusions based upon this particular analysis.

METHODOLOGICAL CONSIDERATIONS

Validity

One question which arises is how successful the questionnaire technique has been in studying the problem of behaviour under stress. A criticism which is often levelled at this method is the fact that it relies on what people say rather than what they do. Clearly in some studies it is possible to check the validity of questionnaire measures by obtaining some direct observations of the phenomenon. However, the elements of the fire situation tend to preclude this kind of check. One way that the validity of the present data was checked was to require the Fire Brigade officers who acted as the interviewers to compare the replies given by the respondents with their own first-hand knowledge of the incident. By operating this check approximately twenty of the returned forms were rejected prior to analysis (less than 1 per cent). An additional check on validity was possible after initial coding by checking and comparing the responses of interviewees from the same incident. The average number of people interviewed per incident was just over two, the highest number being ten, so it was quite feasible to examine a fairly large number for possible anomalies. In addition to these procedures, we feel that the actual nature of the responses lends

weight to their veracity. There certainly appeared to be little attempt to deliberately put actions in a good light. It seems likely that many of the people who
confessed to an inappropriate action were ignorant of the fact that they were so
doing.

Sample bias

It should be pointed out that the sample may be somewhat biased in that people interviewed tended to be those immediately involved with the fire. This inevitably
follows from the use of fire brigade officers, who have as data gatherers only a limited time at the scene of the incident. If the fire occurred in a large building they
could not possibly hope to interview all the people who were aware of the fire, and
therefore not surprisingly elected to obtain information from those closest to the
actual scene. However, even a large team of independent interviewers would be unlikely to obtain a complete picture of the incident, as the time demanded by such a
an exercise in, say, a factory containing 200 people, would be quite unacceptable.
This possible source of bias should not be overemphasized since in many incidents,
particularly fires in houses, all the people in the building were interviewed.

In retrospect, perhaps the least satisfactory aspect of the questionnaire lay in the
unstructured questions relating to the courses of action. Although these were arranged to provide sequential responses, the difficulty when analysing the data arises
from having no knowledge of the time-scale occupied by each action. The length or
brevity of the recorded comments did not appear to be related to the duration of
the actions, rendering it difficult to assess over what period of time a person continued to pursue any specific action. This was particularly so when the course of
action was a general one, directed mainly towards one end, for instance firefighting.
It may be that the person was fighting the fire for several minutes, but superficially
he appears to be less 'active' than someone who did several specific things, which
may well have occupied less time. Any future attempt to study stress situations
which are not of very short duration should include some measure of the period of
time involved, to provide a more valid basis for comparison between types of behaviour.

Unexpected findings

A number of aspects of the results were not anticipated.

The popularity of firefighting as a course of action, for instance—it certainly
seems that the general public are willing to attack the fire more frequently than is
popularly supposed. It is clear, however, that the decision whether or not to immediately fight the fire was based to a certain extent upon the person's judgement
of the seriousness of the fire, the proportion of people firefighting being inversely
related to the seriousness judgement. This seems to indicate that people are more
likely to attempt to fight the fire if it was not thought to be very serious, and they
thus judged that they had a good chance of extinguishing it. However, since our
data all relates to fires to which the fire brigade were called, presupposing a certain

level of seriousness, this may also be a comment on the lay-person's misjudgement of fire dangers. Unfortunately, it was not possible to discover at what stage in the proceedings the Fire Brigade was called, which would shed light upon this problem.

It would certainly seem likely that there is some threshold value of 'fire-severity', as judged by the person, beyond which people call the fire brigade. It may be that this limen is different for different people, depending on age, sex, presence of others and other individual differences. In view of other widespread public misconceptions concerning fire, it would be extremely interesting to discover what factors or combination of factors actually determine the calling of the fire brigade.

Frequency of firefighting was also closely associated withe the type of building, accounting for almost one-quarter of the first actions taken in factories compared to only one-tenth in dwellings. This being so, one would anticipate some correlation with frequency of training, but surprisingly this did not emerge. The absence in some cases of a clearcut relationship between training frequency (and to a lesser extent, familiarity with the building), was to the author one of the major surprises of the investigation. Both the above variables were expected to be key variables and we feel that further investigation, especially concentrating on the nature and type of training, is urgently needed. This should not be interpreted as an indictment of all training, since the present investigation merely studied one aspect of it, its frequency. It is suggested, however, that little is known about the effectiveness of various types and frequencies of fire training. In view of the large sums of money which are invested in buildings to provide fire escape and fire-protection facilities, one must wonder at the small amount of time and effort which appear to be devoted to ensuring that they are used properly.

One of the most interesting aspects of the analysis was the sex difference which emerged. In terms of general actions women appear to be more concerned with the safety of people, themselves included, in that they were more likely to warn others and evacuate family, in addition to being more likely to immediately leave themselves. In contrast, men seem more situation-orientated, being more likely to attempt firefighting or minimize the risk. With regard to leaving the building over the course of the incident, women were again more likely to leave, contrasting with aircraft emergency experience which in general has shown that a greater percentage of men escape (Mohler, 1964). However, in this case the alternative actions are of course extremely limited, the superior strength and size of men appearing to be a determining factor. Frequency of returning into the building and moving into smoke also demonstrated sex differences, the proportion of men being significantly greater in both cases.

A surprisingly large percentage of people move through smoke, an action which we were led to believe was fairly rare. Having discovered that this behaviour is actually fairly frequent, one must ask to what end people moved into the smoke. It is demonstrated that moving through smoke is not associated with leaving the building, nor with knowledge of emergency escape, and one can only assume that it is undertaken whilst performing some other activity, such as fire-fighting or warning others. However, there is certainly room for further research into the reasons for and the conditions under which people move into smoke.

Previous involvement in a fire incident appears to reduce some of the stressful elements of the fire, since people who had been previously involved were less likely to leave immediately and more likely to firefight in addition to moving further through smoke. They were not, however, any more likely to behave in a 'correct way', such as contacting the fire brigade, and indeed were more likely to return into the building.

The high percentage of people who returned into the building must be a comment—not only upon the ineffectual nature of fire-safety propaganda which invariably preaches the folly of such behaviour, but also the difficulty many people experience in apparently 'doing nothing' when their possessions and property are threatened. Having left the building, a certain proportion of people seemingly feel motivated to return in to check 'the progress' of the fire even if they do not actually do anything about it. If this type of behaviour is liable to occur at all types of fire, then one must have some reservations concerning current thinking on means of escape, which seems to be moving towards a policy of either moving people to a safe place, or persuading them to remain in a safe place without actually leaving the building (Silcock, 1969). If, as we have seen, a certain percentage of people will return into the building for apparently not very rational reasons, then perhaps a similar percentage will wish to leave the 'safe area' within the building, rather than remain passive and inactive. This aspect appears to require urgent investigation, initially to see if it is a general problem. If it proves to be so, then one would attempt to identify how large a percentage of people will react in this way, what sort of conditions determine its occurrence, and how such behaviour might be altered or redirected. It undoubtedly seems to be expecting rather a lot of people to merely direct them to a safe place, and then in the face of an extremely stressful, unusual situation, require them to be inactive. These needs are recognized in other forms of disaster or emergency planning, by giving people something to do, even though they may know on a rational level that their actions may not alter a particular outcome. The mere fact of doing something not only lowers the level of threat, but may also divert potentially dangerous investigative behaviour.

CONCLUSION

From the analysis of the present data we may construct a 'probable' picture of people who take certain courses of action.

(1) The person who, as soon as they become aware of the fire, immediately leaves the building, will more frequently be female; (rightly) consider the fire to be extremely serious; and not have been previously involved in a fire incident.

(2) For firefighting the person is more likely to be male; between 30 to 59 years old; in the working environment; and previously have been involved in a fire.

(3) Over the course of the incident, in evacuating the building it is probable that the person will again be female, not know a means of emergency escape; not have been previously involved; and never have received any training.

(4) A similar picture for returning into the building reads, men; between 20 and 39; completely familiar with the building; previously involved; smoke being present in the incident.

(5) In moving through smoke the person is more likely to be male; completely familiar with the building; consider the fire extremely serious; in the home environment; the incident occurring at night.

It is clear that no one factor is acting consistently on all the behavioural variables. However, for specific types of action, it is possible to isolate and examine which variables were important in determining the behavioural responses.

The process of encoding and simplifying that data for computer analysis unfortunately renders examination of some aspects more difficult. One such aspect is the effect of people's actions and communications upon each other, that is, social effects. We have, of course, recorded and sorted by the presence and relationships of people in the building; however, to obtain a more detailed scrutiny of this factor a sample of questionnaires relating to incidents at which three or more people were interviewed was inspected. Perhaps the most interesting impression from this analysis was the essential similarity between behaviour at work and in the home. The cooperative nature of actions in the work environment was often a feature, tasks being allocated and undertaken, assistance to workmates rendered and people being generally helpful. Most of the larger incidents in fact involved factories, and in the relatively small number of fires at which six or more people were interviewed this active, helpful and cooperative behaviour is shown to a marked degree. In the home, although there were naturally individual differences, this type of behaviour is repeated, albeit on a smaller scale. For instance, where families are involved, the roles of husband and wife seem to become fairly sterotyped, in so far as the husband became responsible for fighting or containing the fire whilst the wife was allocated the job of contacting the fire brigade and evacuating other family members. Often the 'operations' are directed by one specific family member, in many cases the husband, although where a young married couple is living with parents it may well be one of the parents. The apparent ease with which tasks are allocated and roles assumed in this situation is perhaps a function of the underlying hierarchial nature of family relationships, and is a reflection of the more formalized relationships of work.

Although not a specific aim of the study, the popular belief that 'panic' frequently occurs in a fire situation led to the results being examined for evidence of this. We might construe that immediately leaving the building represents a 'panic-type' response. This type of behaviour was associated with high levels of seriousness-rating, our measure of seriousness being intended to indicate the threat level to the person involved. There is some evidence that people did not rate the fire as inappropriately serious, since seriousness is correlated with some of the objective measures of fire severity, such as high levels of smoke spread and density. High seriousness ratings were also associated with fires which occur in the home, although this was not reflected in the proportions of people immediately leaving. It may be that people rated seriousness in terms of threat to property rather than threat to life. Other factors which we would expect to affect our self-defined panic-type responses, such as familiarity with the building, and training frequency, did not in fact do so, although people who were trained very frequently, (at least once per month), were significantly less likely to leave immediately. It is difficult to draw firm conclusions about the incidence of a panic-type response largely because this study was descriptive rather than analytical. Any *post hoc* interpretations of actions involve assumptions whose validity cannot be checked. What we are really lacking, and what obviously follows from this research is some attempt to gain insight into the decision processes which lead to certain courses of action. In the present research we have simply asked people what they did without reference to what other courses of action were considered and rejected. Clearly we cannot hope to make predictions about such behaviour or attempt to alter it, if we do not have some evidence as to *why* people did one thing rather than another. Such more intensive studies will have to look at people's attitudes, knowledge and beliefs concerning fire, in addition to the measures recorded in the present study.

COMPARATIVE DATA

The present research, which was undertaken in Great Britain, was completed in 1972. An essentially similar study was undertaken by Professor John L. Bryan during 1975–76 in the United States (Bryan, 1977). In this case 584 people were interviewed at the scene of 335 fire incidents.

Although the methodology of the two studies was very similar, many differences emerged in comparing the results. In terms of occupancy, that is the 'type' of building involved, the US study has an even heavier preponderance of 'home' fires (63.6 per cent in dwellings and 20.9 per cent in apartments), with a very low (0.6 per cent) proportion of industrial incidents. This may partly serve to explain the statistically significant difference in the sex of the two study populations illustrated in Table 6.5. In addition, the British population were slightly older. In both studies, a surprisingly high proportion of the participants had previous fire experience (24.8 per cent British, 28.3 per cent US).

In terms of the general actions taken, Bryan selected five categories of behaviour on which to compare the two studies. This comparison is illustrated in Table 6.6. Only in the proportion of people moving through smoke are the differences non-significant.

Perhaps the most interesting points are the very high proportion of people leaving the building in the US study, and similarly the high proportion who re-enter the building in the British study. In both studies a large number of people stated they moved through smoke despite the incidents in the US study having generally more

Table 6.5 British/American statistics, male/female

Sex	British study Number	(per cent)	US study Number	(per cent)
Male	1293	56.5	263	45.2
Female	954	43.4	319	54.8

Table 6.6 British/American statistics, behaviour

Behaviour	British study (per cent)	US study (per cent)
Evacuation	54.5	80.0
Re-entry	43.0	27.9
Fire fighting	14.7	22.7
Moved through smoke	60.0	62.7
Turned back (in smoke)	26.0	18.3

extensive smoke spread. Visibility estimates from those moving through smoke tend to indicate, however, that the smoke was denser in the British incidents.

Although the results from the two studies show many significant differences, the majority of these may be due to the inevitable (and uncontrollable) differences between the occupancies involved in the fire incidents, and the subsequent difference in the sex of the participant populations. Future comparative studies may need to select only a restricted range of occupancies for study, tall buildings with ten or more floors, for example.

REFERENCES

Bryan, J. L. (1977). *Smoke as a Determinant of Human Behaviour in Fire Situations.* Report from Fire Protection Curriculum, College of Engineering, University of Maryland, College Park, Maryland. Prepared under support from the Center for Fire Research, National Bureau of Standards. Program for Design Concepts (mimeo).

Mohler, S. R. (1964). *Human Factors of Emergency Evacuation.* Washington: Office of Aviation Medicine, Federal Aviation Administration AM-65-7.

Silcock, A. (1969). *A Critical Look at Some Current Problems of Escape Route Planning.* Borehamwood: Joint Fire Research Organisation. Fire Research Note No. 753.

Fires and Human Behaviour
Edited by D. Canter
© 1980 John Wiley & Sons Ltd.

CHAPTER 7

Fat Fires: A Domestic Hazard

CLAIRE WHITTINGTON AND JOHN R. WILSON
Institute for Consumer Ergonomics
Loughborough University of Technology

INTRODUCTION

In 1974 there were 19 399 domestic fires originating in cooking appliances which were reported to Fire Brigades in the United Kingdom (Home Office, 1977). These fires, of which approximately 80 per cent were fat fires comprised over one-third of the total number of domestic fires reported that year. These figures indicate a clear need for research to be carried out in the area. As a preliminary, this chapter examines available data relating to incidents involving hot fat and associated cooking facilities, in an attempt to provide a basis for such research.

OVERALL EXTENT AND NATURE OF THE PROBLEM

With respect to fires involving domestic cooking facilities comprehensive national data only exist for those incidents which result in a call to the fire brigade. However, it must be borne in mind that evidence suggests that large numbers of cooking fires are dealt with by the individual householder. For example, an evaluation of a recent Home Office chip-pan fire prevention advertising campaign revealed that under 5 per cent of respondents who had claimed to have had a fat-pan fire actually called the fire brigade (Research Services Ltd., 1976). This result is supported by *ad hoc* evidence from a small-scale Electricity Council study which showed that of seventeen cases reported to a local area board, only one involved the local fire brigade (Parker, 1976).

The large contribution made by fires originating in cooking appliances to the total number of fires within dwellings has already been mentioned. It is also apparent from the national statistics discussed in Chapter 2, as summarized in Figure 2.3. Table 7.1 also presents information concerning the principal sources of ignition of fires in dwellings for the years 1969–74. This table confirms that over the time period considered, cooking appliances were responsible for a major proportion of all fires occuring in dwellings. Two other major points emerge from the table.

Fires and Human Behaviour

Table 7.1 Principal sources of ignition of fires in dwellings, 1969–74

Source of ignition	Year					
	1969	1970	1971	1972	1973	1974
Cooking appliances (electric and gas only)	11 432	12 468	13 754	15 984	18 354	18 579
Children with fire	3138	3111	3388	4224	4290	4624
Smokers' materials	3372	3408	3307	3554	4028	4619
Electric wiring installation	3252	2990	2951	3202	3250	3304
Chimney stove pipe, flue	3078	2379	2235	2202	2116	1691
Oil space heating	2468	2316	1862	2182	1844	1493
Electric space heating	1896	1757	1551	2021	1996	1871
Radio and television	1432	1711	1910	1946	2272	2332
Fire in grate	2846	2306	1830	1921	1828	1537
Malicious ignition	964	895	1211	1690	1968	2140
Electric blanket	1624	1432	1333	1499	1470	1521
Other known sources	7500	7462	7589	9215	8740	8158
Unknown	2870	3070	3043	3228	3326	3272
TOTAL	45 872	45 305	45 964	52868	55 482	55 141

Source: United Kingdom Fire and Loss Statistics for 1973 (Department of the Environment, 1976), and United Kingdom Fire Statistics for 1974 (Home Office, 1977).

First, during the six years 1969–74 the number of fires attributable to cooking appliances increased from 11 432 to 18 579, an increase of 9269 (20.2 per cent) in total reported fires in dwellings over the same period. Figure 7.1 presents a more detailed account of information also shown in Figure 2.3. It shows the trend for some major causes of fires in dwellings for 1969–74.

The second major point to emerge is the increase in the contribution of cooking appliance fires to all domestic fires. This is emphasized in Table 7.2. It can be seen that the percentage of domestic fires which originate in electric and gas cooking appliances has increased from 24.9 per cent in 1969 to 33.7 per cent in 1974.

The statistics presented so far have related to all fires associated with cooking appliances. However, the emphasis of this chapter is upon those cooking fires in which the material first ignited is fat. Unfortunately, only limited statistics relating to fires in dwellings (as against fires in all buildings) are available in the annual fire statistics, which were produced by the Department of the Environment (DoE.) until 1973 and by the Home Office since then. For the purpose of this review, it was therefore necessary to request more specific analyses of available data to be carried out by the Fire Research Station. The tabulations provided relate only to incidents occurring during 1973 (Fire Research Station, 1976). Hence, in the remainder of this chapter some data are presented for 1973 only. It should be noted that these specific analyses were carried out prior to the ratification of the final published statistics for 1973. Therefore, throughout this chapter there is a slight discrepancy between the total figure for 1973 presented in the two sets of data. Table 7.3 gives the distribution of electric and gas cooking appliance fires by the material first ignited for 1973.

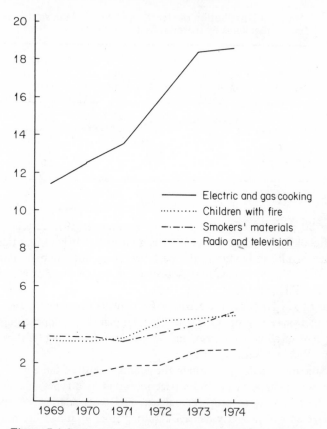

Figure 7.1 Some major causes of fires in dwellings 1969–74
Source: United Kingdom Fire and Loss Statistics for 1973
(Department of the Environment, 1976) and United King-
dom Fire Statistics for 1974 (Home Office, 1977)

Table 7.2 Proportionate contribution of electric and gas cooking appliance fires to all domestic fires, 1969–74

Year	Fires originating in cooking appliances (A)	Total number of domestic fires (B)	$\frac{A}{B} \times 100$ per cent
1969	11 432	45 872	24.9
1970	12 468	45 305	27.5
1971	13 745	45 955	29.3
1972	15 984	52 868	30.2
1973	18 354	55 482	33.1
1974	18 579	55 141	33.7

Source: United Kingdom Fire and Loss Statistics for 1973 (Department of the
Environment, 1976); and United Kingdom Fire Statistics for 1974 (Home Office,
1977).

Table 7.3 Distribution of electric and gas cooking appliance fires in dwellings by material first ignited, 1973

Material first ignited	Absolute frequency	Relative frequency (per cent)
Fat	14 574	79.7
Food—not fat	2060	11.3
Other	1654	9.0
TOTAL	18 288	100.0

Source: Fire Research Station (1976).

Nearly 80 per cent of cooking fires in dwellings in 1973 resulted from the ignition of cooking fat. Hence, since there is no reason to believe that this proportion has altered significantly during the past few years, the dramatic increase in the number of cooking appliance fires is assumed to reflect a similar increase in the incidence of cooking fat fires.

Of the 14 574 fat fires on electric or gas cookers occurring in 1973, 13 434 involved a pan on a ring or hot plate. Thus fat-pan fires comprised approximately 73.5 per cent of all cooking fires, and 92 per cent of cooking fat fires occurring in 1973 (Fire Research Station, 1976).

Few data are available concerning the economic cost of these incidents. Leaving aside the actual drain on the fire service resources, two types of cost can be considered: costs relating to property damage, and costs relating to personal injury. The average cost of property damage resulting from a fat-pan fire has been estimated at £150 (at 1973 prices). This is less than the average cost of property damage in all domestic fires which was estimated at £250 in 1973 (Rutstein and Butler, 1977).

Table 7.4 shows the numbers of casualties which resulted from fat-pan fires in 1973, compared to all other domestic fires. It can be seen that the incidence of casualties per 1000 fat-pan fires was approximately two-thirds the comparable rate for all other domestic fires. However it needs to be stressed that, per year, the overall number of casualties associated with fat-pan fires is greater than for any other source of ignition. For example, in 1973 the total number of casualties known to have resulted from fires involving 'smokers' materials' was 611, from 'children with fire' 412, and less than 350 from any other single known source, as compared to a total of 940 for fat-pan fires (Department of the Environment, 1974; and Rutstein and Butler, 1977). In terms of the total number of injuries and their attendant economic, social and medical costs, therefore, fat-pan fires must be seen as a significant problem.

CHARACTERISTICS OF FAT-PAN FIRES

Having considered the extent of the problem of domestic fat-pan fires, the discussion will now be extended to an examination of the events leading up to such

Table 7.4 Casualties caused by fat-pan fires and other domestic fires* in 1973

	Domestic fat-pan fires on ring or hotplate	All other domestic fires
Number of fires	14 010	41 472
Number of fires with casualties	834	2884
Number of fatalities	17	748
Total number of casualties (fatal and non-fatal)	940†	3993
Number of casualty fires per 1000 fires	60	70
Number of fatalities per 1000 fires	1.2	18
Total number of casualties per 1000 fires	67†	96

Source: Rutstein and Butler (1977).
* The figures refer to fat-pan fires associated with all cooking appliances (that is, gas, electric, solid and other fuels).
† This is an estimated figure. Most fat-pan fires involved only one casualty.

fires. This will be done first in terms of the behaviour of the person(s) concerned and second in terms of the characteristics of the cooking appliances and associated equipment. For reasons of simplicity these factors are considered separately. However, it will become apparent that such a division is purely artificial. The accident sequence is initiated and sustained by a complex interaction of behavioural, product and environmental characteristics.

Behavioural characteristics of the person(s)

Cooking with hot fats and oils using traditional kitchen equipment is an inherently hazardous task. However the risk of an accident may be substantially increased by certain types of behaviour on the part of individuals. This issue will be discussed in this section.

The official fire statistics contains no classification of the causes of cooking fat fires. Moreover even the specific analyses carried out for the authors by the Fire Research Station provide only a limited classification (see Table 7.5). Unfortunately no further breakdown is given within the primary category identified; that is, no indication is given as to whether the power was turned on accidentally or left on, or whether the cooker was left unattended. The only data available which provide further insight into these causes are restricted to the results of small *ad hoc* regional studies. During a study commissioned by the Scottish Special Housing Association, Andrews (1969), using mail questionnaires, obtained data from 66 households which had experienced a fat-pan fire and from 83 control households. Of the fire sample, 88 per cent claimed not to have been present in the kitchen at the time of ignition (Andrews, 1969).

> Absence from the kitchen when the fire commenced followed an unexpected demand for attention in 35% [of the cases]. A further 31%

Table 7.5 Distribution of all cooking fat fires* by cause of ignition, 1973

Cause of ignition	Absolute frequency	Relative frequency (per cent)
Power accidentally turned on or left on, unattended	14 968	98.1
Electrical fault	30	0.2
Flared up or in draught	2	–
Other causes	258	1.7
Total	15 258	100.0

Source: Fire Research Station
* The above figures refer to fires associated with all cooking appliances (that is, electric, gas, solid and other fuels) and with all parts of the cooking appliances.

[of the respondents] had made a planned absence. In other cases (34%) the fire occurred after cooking had ceased or had been thought to have ceased. Nearly 66% of the absentees had forgotten that they had a chip pan being heated. This points to a failure of short term memory. The other absentees seemed to misjudge the length of safe period, delaying their return to the kitchen too long.

During an evaluation of the Home Office advertising campaign on chip-pan fire prevention all respondents in the three selected geographical areas were asked 'to explain how it is that a chip-pan can go up in flames?' (Research Services Ltd., 1976). Before the campaign 'inattention' or 'leaving and forgetting the pan' were each given as likely reasons for a fire by only about 10 per cent of the respondents. However, those respondents who had actually had a fat-pan fire were asked what they perceived to be the cause. Their responses before the campaign are shown in Table 7.6. It can be seen that approximately 50 per cent of the respondents mentioned 'inattention' as a primary factor and about 10 per cent mentioned 'forgot to turn heat off.

This finding is supported by the results of two small-scale studies. The first carried out by the Electricity Council (Parker, 1976), concluded that 'the fires [almost] invariably occurred when the cook was momentarily out of the room'. Examples of reasons given for such absence were as follows:

(1) Neighbour in accident called for help
(2) Child fell and cut himself badly
(3) Visited lavatory
(4) Fetched washing in as rain started
(5) Husband (cooking) went to speak to wife ill in bed

Although only a small sample of fire cases was studied (seventeen cases in all) the unpredictable nature of the unplanned absences is worth noting. Such is the nature

Table 7.6 Reported causes of informants' chip-pan fires, before the campaign

	Area 1	Area 2	Area 3
Base: those who have had a chip pan fire	78	79	84
Weighted	72	91	86
	per cent	per cent	per cent
Inattention	56	45	55
Overheating	24	34	24
Forgot to turn off heat	13	12	9
Fat boiled/splashed over	7	12	18
Overfilling	3	5	4
Wet chips	7	7	11
Oil on outside of pan	4	7	3
Electrical/cooker fault	6	–	–

Source: Abstracted from Research Services Limited (1976).

of the unexpected external demands on the person cooking that they are often unable to make safe contingency plans should their return to the kitchen be delayed.

The second of these small-scale studies was a pilot survey conducted by the Institute for Consumer Ergonomics for the Department of Prices and Consumer Protection in 1976. This involved conducting home interviews in households in which either a fire or personal injury involving cooking facilities had occurred. A total of 51 cases were followed up during a three-month period. Of the eleven cases in which fat being heated on the burner or hot plate ignited, ten occurred when the person had vacated the kitchen. Of these absences three were unplanned. Such an example, taken from the interview details, is briefly described below:

A woman aged 25–30 put a chip-pan of fat to heat on a back ring of an electric cooker. She subsequently felt unwell, left the kitchen and went to lie down in the lounge (shutting the lounge door). The first thing she remembered was hearing a crackling noise and seeing smoke appearing through the gaps in the door. Both kitchen and upstairs were filled with smoke. She managed to turn the cooker off and carry the burning pan through the back door of the house to the outside. A neighbour alerted the fire brigade.

Another three absences were planned by the persons concerned; however, they had misjudged the length of time that had elapsed since the fat had been placed on the cooker. A typical example of this type of behaviour is given below:

A family was preparing to go on holiday. The wife was alone in the house and had decided to cook chips for the family lunch; she had put the chip-pan on an electric cooker, left the kitchen, and went upstairs to continue packing. She recalls she was rushing to complete the packing, when she smelt burning. On going downstairs she saw the kitchen full of smoke. Unfortunately, she failed to extinguish the fire and in trying to do so delayed calling the fire brigade. The resultant damage extended to the whole of the kitchen, dining and conservatory area at the rear of the house.

Characteristics of the appliances and associated equipment involved in the incidents

In this section details of the appliances and associated equipment involved in cooking fires are considered. These comprise a group of interrelated products: the cooker (its power and controls), the cooking utensil and the cooking fat. Each of these will be considered in turn below.

The cooker

The most obvious distinguishing feature of domestic cooking appliances is their fuel or power. The numbers of fires reported to the fire brigade for 1969–74, in which the source of ignition was an electric or gas cooking appliance, are shown in Table 7.7. These figures represent all electric and gas cooking appliance fires, regardless of the material first ignited. Comparable figures for fat fires only are given in Table 7.8. This table however only refers to the year 1973 since the breakdown of type of cooking appliance by the material first ignited, for incidents occurring in dwellings, is not available in the annual published fire statistics.

It can be seen from Table 7.7 that the total number of fires reported each year involving electric cookers is at least twice the number reported for gas cookers, and that the ratio is increasing (1969–2.1:1.0; 1974–2.7:1.0). As is seen in Table 7.8 this ratio is even higher when fat fires only are considered (1973–3.8:1.0). These ratios must be considered in the light of related exposure figures. In 1972, the number of households using electricity for cooking was 7.5 million and those using gas numbered 10.8 million. The associated incidence of fat-pan fires per 100 000 households are given in Table 7.9. The rate, therefore, of fat-pan fires involving electric cookers is over four times that for gas cookers. Several reasons for this discrepancy have been advanced; the major ones are listed below, mainly drawn from three reports: Hogg (1963), Andrews (1969) and Department of Prices and Consumer Protection (1976).

(1)*Visibility of flame*–'the user is more likely to be aware of danger from the visible flame of a gas ring than from the surface of a hot plate' (Hogg, 1963).

(2) *Audibility*–'the user can perhaps subconsciously be aware of the sound of a gas burner whilst the silence of the electrical appliance gives no indication that it is on' (Hogg, 1963).

(3) *Control of heating rate*–'gas cookers have a much higher degree of control over the heating rate' (Andrews, 1969).

Table 7.7 Fires in dwellings associated with gas and electric cooking appliances 1969–74

Source of ignition	Year					
	1969	1970	1971	1972	1973	1974
Electric cooking	7696	8507	9512	11 305	13 278	13 568
Gas cooking	3736	3961	4233	4679	5076	5011
Total	11 432	12 468	13 745	15 984	18 354	18 579

Source: United Kingdom Fire and Loss Statistics for 1973 (Department of the Environment, 1976); and United Kingdom Fire Statistics for 1974 (Home Office, 1977).

(4) *Thermal capacity*—'the electric hot plate (and radiant ring) remaining hot after use is liable to be more hazardous' (Hogg, 1963).

(5) *Unfamiliarity with heating rate*—'unfamiliarity with the speed of the heating response of a new [radiant ring] electrical cooker compared with the older [solid plate] ones' (Andrews, 1969).

(6) *Control confusion and malfunction*—this appears to be exclusive to electric cookers.

(a) 'some control knobs on electric cookers permit 360° rotation in both clockwise and anticlockwise directions. When there is no positive stop at the 'Off' position . . . when one is under stress or in a hurry it is possible to switch or leave 'On' accidentally' (Andrews, 1969).

(b) 'the facility of independent control (mains switch and individual heating element control) on electric cookers can lead to individual switches being inadvertently left in an 'On' position' (Andrews, 1969).

(c) 'faulty electrical contacts on the cooker on mains switch whereby power is still supplied when the switch has been turned to the 'Off' position' (Andrews, 1969).

Table 7.8 Distribution of cooking fires in dwellings by source of ignition and material first ignited, 1973

Source of ignition	Material first ignited			
	Fat	Food—not fat	Other	Total
Electric cooking	11 534 (87.2)*	1080 (8.2)	614 (4.6)	13 228 (100.0)
Gas cooking	3040 (60.1)	980 (19.4)	1040 (20.5)	5060 (100.0)
Total	14 574 (79.7)	2060 (11.3)	1654 (9.0)	18 288 (100.0)

Source: Fire Research Station (1976).
* Figures in brackets are percentages.

Table 7.9 Rate of occurrence of fat-pan fires by source of ignition, 1971–74

	Fat-pan fires* (in dwellings)	Number of households in England Scotland and Wales	Rate of fat-pan fires (per 100 000 households)
Gas users	3500†	10.8 million††	32.4
Electricity users	10 000†	7.5 million††	133.3
Total	13 500	18.3 million**	74.2

Source: Mytton and Butler (1976).
Notes. * Average for 1971–74
      † Unknown split proportionately between gas and electricity
      ** 1971 Census data
      †† Information from Gas Board

(7) *Fat ignition after spillage*—'if fat spills or spits over when being heated by a hot plate (or to a certain extent by a radiant ring) it falls directly upon the surface from which the heat is emanating, whereas when a gas ring is used the fat can fall through the flame without igniting' (Hogg, 1963).

(8) *Area of heating surface*—'chip-pans (often) do not entirely cover radiant rings or (solid plates), whereas it is natural to adjust a gas flame not to extend further than the base of the pan' (Department of Prices and Consumer Protection, 1976).

Of the above reasons, only those numbered (1) to (5) inclusive may have some influence on the behavioural factors discussed earlier—that is, inattention and misjudgement of heating time. Even in total it is felt that all the above reasons do not adequately explain the large discrepancy between the rates with which cooking fat fires occur on the two types of cooker. The authors consider that a further explanation of this discrepancy appears to have been overlooked in the literature. Of necessity all comparisons between the numbers of fat fires on the two types of cooker have been limited to incidents requiring a call to the fire brigade or medical services. There is no evidence on which to base a true comparison of the actual incidence of such fires—that is, to include those successfully extinguished by the householder. There may be little or no difference in the probability of the occurrence of a cooking fat fire on a gas or electric cooker. However those incidents occurring on gas cookers may be easier to cope with than those on electric cookers for reasons already indicated in the listing above, for example, the residual heat of electric hot plates and rings after switching off and also the more inaccessible controls often found on electric cookers.

The issue of cooker controls is worth expanding. In studies of fat-pan fires, a small proportion of the incidents have been attributed to misunderstanding, misuse or breakdown of cooking controls. The reasons for the increased likelihood of these types of incidents occurring in electric cookers have been discussed. Andrews (1969) found that of the 10 per cent of fires which occurred after the cooking stage had been completed, the majority were largely due to control error. In three of the seventeen cases identified by the Electricity Council (Parker, 1976) the cook thought she had switched 'Off' or 'Low' but had apparently turned up the heat instead. However, during the Home Office 'chip-pan fire prevention advertising campaign evaluation' (Research Services Ltd., 1976) only six (1 per cent) out of 467 respondents who had had a chip-pan fire attributed it to malfunctioning of the cooker or its controls. However, there is no indication of the number of fires caused by misuse or misunderstanding of controls.

Cooking utensil involved

As has already been discussed, spillage of oil or fat does not seem to be a major cause of cooking fat fires. Where spillage does occur it must be assumed that this is due to a less than optimum combination of pan size and depth of oil.

In relation to those incidents due to lack of attention Andrews (1969) concluded that 'contrary to popular belief shallow fryers may be safer to use'. The explanation he offers for this is that the shallower the pan the faster will be the heating up time for the oil; fewer persons may therefore be prepared to risk leaving the kitchen.

A further point about pans was highlighted by the Electricity Council study (Parker, 1976). They considered that lightweight pans liable to buckle with time, might develop 'hot spots' (that is, when heat is not dissipated evenly across the base of the pan). This could present a hazard on electric cookers, although less so on gas cookers. Finally, Andrews (1969), during laboratory experiments conducted in parallel with his survey, found that the condition of pan exteriors in terms of solidified fat made no difference to the potential hazard of the cooking task.

However, overall it is felt that no reliable conclusions can be drawn about the contribution of the pan to the cooking fat fires without further data from field, laboratory and accident studies.

Cooking fat

The emphasis of this discussion has been on the hazardous consequences of inattention when heating cooking fats. These consequences are the overheating and subsequent spontaneous ignition of the oil or fat. The process is illustrated in Figure 7.2. The three major points of note on the temperature scale are: first, the maximum recommended cooking temperature for chips; second, the temperature at which visible evidence of overheating first appears; and third, the temperature above which spontaneous ignition will take place.

Technical data on the properties of cooking fats have been produced by Andrews (1969) and the Fire Research Station (Stark and Mulliner, 1965). The most relevant points are summarized below. It must be borne in mind that recent developments in cooker design may have invalidated some of these results.

(1) Given a chip-pan half to two-thirds full of oil or fat the time taken to reach 205°C (correct cooking temperature) from cold is between eight and sixteen

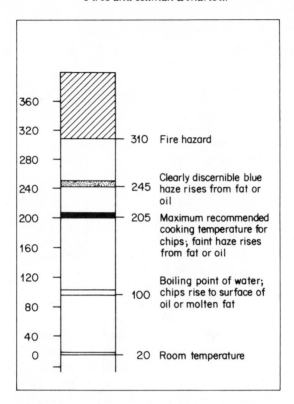

360
320 — 310 Fire hazard
280
240 — 245 Clearly discernible blue haze rises from fat or oil
200 — 205 Maximum recommended cooking temperature for chips; faint haze rises from fat or oil
160
120 — 100 Boiling point of water; chips rise to surface of oil or molten fat
80
40
0 — 20 Room temperature

Figure 7.2 Stages in the process of heating fat or oil to the temperature necessary for spontaneous combustion. Source: Fire Protection Association (1970)

minutes on a gas cooker (depending on heating rate) and between seventeen and twenty-one minutes on an electric *hotplate* (set full). Furthermore, the time required to heat the fat from this temperature to that of spontaneous ignition is between six and ten minutes on a gas cooker and from nine minutes upwards on an electric *hotplate* (Andrews, 1969). Unfortunately, comparable data for electric radiant rings are not available.

(2) There is little difference in the spontaneous ignition temperature of various types of oils or fats. The maximum percentage difference in this temperature between six types of oil or fat was only 4 per cent (Stark and Mulliner, 1965).

(3) There is only a small difference in ignition temperature between old (used) fat and new fat. This is true for all types of fat or oil (Stark and Mulliner, 1965).

(4) The presence of small pieces of food (from previous cooking) makes no difference to ignition temperatures (Andrews, 1969).

(5) Spontaneous ignition occurs at about the same temperature when chips are present in the fat as with fat alone, although the time taken to reach this temperature is greater when the pan is full (Stark and Mulliner, 1965).

Table 7.10 Actions taken to put out fire—reported on survey before campaign

	Area 1	Area 2	Area 3
Base: those who have had a chip pan fire	78	79	84
Weighted	72	91	86
	(per cent)	(per cent)	(per cent)
Turned off heat	32	22	28
Took/threw outside	27	27	45
Covered with damp cloth	23	15	19
Covered with wet cloth	10	4	13
Covered with dry cloth/ cloth unspecified	7	7	5
Covered with lid	5	11	11
Covered with something else	5	6	3
Put soil/sand in pan	3	8	2
Threw on water/put under tap	1	6	8
Left to cool	10	4	9
Opened door/window	3	3	4
Called fire brigade	4	2	—
Did nothing/left to burn	4	—	—
Panicked	7	10	7

Source: Abstracted from Research Services Ltd. (1976).

POST-IGNITION SITUATION

Action taken to put out the fire

As has been previously mentioned it is highly probable that only a small proportion of fat-pan fires result in a call to the fire brigade. Few data exist to compare the behaviour of those persons who called the fire brigade with those who did not. The only substantial data available which do not emanate from fire brigade records are those collected by Research Services Ltd. (1976). The resultant actions taken by all respondents who had experienced a chip-pan fire before the publicity campaign are summarized below in Table 7.10. Because the categories chosen to classify the behaviour patterns are rather unsatisfactory and are not mutually exclusive, interpretation of the information is difficult. The three most reported actions were 'turned off heat', 'took/threw outside', 'covered with damp cloth'. However, it is possible that these actions (certainly the first of the three) could have been taken in conjunction with one or more of the others detailed in the table.

An attempt to provide comparable information for a sample of cases where the fire brigade was called was made by Rutstein and Butler (1977). They obtained their information from an examination of K433 forms (standard reporting forms)

Table 7.11 Actions taken by householders to put out fire—reported on K433 forms before campaign

	Smothering* with towel, cloth, etc (per cent)	Removal outdoors (per cent)	Water from bucket, etc. (per cent)	Other action** (per cent)	Swith off only—leave alone† (per cent)
Area 1	47	16	7	9	21
Area 2	37	19	7	9	28
Area 3	25	19	5	9	42
Area 4	31	12	5	12	40
Area 5	32	25	7	10	26
Area 6	37	15	10	16	22

Source: Abstracted from Rutstein and Butler (1977).
Note: *Only the principal action is considered, such as 'switch off' and 'smother with towel' is coded as smothering
         †Unknowns have been excluded
         **'Other actions' include use of sand, water, pan lid, etc.

completed by the fire brigades in three study areas. Their results are shown in Table 7.11. Unfortunately it is difficult to make valid comparisons between these two sets of data, because of the different ways in which the data were collected.

Injuries associated with the incidents

It has already been pointed out that the number of non-fatal casualties associated with all cooking appliance fires (the vast majority of which are fat fires) is greater than for any other source of ignition. It may be postulated that the probability of personal injury is directly related to the action taken by the individual following the ignition of the fat. A report from the Fire Research Station (Chambers, 1967) provides some interesting data in relation to this; these are summarized below in Table 7.12.

It can be seen that the more inappropriate methods (1, 2, 3, 8) used by the public to extinguish the fires are associated with higher average losses (see final column of Table 7.12). In cases where an attempt was made to extinguish the fire with water or sand (methods 1, 2 and 3), as well as with the more appropriate methods (4, 5, 6 and 7), the losses are largely accounted for by property damage. However, for 'removal outside' (method 8), although the probability of extinguishing the fire is high (fourteen out of fifteen cases) the proportionate contribution of the cost of personal injury to the average losses is extremely large. The implication of these figures is that the chance of personal injury is greatly increased when the pan of hot fat is handled directly. Chambers (1967) reported that 'the action that is often taken, and [which] stands out as completely inappropriate, is removal of the pan out of doors. Although the fire is usually prevented from doing much damage to the room, fifteen fires dealt with in this way resulted in six serious injuries [out of only eight injuries identified on the whole study].' Further evidence to support this

Table 7.12 Fat-pan fires, Surrey 1964. Success of 'First aid' fire fighting by method employed

Method employed	Number of fires	Number out on fire brigade arrival	Percentage out on fire brigade arrival	Average loss excluding injuries	Average loss including injuries
1. Water from garden hose	3	–	–	£117	£117
2. Smothering with sand, earth, etc.	5	–	–	£ 66	£106
3. Water from buckets, bowls etc.	26	19	73	£ 57	£ 65
4. Extinguisher (any type)	7	3	43	£ 50	£ 50
5. Switch off, leave alone, etc	34	12	35	£ 37	£ 37
6. Smothering with lid	2	2	100	£ 32	£ 32
7. Smothering with towel, cloth, etc.	8	8	100	£ 10	£ 10
8. Removal outside	15	14	93	£ 23	£103

Source: Chambers (1967).

view, that the risk of injury in a fat-pan fire is related to firefighting action is provided by Rutstein and Butler (1977). From an analysis of fire brigade K433 forms, they found the number of injuries associated with 'removing the pan outside' to be 111 per 1000 fires. The equivalent rate for 'no action' was 32 per 1000 fires and for 'smothering', 50 per 1000 fires.

Data presented in the annual fire statistics do not provide a detailed breakdown of the injuries associated with cooking fires. However, a recent Home Office analysis of K433 forms gives an age-by-sex distribution of such injuries for the year 1973. Unfortunately, these relate to cooking fires involving all materials (that is, not just fat) and all buildings (that is, not just dwellings). This distribution is illustrated in Figure 7.3. Bearing in mind that the vast majority of these injuries are associated with domestic incidents it is interesting to note that whereas the age breakdown seems to reflect the expected pattern of exposure to the hazards of cooking with fat, this is not so with the sex distribution. In all age categories other than '65 plus' the frequency of injury is slightly greater for males than for females. These figures bear a number of possible interpretations. First, that there is no difference in the level of exposure to the cooking task, and in the risk of injury when exposed, between males and females. Second, that although males have a lower exposure level, due to inexperience in the task they run a higher risk of starting a fire and sustaining injury. Third, that when a male is present in the dwelling at the time of ignition (even though not directly involved in the cooking task) he is frequently called upon to take suitable counteraction.

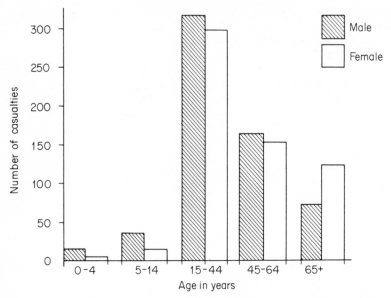

Figure 7.3 Non-fatal casualties resulting from fires in cooking appliances in *all* buildings, 1973. Source: Home Office Analysis (personal communication, 1976)

SUMMARY OF THE MAJOR CHARACTERISTICS OF FAT FIRES

Before discussing practical approaches to the problem, it will be useful to summarize the major characteristics of fat fires.

(1) Cooking appliance fires comprise approximately one-third of all domestic fires.

(2) Eighty per cent of cooking appliance fires in dwellings result from the ignition of cooking fats.

(3) Over 90 per cent of cooking fat fires occur on a ring or hot plate.

(4) The rate of occurrence of fat-pan fires (reported to fire brigades) on electric cookers is over four times that for gas cookers.

(5) The correct cooking temperature for oils or fats is approximately 205°C and spontaneous ignition occurs at temperatures above 310°C.

(6) There is little difference in the above temperatures and in the time required to reach them between different types, ages or conditions of the fat or oil in use.

(7) The majority of fat fires occur prior to or during the cooking stage.

(8) The majority of fat fires result from the inattention of the person responsible for the cooking. This inattention can be associated with unexpected demands, misjudgement of elapsed time, underestimation of heating rates, and memory failure.

(9) The proportion of cooking appliance fires which involve a casualty is lower than for other major sources of domestic fires. However, the overall number of

non-fatal casualties associated with the former is greater than for any other source. (10) Such casualties as do occur seem to be mostly associated with handling the ignited pan during attempts to remove the hazard.

PRACTICAL APPROACHES TO THE PROBLEM

The successful introduction of measures designed to reduce the number or severity of any type of accident usually depends upon a thorough understanding of the components of that accident situation. The purpose of this chapter was to collate such information as is available on the problem of fat fires. However, as has been indicated, published data, particularly the annual fire statistics now produced by the Home Office, are often incomplete, inappropriately classified or lacking in sufficient detail, especially with regard to events prior to ignition of the fat. There are two reasons for this. First, the published fire statistics have to be relevant to a wide range of users, and therefore cannot provide detailed information about certain types of fire. This problem can be partially overcome by requesting specific analyses of the fire statistics. This was done by the authors, for fat fires in dwellings. The second reason is that the official statistics are compiled from the fire brigade K433 forms. The information recorded on these forms about events before the fire brigade arrived is naturally lacking in detail (for example, see Table 7.5 of this chapter).

Notwithstanding the deficiencies of existing data, a number of approaches to reducing the frequency or severity of fat fires and injuries have been proposed in the literature. The following approaches are considered, by the authors of this chapter, to be the most feasible in terms of potential effectiveness, cost, durability, reliability, simplicity and public acceptance. These approaches can be divided into two types. First, modification of traditional cooking equipment to reduce the hazard associated with the task. Second, modifications to the behaviour of the general public by increasing their awareness of both the cause and prevention of cooking fat fires and associated injuries.

Thermostatically controlled deep fat fryers

Several such models are at present commercially available. Whilst these seem to provide the best available solution to the problem, their capital cost may prove prohibitive to all but a segment of the market. 'The main advantage of an electric deep fat fryer over a conventional chip-pan is that the oil is always kept at the right temperature. So there is less risk of the oil burning and causing a fire and the oil lasts longer. . . They also worked out cheaper to run but are a bit expensive to buy' (*Which?*, 1976).

Thermostatically controlled cooker rings or burners

Several models of cookers have already been fitted with 'simmer rings' designed mainly to prevent milk boiling over. It is suggested that such devices could be adapted for use with cooking fats (which require a much higher cooking temperature).

The technical feasibility of the approach (with its attendant fail-safe devices) has not however been established. Also it has been suggested (Parker, 1976) that a thermostatically controlled hob has not gained general (public) acceptance because its mode of operation is not understood, the rate of heating being less than with the simple energy regulator. Nevertheless, it is felt that publicity regarding the hazards of fat frying and the advantages of cooking at correct temperatures (in terms of better food and longer-lasting oil) might well overcome this resistance.

Temperature controlled warning devices

It is considered possible that an audible warning device which comes into operation at a predetermined level above cooking temperature, for instance at 240°C, could be marketed either as an integral part of new chip-pans or as a separate device to be fitted to existing pans. This device would perhaps have more scope for greater market penetration in the short term than those mentioned in the two sections above. However, unlike the former devices, an audible warning system would require manual correction of the heating rate which implies proximity, a degree of attention and a quick response by the individual.

Education of the public

The importance of the behaviour of the individual has been much discussed earlier in this chapter. It might therefore be presumed that a significant reduction in the extent of the problem could be achieved by a comprehensive campaign of public education. The most recent such campaign was that conducted by the Home Office during 1976 and 1977. Two main conclusions have been drawn from the evaulation of this television advertising campaign. First, that the 'campaign has been extremely successful in promoting knowledge of how to safely extinguish a chip-pan fire. It has also, but to a lesser extent, educated housewives in the causes and prevention of these fires.' (Research Services Ltd., 1976). The second main conclusion was that 'there was a 30% drop in the number of fat-pan fires to which the Fire Brigade were called, and a 40% drop in the number of fat-pan fires which then had to be extinguished by the brigades' (Rutstein and Butler, 1977).

However, it looks as though the relative success of this occasion was comparatively shortlived (Home Office, 1977). About six months after the campaign had ended, the number of calls to fat-pan fires were back to their previous level. Also, since a major part of the campaign was to instruct the public in the extinguishing of fat fires it is felt by the authors of this review that a reduction in calls to the fire brigade may well have been accompanied by an associated increase in injuries resulting from such behaviour. Unfortunately, the study did not extend to the usage of medical facilities within the campaign areas. therefore no data are available to refute or substantiate this hypothesis.

Finally, it is considered that, over and above any 'blanket' publicity campaign, the introduction of any practical preventive measures, such as discussed in the sections above, should be accompanied by relevant and specific public education.

REFERENCES

Andrews, C. J. A. (1969). SSHA. *Chip-pan Fires Research Project.* Unpublished report of the Napier College of Science and Technology, Edinburgh.

Chambers, E. D. (1967). *Behaviour When Faced With Fat-Pan Fires.* Fire Research Note no. 654; Joint Fire Research Organisation.

Consumers' Association (1976). 'Electric deep fat fryers.' *Which* (June 1976).

Department of the Environment (1974). *United Kingdom Fire and Loss Statistics for the Year 1972.* Building Research Establishment and Department of the Environment.

Department of Prices and Consumer Protection (1976). *A Commentary on Electrical Fatalities in the Home.* (London: DPCP).

Fire Protection Association (1970). 'Out of the frying pan', *FPA Journal,* **86,** 18–23.

Fire Research Station (1976). Personal communication. Requested analyses of Fire and Loss Statistics for 1973.

Hogg, J. M. (1963). *Fires Associated with Electric Cooking Appliances.* Fire Research Technical Paper No. 9. (London: HMSO).

Home Office (1976). Personal communication.

Home Office (1977). *United Kingdom Fire Statistics for 1974.*

Mytton, M. G. and Butler, A., (1976). 'Measuring the effects of a fat-pan fire publicity campaign; an interim report', Memorandum No. 9/76. Home Office Scientific Advisory Branch.

Parker, L. C. (1976). Personal communication from the Electricity Council.

Research Services Ltd., (1976). *Chip-Pan Fire Prevention Advertising Campaign Evaluation* (London: Research Services Ltd.).

Rutstein, R. and Butler, A. (1977). 'Measuring the effects of a fat-pan fire publicity campaign: a final report'. Home Office Scientific Advisory Branch.

Stark, G. W. V. and Mulliner, W., (1965). *The Fire Properties of Cooking Fats.* Fire Research Note No. 610, Fire Research Station.

Fires and Human Behaviour
Edited by D. Canter
© 1980 John Wiley & Sons Ltd.

CHAPTER 8

Domestic, Multiple Occupancy, and Hospital Fires

DAVID CANTER, JOHN BREAUX AND JONATHAN SIME
Department of Psychology, University of Surrey

INTRODUCTION

This report gives a broad view of the first phase of a project established to elaborate patterns of behaviour of people involved in buildings on fire.

Effective fire safety regulations are, of necessity, based upon assumptions as to what people will do in the event of a fire. In a previous report Canter and Matthews (1976) argued that assumptions about human actions, on which the fire regulations are built are, in the main, unvalidated. They also reviewed relevant research then available and showed that much of it questioned the assumptions then prevalent. Existing research was extremely limited and whilst the quality and quantity of research is growing internationally (as revealed by the present volume) the available sources of information, discussed in the introductory chapter, on what actually happens in a fire are still severly restricted.

The studies to be reported support the view that fire is experienced as a complex, rapidly changing event, which, in its early stages at least, is usually highly ambiguous, providing little positive information to act upon. In other words any fire can take on a variety of different forms and change from one form to another very quickly. From the point of view of the person who experiences the fire this means that a lot of information is necessary in order to understand the situation fully and hence to recognize the appropriate role he should take and to understand the appropriate rules (and related activities) that he should follow.

As a consequence central research questions are the extent to which behaviour in fires can be explained in terms of (a) appropriate understanding of the fire situation by those affected; (b) accurate recognition of their role in the event; (c) effective understanding of the relevant rules for their roles; (d) efficient performances of actions relating to their interpretation of the rules. In all four aspects it is necessary to discover the contribution of the building and its equipment to the outcome.

We are grateful to the Department of the Environment who, through the Fire Research Station at Borehamwood, supported the studies from which this paper was derived. We are especially grateful to Ian Appleton and Brian Pigott at FRS for the way in which they have encouraged and guided our research efforts.

In order to answer these questions detailed case studies were carried out. This enabled the researchers to obtain clear information on the sequence of actions which occur in a fire, as well as details of the crucial early stages of fire development.

DATA SOURCES AND MODE OF ANALYSIS

Information was received from the fire brigades assisting in the study on the occurrence of all fires. Those which involved people within a building were followed up. Initial contact was made either with the fire brigade officer concerned, or when he was not available, with any member of the public who could be located who had direct knowledge of the fire. In a number of instances, in addition to information received from the brigades, press reports were drawn upon to make contact with participants. In all cases an attempt was made to obtain information from every individual who had had direct experience of the fire. In some cases police witness statements provided additional information to complement that obtained by other means.

Once participants in fires were contacted, be they members of the public or of the fire brigades, an interview was arranged at a convenient location, usually the site of the fire. At this interview the aims of the research were described, as well as the use to which the interview would be put. The participant was then asked to give as detailed an account as possible of everything which had happened from the time they were aware that something out of the ordinary might have been occurring. An interviewing technique was used whereby the interviewer interrupted the flow of the account as little as possible. Once the respondent had given as full an account as he felt possible he was then questioned on specific issues. These issues were derived from the interview agenda (Table 8.1) and from the need to obtain reference points for the actions reported, both to relate to the growth of the fire and to cross-reference with the reports of other participants to provide corroboration of the acts described. This whole interview was tape-recorded. If further information was volunteered with the tape-recorder turned off notes were taken of this.

The tape-recordings were typed verbatim onto transcripts and these transcripts were then used as the primary data source. The first stage of analysis was to break the transcripts down into single acts, and a note was taken of the position in the sequence of events and physical location of each act. These acts were then reduced for each fire to a standard 'dictionary' of all acts carried out by each person in any given fire. Cross-checking between research assistants who carried out the coding led to a standard composite list for each building occupancy type. It was then possible to relate each entry of the dictionary to the original breakdown of acts and ascertain which act had followed each of the other acts in the dictionary.

A matrix was prepared with the same list of acts as both its rows and columns. Into the cells of this matrix were entered the frequency with which the acts in the columns followed those in the rows. This is known as a 'transition' matrix because it shows the likelihood of each act giving rise to every other act. This transition matrix provides the essential numerical description of the fire for statistical analysis.

The next stage in the analysis is to examine the transition matrix to determine

Table 8.1 Agenda for interview

Recognition of fire
Location of occupant
Ongoing behaviour
Sequence of actions
Perception of the situation including:
  other people's behaviour
  physical circumstances of fire
  time estimates
Related past experience
Background details, such as:
  role of person
  layout of building
  fire damage

---

those acts which are as likely to follow as to precede each other at a point in the overall sequence. The acts are collected together into a group called an 'equivalence' group. It is then possible to determine the sequence in which the equivalence groups occur. The greater the intricacy of the behaviour the greater will be the number of acts in the equivalence groups. Technically this process is known as 'decomposing' the transition matrix. The resultant set of related act sequences may be represented in the 'decomposition diagrams' illustrated (Figures 8.1 to 8.5).

In these decomposition diagrams the act equivalence classes are represented as circles. Dashed circles indicate acts which occur with a lower frequency and are included to give the reader an idea of behavioural relationships. It was felt that if only the principal acts were emphasized (that is, those acts which occur more often than would be expected by chance) then the resulting skeletal picture might be confusing. The relationships between acts (and their typical sequence) are indicated by arrows. At times, one type of act may be repeated; in a sense this is an act following itself, as, for example, successive acts concerned with fighting the fire. In this case the 'fight fire' circle would, apart from any other links, have a looped arrow coming back on itself. The numbers next to an arrow linking any two acts refer to the strength of the association between the two acts. The higher this number, the greater the strength of the association, that is, the more likely it is that given the occurrence of one act, the one specified will follow it. For example, if every time anyone encounters smoke they immediately leave the area then one would expect to see a higher number than if, say, having encountered smoke some would leave the area, others ring the fire brigade and yet others investigate from where the smoke was coming.

The diagrammatical summary of the decomposition of the act matrix is valuable in that it summarises in a visual form what may be a complex series of events. It is possible to test theoretical predictions directly by placing an index of the likelihood of actions following each other on the links in the diagram. More particularly tests can be carried out by making separate decompositions for acts performed by a (particular) group such as men or women. It is also possible to examine the frequency of occurrence of particular acts and to attempt direct tests on the statistical difference between these frequencies.

RESULTS

In this results section each of the three occupancy types are in preparation for publi-
Further reports associated with each occupancy type are in preparation for publication. In the present summary only the general trends for that occupancy will be
reported. The order of discussion of the occupancies is intended to reflect the increasing complexity of the fire situation and the resultant need for an increasingly
elaborate model to explain behaviour in fires in those settings.

Domestic fires

Table 8.2 gives the act dictionary which was produced from studying fourteen fires
and the accounts of 41 people involved. It summarizes the 1189 acts which occurred in all those fires. Figure 1 is the decomposition diagram for all cases. It shows
the departure from present activity as a cluster of interrelated acts. These consist of
noticing cues but finding them ambiguous, often hearing strange noises, misinter-
preting or ignoring these, or discussing them with anyone present, and, if the cues
persist, investigating. The only variation from this initial group of actions is if
smoke or the fire is encountered directly. Furthermore, if investigation follows the
early ambiguous cues, then it inevitably leads to an encountering of smoke, either
within the room of fire origin or outside this room if the smoke is spreading. In this
case people are still likely to enter the room where the fire is.

It can be seen from Figure 8.1 that a direct encounter with the smoke or fire
generates variability in the likely response sequence. Clearly this variability is a
function of the stage of fire growth and location of the fire at the time when it is
encountered. Much of the variability present in Figure 8.1 can be accounted for by
differences between men and women, occupants and neighbours. Figures 8.2 and
8.3 make comparison of the male/female groups possible by presenting separate decompositions for each of them. The main differences exist at the initial interpretation stage and the behaviour following investigation and encountering the smoke.

While both males and females tend to misinterpret ambiguous cues, males are
more likely to do so and delay investigation. The response of a female may be delayed by interaction with a male if present. Eventually one of them initiates investigative activity. Both males and females are likely to investigate. If informed by
someone who has returned to say there *is* a fire, the tendency is to check this information for oneself. The indications are that this may be more likely if males initially receive a warning from females than vice versa. This tendency to continue investigation after being informed, is particularly characteristic of domestic as opposed to other building/occupancy types. It is apparently related to the role of the
individual in his or her own home as well as the proximity of a fire. More responsibility may be felt for the safety of others who are likely to be present and for the
prevention of damage.

The variability of the actions which follow the encountering of the smoke and
fire itself is explained by male/female differences. Females are more likely to warn
others and wait for further instruction (for example, if husband and wife are both
present). Alternatively they will close the door to the room of fire origin and leave

Table 8.2 Act dictionary for domestic fires

Code no.	Code	Frequency	Overall frequency of sub-categories
(1)	Pre-event actions	42	42
(2)	Perception of stimulus (ambiguous)	40	
	Alerted/awoken (ambiguous)	20	76
	Note behaviour of others (ambiguous)	16	
(3)	Perception of stimulus (associated with fire)	12	
	Note fire (development)	31	78
	Encounter smoke	35	
(4)	Interpretation (incorrect)	23	38
	Disregard/ignore prior stimulus	15	
(5)	Receive warning/information/instruction	73	89
	Ask advice/request information	16	
(6)	Search for people (in smoke)	18	
	Encounter person in smoke	12	36
	Check state of victim	6	
(7)	Observe rescue attempt	19	19
(8)	Advise/instruct/reassure	51	56
	Note agitated state of person	5	
(9)	Feel calm/unconcerned	11	11
(10)	Experience negative feelings	16	
	Experience uncertainty	15	49
	Feel concern about occupants	16	
	Request assistance (urgent)	2	
(11)	Fire equipment faulty/unable to work	8	13
	Struggle with environment	5	
(12)	Seek information/investigate	76	76
(13)	Realize door to fire area open	6	
	Prevent fire spread	29	45
	Ensure accessibility	10	
(14)	Indirect involvement in activity	27	49
	Wait for person/action to be completed	22	
(15)	Rescue	42	42
(16)	Go/gain access to house with fire	40	52
	Go to neighbour's house	12	
(17)	Dress, gather valuables	20	20
(18)	Fetch things to fight fire with	22	47
	Fight fire	25	
(19)	Evasive	33	84
	Leave immediate area	51	
(20)	Forced back by/breathing difficulties due to smoke/flames	48	
	Cope with smoke	15	77
	Struggle through smoke	8	
	Injured	6	
(21)	Pass through/enter fire area (investigate etc)	35	35

[Continued on page 122

Table 8.2 Act dictionary for domestic fires *(Continued)*

Code no.	Code	Frequency	Overall frequency of sub-categories
(22)	Warn	34 }	41
	Phone for assistance	7	
(23)	Rescued/assisted	13 }	17
	Rescued from window	4	
(24)	Note/wait for fire brigade arrival	45	45
(25)	Enter area of minimal risk	52	52
	Total number of acts =		1189

the house. In both cases females are more likely to seek assistance from neighbours. Male occupants are most likely to attempt to fight the fire. Male neighbours are more likely to search for people in smoke and attempt a rescue.

With reference back to the postulated role/rule model and the 'panic' model, it can be seen that there is no need to draw upon the panic explanation in order to give an overview of what happened in these domestic fires. Although confusion can be seen in the act sequences the inherent logic of them is apparent. Futhermore the effect of the initial interpret/investigate stage is graphically represented. The pattern of behaviour which has emerged has role/rule implications. The active role for males is characterized by firefighting or rescue, of females by warning others. Where males and females are present, the latters' role tends to be more passive and/or supportive. The success with which the role is performed depends upon the stage of the fire growth and the skills of the individual. The role and rules followed may be determined both by who is present, as well as socially determined pre-existing sex roles.

The goal of the pilot research is only to clarify issues, yet it is helpful in explaining these findings to indicate some of their implications. The first implication is to question the extent to which any roles and related activities which occur in a fire are particular to that fire situation or are determined by general pre-existing role definitions. If the latter, then the possibilities for predicting behaviour in fires is likely to be very great.

Beyond the theoretical implications are the practical ones to be considered. Clearly education, and general fire prevention propaganda could be much more effective if based upon models such as those illustrated in Figures 8.1, 8.2 and 8.3. Given the high proportion of deaths which occur in domestic fires, the potential value of research such as this should not be minimized. Clearly the current data base is limited but data from domestic fires would be the easier to enlarge in future research than from the other building occupancies.

The implications of an approach such as the present one depend of course on policy decisions about legislation for domestic premises. The significance of misinterpretation and investigation will be considered again later, but clearly has relevance for any form of domestic alarm/detector system which is recommended.

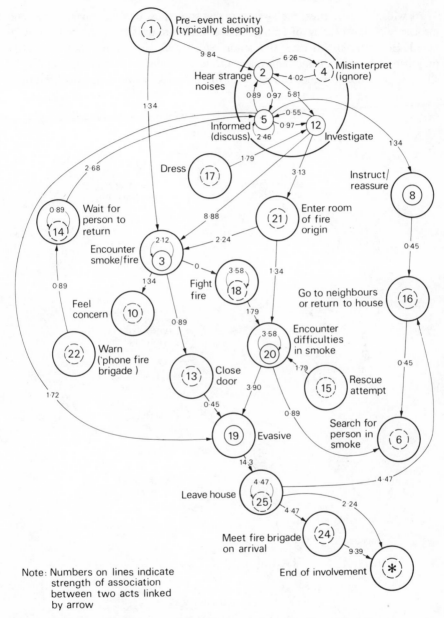

Figure 8.1 Decomposition diagram—domestic fires—all cases

Multiple occupancy fires

Table 8.3 presents the act dictionary which was produced from studying eight multiple occupancy fires and the accounts of 96 people involved. The dictionary summarizes the 1714 acts which occurred in all those fires. Figure 8.4 presents the decomposition diagram for all cases.

As with domestic fires, the awareness that something unusual is happening commences with the hearing of strange noises which are usually misinterpreted or ignored. However, in this case, the action which normally follows these cues after they persist forms a distinct equivalence class associated with investigation. Typically this gives rise to direct contact with fire or smoke and a consequent return to

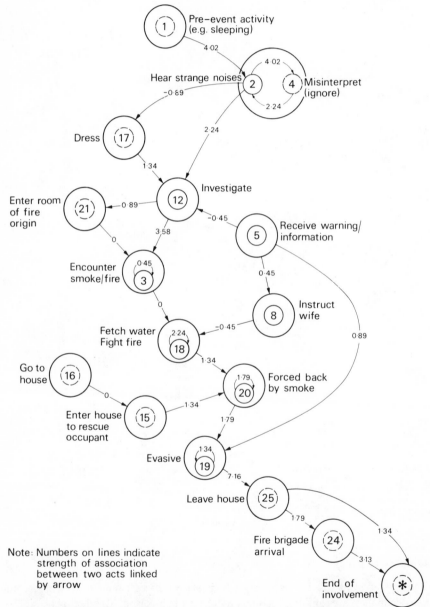

Note: Numbers on lines indicate strength of association between two acts linked by arrow

Figure 8.2 Decomposition diagram—domestic fires—males

the original room where the individual was. The characteristic sequence which
follows from this then relates to the individual going to the window, shouting for
help and being rescued. Clearly the multiple occupancy case, especially that occurring in hotels, produces a pattern much more complex than that for domestic fires.
This complexity is most apparent in the number of significant outcomes of any

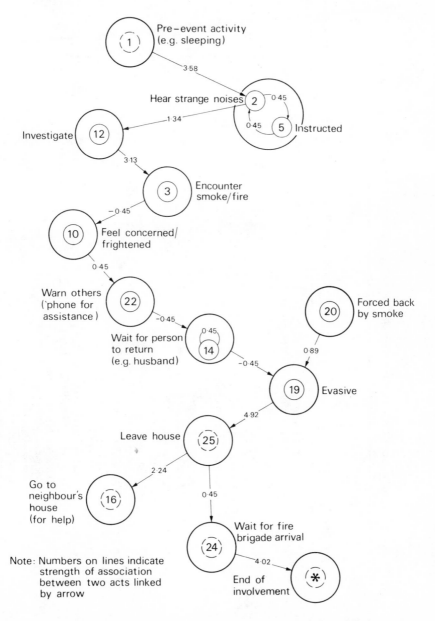

Figure 8.3 Decomposition diagram—domestic fires—females

Table 8.3 Combined act dictionary for multiple occupancy and hospital fires

Code no.	Code	Multiple occupancy		Hospital	
		Frequency	Overall frequency	Frequency	Overall frequency
(1)	Pre-event actions	100	100	56	56
(2)	Perception of stimulus (ambiguous)	112	130	61	68
	Note behaviour of others (ambiguous)	18		7	
(3)	Perception of stimulus (unambiguous)	83	172	46	96
	Note worsening of immediate situation	43		25	
	Note fire development	46		25	
(4)	Correct interpretation	19	41	11	29
	Arrive at conclusion	22		18	
(5)	Incorrect interpretation	45	61	25	31
	Disregard/ignore prior stimulus	16		6	
(6)	Seek information and investigate	163	180	79	108
	Approach fire area	17		29	
(7)	Disseminate warnings/information	59	91	27	60
	Receive query	7		3	
	Raise the alarm	19		19	
	Inform superiors	6		11	
(8)	Rescue related	4	14	11	68
	Ancillary rescue actions	10		16	
	Evacuation with/limited transfer	0		41	
(9)	Dress/gather valuables	83	83	38	38
(10)	Evasive	86	86	28	28
(11)	Coping (self-related)	68	68	9	9
(12)	Securing environment	43	47	25	38
	Check security of others	4		13	

(13)	Fetch things to combat fire	4 ⎫		7 ⎫	
	Combat fire	5 ⎭	9	12 ⎭	19
(14)	Give instructions	19 ⎫		16 ⎫	
	Receive instructions	33 ⎬	55	41 ⎬	63
	Intervene	3 ⎭		6 ⎭	
(15)	Give assistance	25 ⎫		9 ⎫	
	Receive assistance	50 ⎪		3 ⎪	
	Note arrival of assistance	32 ⎬	139	24 ⎬	41
	Seek assistance	32 ⎭		5 ⎭	
(16)	Experience movement/breathing difficulties	47 ⎫		24 ⎫	
	Overcome hindrance	7 ⎬	54	2 ⎬	38
	Experience negative feelings	0 ⎭		12 ⎭	
(17)	Note persistance of stimulus	17	17	11	11
(18)	Receive information (verbal)	56 ⎫		48 ⎫	
	Note behaviour of others (unambiguous)	58 ⎬	138	25 ⎬	90
	Note people who need to be rescued	24 ⎭		17 ⎭	
(19)	Alter plan (self-initiated)	4 ⎫		4 ⎫	
	Model behaviour	0 ⎭	4	1 ⎭	5
(20)	Duty related	27	27	39	39
(21)	Experience uncertainty	15	15	8	8
(22)	Note nothing unusual	5	5	5	5
(23)	Enter area of minimum risk	88	88	53	53
(24)	Leave immediate area	72	72	72	72
(25)	Approach/enter duty area	0	0	12	12
(26)	Encounter colleague/superior	6 ⎫		17 ⎫	
	Encounter subordinate	1 ⎭	7	2 ⎭	19
	Total number of acts =		1714		1104

given act and the number of routes through the diagram. This means that the typical route described in the previous paragraph is far less common than are the analogous typical routes described for domestic fires. The increased complexity appears to be a function of the range of potential sources of information about the fire and about appropriate actions. Where actions in the domestic situation could be

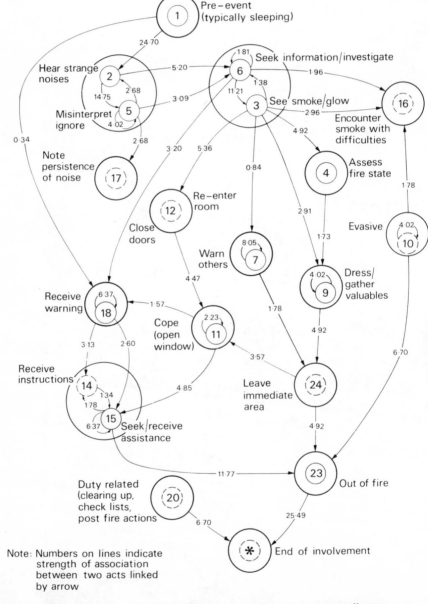

Note: Numbers on lines indicate strength of association between two acts linked by arrow

Figure 8.4 Decomposition diagram—multiple occupancy—all cases

related to the roles of husband, wife or neighbour, in the multiple occupancy setting a person can be at a loss as to whether they are the prime discoverer of the fire or one of many individuals with similar experience. This added complexity can be highlighted by the emergence of the action cycle associated with receiving a warning, receiving instructions, seeking assistance and making attempts to cope with the increasingly dangerous situation before again interacting with others who can possibly help.

The fact that people are aware of the likelihood of meeting others is demonstrated most directly by the way in which their assessment of the fire state is typically followed by their dressing and gathering valuables. Surely these acts which waste time would feature less if the individual did not anticipate that there were others in the immediate vicinity. There is no clear indication of any parallel acts in the domestic situation.

Returning to the opening acts, once again, these are associated with the reception of ambiguous cues, usually odd noises, followed by a process of misinterpretations. The misinterpretation seems to be of more danger in the multiple occupancy fires studied than in the domestic fires. In the multiple occupancy situation the options for leaving the building appear to reduce more rapidly because the initial time loss militates against later unassisted escape. It is also of interest that the early cues to the fire in domestic settings may more often be olfactory or auditory as opposed to the auditory cues of the multiple occupancy fire. A further distinction from the domestic setting is the absence of any attempt to fight the fire. Indeed, people rarely get to the room of origin of the fire as they do in a domestic setting. The pattern, then, is much more one of people becoming vaguely aware of something happening at some distance, realizing its danger, then drawing upon whatever assistance they can from others in order to get away from the dangerous situation by whatever means they can. The means of escape employed will be discussed later, but for the moment it is to be noted that many individuals return to their rooms. Looked at from a role/rule perspective this is intriguing because it suggests that many hotel guests may well associate the role of being a guest with expected actions linked to their hotel room. The room may be regarded as the only setting that is unique to the guest, and private to them. Is it perceived as a natural place to escape to from danger? If it were, the location of escape routes in relation to hotel rooms would need reconsidering.

Hospital fires

Table 8.3 presents the act dictionary which was produced from studying six hospital fires and the accounts of 61 people involved. The dictionary summarizes the 1104 acts which occurred in all those fires. Figure 8.5 presents the decomposition diagram for all cases. The case studies covered a variety of hospital types (geriatric, psychiatric, general) yet a number of patterns are revealed from the overall decomposition.

The general decomposition is now very intricate, this being revealed most clearly by the very large equivalence class. It is therefore necessary to describe typical event sequences.

Detection and investigation of the fire takes place relatively early in fire development as compared with multiple occupancy incidents, possibly because of the
higher spread of people in the building. Once detected, the transfer of information
concerning the fire is highly specified, with senior nursing officers tending to be initial recipients. Thus investigation is typically first carried out by these people.
Shortly thereafter they relay information to their junior colleagues. Although jun-

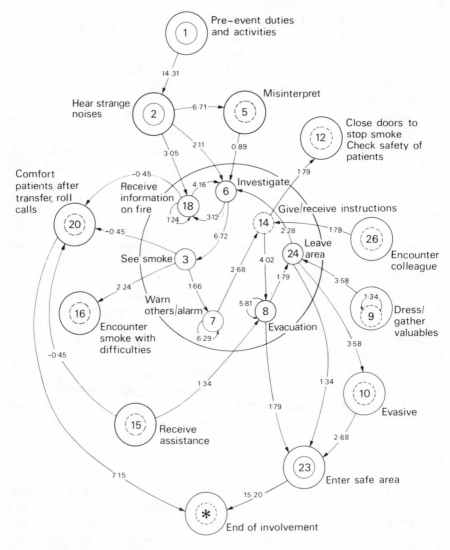

Note: Numbers on lines indicate
      strength of association
      between two acts linked
      by arrow

Figure 8.5 Decomposition diagram—hospitals—all cases

ior staff are likely to receive early warning that there is a fire, there is a great demand on their part for information concerning location and intensity which they will subsequently need to know in planning patient evacuation. Interviews suggest that from their point of view this information is often late in arriving.

The nature of organizational hierarchies is particularly evident in the way an evacuation or transfer of patients transpires. Ultimate destinations and routes are usually specified by senior staff. Action by junior staff (except for preparing patients) is guided by prior instructions, both through previous training and orders received during the incident. The act of evacuating patients is often related to several other actions and processes which when viewed *in toto* reflect greater behavioural complexity than is the case for multiple occupancy fires. However, due to greater organizational sophistication this higher action complexity (as evidenced in the central equivalence class) does not appear to be strongly related to increased threat. Evacuation and movement through smoke does occur. It would seem to be due to inadequacies in the building structure (exits not wide enough to accept beds leading to slower or delayed movement; ventilation systems contributing to smoke spread) or delays in information reaching junior staff.

In the hospital situation the role aspects, then, become more sophisticated than in the earlier two cases. Now the existing organizational structure imposes a pattern with differing roles carrying out different aspects of the 'investigate' and 'assist' sequences. Furthermore, the dangers in the situation are more clearly a function of the effectiveness of the communication between different role groups. Parallels here between residents and neighbours, or between hotel guests and staff are of interest because in these two cases there is no strong evidence for effective communication in the present case studies. However, in hospitals the effective communication between 'investigators' and those who 'give assistance' is typically present.

The implications of this pattern for fire regulations is that if they are framed independently of the existing organization then confusions in the communications system may be aggravated. Perhaps even more significant is that the senior individuals should know how the building is to be used during a fire and agree with that proposed use. The details of this building use may well be crucial because of the time delay involved in issuing instructions and then changing them. However, the great variability amongst buildings does not make the current data very useful for detailed comment on design.

GENERAL TRENDS ACROSS FIRES AND PROPOSALS FOR FUTURE RESEARCH

In previous sections it has been demonstrated that characteristic patterns of behaviour do occur in each occupancy type which has been examined, even though there are notable variations within these types. It is fruitful to take the possibilities for generalization a stage further and to explore the patterns which might be derived across all the fires investigated. The main value of such an exercise is to generate a broad model of behaviour in fires; any given fire can then be viewed as a particular variant of the general trend. Consequently the overall trends can be used both for indicating the key areas in which fire regulations could have a substantive impact, and for structuring the development of future research.

As a first step in producing a summary general model, each of the overall sequence decompositions for a particular occupancy was examined and those act sequences which occurred a statistically significant number of times were listed. Logically similar act sequences were then combined to produce the flow chart. Such a chart serves mainly to show the potential complexity of people's behaviour in a fire; in order to reveal the general patterns a simplification of it is produced in Figure 8.6.

The first thing to note from this general model of behaviour in fires is that a number of acts recur, through the model, at different stages in the sequence. This has been a regular finding of the present research. In essence it implies that particular acts and act sequences derive their behavioural significance from the position in the overall sequence at which they occur. In other words, survival implications of any given act can only be determined in relation to the position in the sequence at which it occurs. In the general model, examples of this implication of location in a sequence can be seen for the acts, 'investigate', 'see smoke', 'return to room', 'warn others' and 'evacuate'.

The relevance of this finding for research activities and reports is considerable. It indicates that any summary of behaviour in fires (Wood, 1972; Bryan, 1977) which is essentially the percentage of occurrence of particular acts can at worse be misleading and at best ambiguous. The relevance of sequence position also demonstrates that any regulations aimed at increasing the frequency of occurrence of a particular behaviour could be counterproductive if it did not also ensure that the behaviour took place at the most effective point in the fire sequence. Thus, for example, fire alarms which generate a late investigative reaction could, in some situations, be more 'dangerous' than early recognition from awareness of smoke. Notably, this would be a situation in which people took little notice of smoke because there had not been any alarm.

Given the importance of the whole sequence of acts it is most useful to examine Figure 8.6 in terms of the points of potential sequence change (the 'nodal points'). There are three of these: the first is immediately after the initial cues at which an 'investigate' or 'misinterpret' sequence can be set in motion; the second comes after seeing smoke, at which one of three 'prepare' sequences can be entered; the third follows the occurrence of the particular preparation, giving rise to choice between 'wait', 'warn', 'fight' or 'evacuate' sequences. To reduce it to its broadest skeleton the structure of behaviour in fires is shown in Figure 8.7, which summarizes a number of recurrent findings. First, it illustrates how the potential actions increase in variety as the sequence of behaviour unfolds. One consequence of this is that whilst general statements about all fires can be made with some confidence for the initial stages of the sequence, at later stages the actions are more likely to be highly specific to a particular occupancy context.

The second point to which Figure 8.7 draws attention is that the role of influences on the sequence should be examined in relation to the three nodal points, whether these be influences derived from the building and its fire-fighting hardware or from the people and their previous training. Thus the questioning of the role of

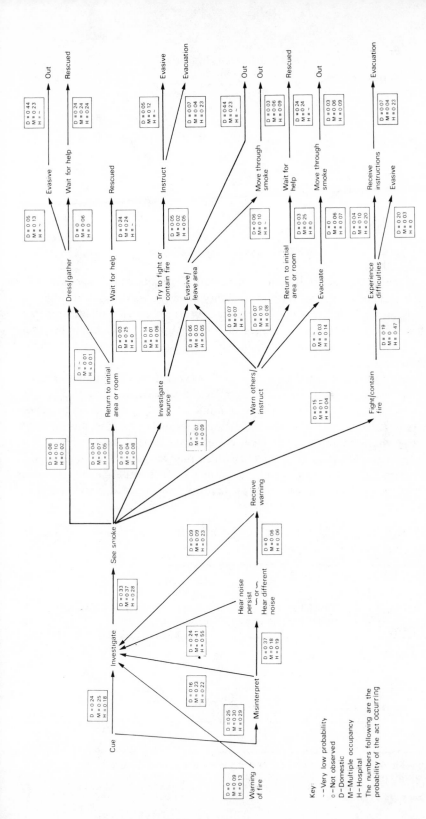

Figure 8.6 Major sequences in fires in general

Fires and Human Behaviour

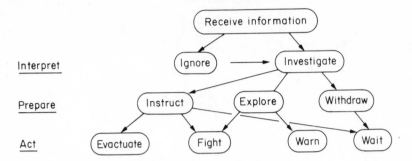

Figure 8.7 Summary of general model

alarm systems in leading to effective interpretation of early cues is raised by this figure. It should be noted that the great majority of early cues which were misinterpreted were acoustic and that therefore there appears to be great potential for misunderstanding the significance of noises in fire situations. Clearly future research should be directed to identifying the frequency and causes of the misinterpretations which take place here.

With regard to the 'prepare' node, the particular occupancy situation is liable to have a great influence. Here, however, the roles and associated rules which people regard as relevant to their situation have an influence on the particular outcome of this stage in the sequence. Consequently, future research exploring the activities at this stage will need to pay particular attention to the roles of the individuals performing the acts and the social context in which they are acting. So, for example, general statistics on rescues from multiple occupancy fires would be of great value if details of who was present, and from where people were rescued, could be collected at the same time.

Table 8.4 Use of fire escape. Number of people for four fires: hotel, hospital, pub and high-rise block

	Used fire escape	Unable to use fire escape effectively	
No attempt to use fire escape		74*	(74)
Attempt to use fire escape	5†	6	(11)
	(80)	85	Total

* Included in this figure are 44 patients in the hospital fire who were evacuated via a lift (under brigade supervision) rather than via the fire-escape.
† In three of these five cases the fire escape was used by members of the public to *rescue* people from a fire.

Table 8.5 Statistics on number of people using a fire extinguisher in thirteen fires. Hotel, hospital and domestic, in which fire extinguishers were directly available

Attempt to fight fire						
Handled fire extinguisher						
Able to work extinguisher		Unable to use extinguisher	Fight fire by other methods	No attempt to fight fire	Total
First extinguisher	Second extinguisher only				
2	5	9	9	106	131

The outcome and sequences of the third 'act' node are likely to be largely a function of the earlier stages, their related roles and the particular situation. Further analysis of current data to relate physical design features to the act sequences of this and the previous stage are in progress.

Some evidence of how effective current provisions are for improving the successful selection of options at this third stage may be gained from a consideration of two sets of hardware provisions specifically designed to increase effective outcome. These are fire escapes (that is, those stairways which are only to be used during a fire) and fire-extinguishers. Table 8.4 provides the figures on use of fire escapes in all buildings where they were present; Table 8.5 provides similar figures for fire-extinguishers. All the *caveats* in earlier parts of this paper concerning the interpretation of act frequencies in isolation are relevant to considering these tables. Nonetheless the figures do demonstrate clearly that the provision of fire-escapes and fire-extinguishers has played a minimal role in influencing behaviour in the fires examined here. It will be of great value if future research can demonstrate how widespread such findings are across the full range of fires.

CONCLUSIONS

This report has demonstrated that a full and coherent account can be given of the sequence of actions which take place in fires, without any need to refer to explanations drawing upon 'panic' or 'irrational' behaviour models. It has been found that behaviour in fires is complex, but that that complexity does not preclude the possibility of providing valid summary accounts of classes of fires and fires in general.

It should be emphasized that in eschewing the prevalent 'panic' framework it has not been suggested that all behaviour in fires is either efficient, or effective, or indeed that it is necessarily even intelligent. Reference has been made to many behavioral sequences which are time-consuming and potentially dangerous. However, these sequences appear to be more a product of attempts to cope with ambiguous, rapidly changing information which also has potentially conflicting implications for behaviour.

It is by considering the role of the person carrying out the acts and attempting to postulate rules which may be associated with those roles that much of the observed variation in behaviour can be accounted for. An example may be useful here—that of the hotel guest. He has already adopted a 'role', that of being a 'guest', for which there are 'rules'. In the event of a fire the best thing for the guest to do may be to leave the building by a certain exit, but he can only do this if he has the information to do so. If that information is not clear to him at the time of the fire then he will behave in accordance with the 'rules' and return to his own room; then he may start to leave the building from that point. In this respect, Figure 8.7 may be regarded as the first step in providing a detailed role/rule model of behaviour in fires.

Finally, the broad implications for fire regulation policy can be indicated. It is suggested that current regulations, in so far as they relate to human behaviour in fires, are designed to minimize the likelihood for 'panic', and to reduce the possibility for people being trapped within the building in the fire compartment. It is possible that the same overall objectives of reducing loss of life and injury could be more effectively achieved, with less disturbance to the building users and probably for the same or reduced capital costs, if the goal or regulations were reorientated to increase the likelihood of informed decisions being made by people in fires. Future research will help to clarify the design potentials of these considerations.

REFERENCES

Bryan, J. L. (1977). *Smoke as a Determinant of Human Behaviour in Fire Situations* (Project People). (University of Maryland, College of Engineering).

Canter, D. V. and Matthews, R. (1976). *Behaviour in Fires; The Possibilities for Research.* (Borehamwood: Building Research Establishment Current Paper CP 11/76).

Wood, P. G. (1972). 'The behaviour of people in fires'. (Borehamwood: Fire Research Station Fire Research Note 953).

Fires and Human Behaviour
Edited by D. Canter
© 1980 John Wiley & Sons Ltd.

CHAPTER 9

Human Behaviour in Fire in Total Institutions; A Case Study

GILDA MOSS HABER
University of the District of Columbia, Washington

INTRODUCTION

The following three case histories of fire were selected out of a number of cases in a study (Haber, 1976) for the following reasons: They represent catastrophe, near catastrophe and success in human response to fire, in total institutions showing a variety of situations and outcomes in fire.

In all cases we are dealing with 'total institutions' as defined by Goffman (1973), that is, 24-hour institutions where the residents or inmates spend their entire day and night. The staff has considerable authority in such institutions, the residents little or none.

Case histories

Case 1: 'Catastrophe' in a fire situation

This case involves a nursing home where a fault in the building design,together with the opening of the (shut) door of the fire, releasing smoke onto incapacitated residents caused 21 deaths. The speed of the fire and deaths is discussed, and design implications and fire protection procedures are explored.

Case 2: 'Near catastrophe' in a fire situation

This case represents a fire in a penal institution. Here we see the speed with which fire and smoke develop and the fact that communications were inadequate when help was needed. No panic behaviour occurred. Although those who came to help were unprepared for a serious emergency, they redirected themselves immediately to cope with the situation.

Case 3: 'Success' in handling a fire emergency

This occurred in a home for the retired aged, and demonstrates that fast action by the staff and good fire protection and drill systems can save patients' lives from fire.

It highlights the unexpected nature and high speed of fire emergencies and also shows an over-reliance on telephones for communication in an emergency. Design implications and fire-protection concepts are suggested at the end of the account.

In each of these cases the fire was deliberately set by a human rather than by faulty electricity or some natural disaster. In all cases, the residents or inmates were dependent upon the decisions of the staff for safety, for protection and escape mechanisms. In case two, some protection mechanisms could not be utilized either because of the lack of familiarity of the helpers with the terrain, or because the staff involved could not handle the fire equipment which was inaccessible or non-functioning. Common to all three cases is the lack of panic. The term 'panic' is defined as a situation where:

(1) There is severe, imminent threat, either physiological or psychological.

(2) Escape is possible but the means is limited.

(3) People overlook possible escape routes.

(4) The situation becomes competitive and invidualistic rather than having the group solve the problem together, therefore,

(5) People involved increase the danger to themselves and others, rather than decreasing it (Turner and Killian, 1972, p. 83).

The fast action by staff, the importance of knowing the area involved, the importance of shutting doors which both contained the fire and protected others from the fire was common to all three cases; so were the feeling of satisfaction by the administration that the best possible effort had been put forth, and the feelings of mixed satisfaction and guilt on the part of the staff directly involved in the fire which found expression in their saying that although they felt they had done their best they 'hadn't done enough, they should have acted faster, done more', and so on. Common to all three cases is the need for adequate communication, and in forms other than the telephone system, which can break down in a fire. Common to all three cases too is a need to know about shutting the door on a fire, having sturdy doors on rooms and having some way of informing visitors or others that the door of the fire should not be opened, otherwise smoke will course out and kill others. Also common to all is the need for adequate fire detectors and alarms such that all people know exactly *what kind* of emergency is on their hands, rather than knowing there is *some kind* of emergency.

United States prisons have the built-in disadvantage of having a high rate of arson. This conclusion was drawn after extensive interviews with wardens, staff and prisoners in eight American prisons, and one each in France, Italy and Egypt. American prisons seem to have at least one fire a month (Haber, in preparation,), whereas French, Italian and Egyptian prisons claim they have more furniture-breaking than fire.

This case study explores the relationship between the social structure of the institution and the fire. The first fire could have been most easily controlled, but was not due to lack of social consensus on keeping closed the door of the room where the fire was. In the second fire there was a built-in contradiction for prisons in that during a fire staff must evacuate prisoners from the fire area, but not let them escape. The last fire could have been the worst but strong staff teamwork handled it.

RESEARCH METHODS USED IN COLLECTING DATA

Before going to the scene of the fires, experts in the fire field were extensively consulted regarding the type of information to be gathered. Experts were consulted at NBS, The National Bureau of Standards; NFPA, The National Fire Protection Association; and Maryland University. Questionnaires which had already been used after fires were read and information valid for this research was extracted. 'Fire' questionnaires examined were those used by Wood (1972), Slater (1976), and Bryan (1977). The author used a focused inverview—this loosely but consistently follows a designed set of questions, but allows the respondent to volunteer information which often turns out to be significant (Selltiz, Jahoda, and Deutsch, 1959). Persons interviewed were known to have all been involved in a particular situation, in this case a fire in a total institution. The interview focused on a definite set of interests of the interviewer, and on the respondents' definition of the situation. The reported material helps test hypotheses, and develop fresh ones from unanticipated responses (Merton, Fiske, and Kendall, 1956). The focused interview followed the same outline for each fire, with each respondent from a fire, so that the information collected would be standard. The technique turned out to be very successful as it collected the desired data but also left room for the respondent to add any statements or opinions which often threw light on thoughts, feelings or observations about the fire. Interviews in cases 1 and 2 were done within 24 hours of the fire; case 3 interviews were carried out four years later.

Each building adminstrator was interviewed first, both for permission to interview and to allay suspicions about the nature of the research. They all cooperated. Information was requested on the number and type of staff RNs (registered nurses), nurse's aides, and LPNs (licensed practical nurses), and the interviewees provided a percentage of patient breakdown regarding age and sex, and the type of treatment given at the particular institution. Interwoven with conversation, information was obtained on the training of the administrator, the number and type of fire drills and/or training. In all cases both the non-fire and fire areas were visited to see the differences, or what the normal area looked like. Data was collected on the history of the neighbourhood, its racial make-up and changes over the past ten years, and the general class, income and type of occupation of the residents in the area. Where the NFPA (The National Fire Protection Association) also visited the scene of the fire, information was exchanged, the NFPA learning more about the behavioural aspects of people in fire as seen by a social psychologist, and in return giving information on building and fire conditions.

In what follows an attempt has been made to put together the varying accounts of each person's story into one composite. Where people gave inconsistent reports of an event, a third and fourth person was asked about the event, and this usually clarified it.

Case 1 'Catastrophe'—a family-owned nursing home

Staff and inmate make-up

There was an average daily total of 82 patients; 75 of the patients were female, and the average age was 83. The type of care offered was long-term care facility, but with no tube fed, intravenous or oxygen cases.

At the time of the fire, the ground floor (Figure 9.1) was entered by a ramp and led into a warm, cosy, carpeted foyer. The fire occurred in January 1976 in bitter cold, -11°C (12°F), with heavy snow and ice and a 25mph wind. Nursing homes are required to maintain a heat of 24°C (75°F). The reception room and administrator's office was on the ground floor, and the kitchen was almost directly opposite her office. The building was mostly fire-resistant. The ground floor was used for administration, security, domestic reasons, and visitors; the other three floors were used for patient care.

The administrator stated that the fire alarm went off at 11.43 am, and the smoke detector alarm at 11.46 am. This was confirmed by the fire department in separate interviews.

Summary

The fire broke out on the third floor in room 306, half way between the social hall and the nursing station. There were 28 patients, a nurse's aide and the social director in the social hall. A second nurse's aide was on the third floor passing 306 when she detected smoke and heat.

No patients were in 306 at the time of the fire, as three were in the social hall, and one was out having her hair done.

The corridor in front of room 306 sustained heavy heat damage to within 3.9m (13 feet), and medium heat damage to within 8.5m (28 feet) of the social hall. Heat damage was heavy in the other direction, 4.5m (15 feet) beyond room 306 to the east, with medium heat damage 8.5m (28 feet) beyond the room.

Apparently the fire started in the area of a wooden clothes wardrobe in room 306. Two wardrobes were side by side, and each was constructed of 12.5mm (¾ in.) plywood with a hinged folding door. The wardrobes were 500mm (20 in.) by 950mm (38 in.) by 2050mm (6 feet 9 in.) high. Mattresses were innerspring units consisting of 69 per cent cotton felt and 31 per cent sisal pad. At about 11.40 am the janitor was on the third floor hallway with his assistant (aged about 20). He mopped just inside the open door of the fire room, 306. Neither he nor his assistant noticed any-

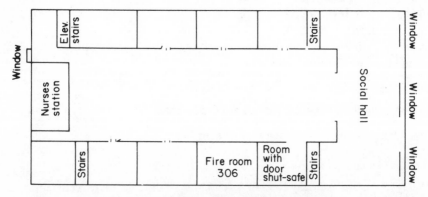

Figure 9.1 Floor plan of Case 1, the nursing home

thing suggesting fire, although (if his timing is correct) the alarm went off three minutes later and the fire was already building up in the closet of 306. The janitor and his assistant took their buckets down to the second floor.

The administrator was sitting in her office; she had just taken a caseworker up to the third floor and returned to her office. On the third floor, a nurse's aide, Miss G, passed by the door of 306, felt the heat and saw smoke. She 'hollered to the social hall for help'. Another nurse's aide, Miss T, who was in the social hall, came and pulled the wall alarm. The fire bell went off, clanging three bells at a time which indicated that the fire was on the third floor. The caseworker shut the door of the fire room, then walked downstairs, informing staff about the fire on her way down.

Miss G, who discovered the fire, ran to the nurses' station on the third floor and called down to the office by telephone saying that fire had begun in the closet behind the door (to the corridor) of 306. The closet door, an accordion type, was slightly open. Nurses's aide Miss T then ran to the social hall where about 28 people were watching television. There was no door at all on the social hall. Only the mobiles on foot or wheelchair cases were up there. People who could not move at all, or with great difficulty, were waiting on the second floor for someone to come down to them and bring them up. Hence another nurse's aide was standing at the second floor staircase waiting for help to come down; when the caseworker passed the second floor and informed those waiting, Miss T pulled the alarm and then ran back to the social hall. Miss T says the social director (male) opened the window in the social hall, although she tried to stop him, saying that this would fan the flames. According to Miss T, this created a draught, with the wind going from the nurse's station end of the floor, where a slightly open window carried a 25 mph wind.

At the sound of alarm the janitor snatched up two fire extinguishers, and with his assistant rushed back upstairs and tackled the fire. The social director, too, was aiming an extinguisher over the top of the fire door, which someone had opened. The administrator checked the panel on the ground floor to locate the fire, and calling to the kitchen staff to follow rushed upstairs with an extinguisher. At least five extinguishers were in use on the third floor—by the janitor (who used two), the administrator, the social director and a nurse's aide.

The janitor, who had been on the second floor, said that the alarm went off within five minutes after he had left the third floor. He thought it was a patient, and said to his assistant, 'some patient pulled the fire alarm' because this had happened once before, 'Let's go down and turn it off'. The janitor and his assistant went down to the first floor in order to turn off the alarm and someone said, 'There's a fire on the third floor'. They therefore rushed back up the stairs, which was faster than using the slow elevators (geared to the aged), feeling 'shocked and nervous'. The janitor arrived at the scene of the fire with two extinguishers picked up from the second and third floors. He said that he saw the social director outside room 306 aiming the extinguisher over the top of the door. The janitor went into the room, and saw smoke, both black and grey, coming out of the closet. The janitor tried to put out the fire with a fire-extinguisher but it gave out such heat he was unable to get closer than within 3m (10 feet) of it. His idea was to open the door of the closet and he was joined by three other people who tried to aim extinguishers over the top of the closet where there was a space of about 25 to 50mm (1 to 2in.). Driven back by heat, the janitor tried to shut the door of the room, but the handle was too hot to touch. He tried to grasp the door around the wooden part and pull it closed but the heat was too intense, so intense in fact that it caused the water in the extinguisher to evaporate. The administrator, her face black with soot, was still trying to put out the fire, but was pushed away by the janitor and the firefighters who had arrived by this time.

The administrator reported that she saw two men with fire-extinguishers, the janitor and the social director, and they were controlling the fire. She also said that the fire was so bad now she was 'coughing up black stuff'. She returned downstairs to man the phones, she called her own ambulance company and organized evacuation of the patients. In the heavy snow and bitter cold, the ambulances came to the alley which ran behind the nursing home. The administrator reported that all nurse's aides were removing patients who could not walk (one in a wheelchair), and a nurse said, 'I collapsed the chair, and put her on the bed and closed the door'. Aides closed the doors and brought patients down initially to the second floor, but due to the firefighters' hose water dripping down from the third floor surviving patients were moved down to the first floor lobby, and given mouth-to mouth resuscitation. Firefighters told the nurse's aides and nurses to spend the minimum time on each with oxygen and get to the next person fast. Meanwhile, Miss T closed the doors of rooms, 308, 307, 305 and other doors, went back to the social hall to close the window there, and then passed out. She woke up in a hospital room about 1.0 pm, having been taken there by a firefighter. Since then, said the administrator, she had been in a nervous state and refused to go up to the third floor again. By now the public had heard the announcement of the fire on the home and car radios on the 12.0 pm news; this caused the problem (administration report) of news media, and patients' families jamming the phones, and thrill-seekers trying to crowd or sneak in, and carry out souvenirs.

The doctor on call to the nursing home came immediately and left within the hour. About 12.45 pm the owner's son, a doctor, arrived about ten minutes after the fire had been put out to find about 75 people standing around fire—bomb, arson and police officials, newpapermen, patients' families and curiosity-seekers. The administrator told him to go and see to a patient on the second floor; patients were in the halls on the first and second floors. The dead had been removed within an hour. The doctor felt that sprinklers, or water, was dangerous for the aged who could slip and break a hip, or suffer pulmonary reactions: 'Moving old people is a trauma itself, and could result in death'.

To summarize: Miss G discovered the fire, called Miss T, who pulled the fire alarm, closed several doors, returned to the social hall to close the window. Miss T attributed her calm behaviour to the 'New York Hospital where I worked. I was trained never to panic and I got this training, namely pulling the alarm from the fire drills in the hospitals in New York City.'

Meanwhile, on the second floor, nurses B and C heard from the alarm and from the caseworker who was coming downstairs, that there was a fire upstairs on the third floor. Nurse C told nurse B to stay out and ran upstairs with an extinguisher, finding on the third floor smoke, black, thick, choking, also a couple of people with fire extinguishers, I counld't see *anything* [this is the five-minute difference between the light and the impenetrable smoke described by the janitor], so I turned around and ran down the back stairs and came up the front stairs back to third floor. When I came back upstairs the front stairs' smoke was so thick I couldn't see. It choked me, it looked like the whole building was on fire. I was feeling frightened, I knew so many people up there in wheelchairs who couldn't move. I knew unless I could save them myself they were trapped; I could hear the coughing. I found a number of patients dead at the entrance to the social hall. The smoke had risen and was above chest level. When I stepped back out the back stairs of the third floor, firemen went up and shortly after one came out and stood on the landing for a minute and yelled; they kept opening the third floor back door to let in equipment and firemen. The firemen were spreading the smoke by keep opening the door of the stairwell.

Nurse C returned to the second floor. Nurse B took walking patients down the front stairs from the second to the first (ground) floor and Nurse C ran back to the third floor. 'The firemen stopped me told me go down and revive those on second [brought from third]. The second floor was fairly safe except when door opened from third.' Miss T knew what to do, and the rest of the nurse's aides also seem to have done all the right things, but they were handicapped by numerous immobile patients trapped behind heavy smoke. Miss T said, 'After I got home, I was dazed; I thought I'd been dreaming. I couldn't believe it had happened, it was unreal; I wasn't nervous, just dazed. After I seen all the people dead I thought I hadn't done enough.'

Asked if there was anyone who acted as a leader, Miss T said the director of nurses directed the action, to which the others agreed. The director of nurses was hospitalized for smoke inhalation longer than other people and unavailable at time of the first visit (4 February 1976). When I returned a few days later the owner refused to let me talk to her, though I saw that she had very red eyes and a distressed appearance. The social director, who is said to have received facial burns, remained incommunicado from the beginning; others took all the telephone calls and consistently refused access to him.

Nurse B was asked her feelings about the fire. She said,

till the smoke alarm went off I thought some little old lady done it. When the smoke detector went off I knew it was no accident. I felt disbelief. A fear of getting burned. In about three to five minutes she [Nurse C] came down and told me where the fire was, so I handed her the big extinguisher that hangs on the wall and she went back up with it. I don't know what happened after that.

Nurse C says that then she ran upstairs to the third floor with a fire-extinguisher and the social director grabbed it from her hand. There was so much smoke she came back down the other stairway. The fire-fighters arrived and put out the fire. A number of patients from the third floor were evacuated to the second and then to the first floor. Some patients on the third floor had been in a room with the door closed, with no ill-effects.

The nursing home was back in working order within hours. All of the above events occurred within about ten minutes. Most patients in the social hall died during the fire, and some later in hospital, a total of 21 patients. A nurse's aide was charged with arson and murder, but found innocent.

Design considerations

(1) The most obvious design consideration here is the necessity of a solid door on all rooms. People in the social hall died largely because there was no door on it and they were overcome by smoke flooding in. The fact that some people remained safe behind a closed door immediately next to the (open) room where others died attests to the importance of heavy closed doors as a safety factor.

(2) Particular care should be taken that escape routes are possible when a number of handicapped or wheelchair patients are involved. These people cannot move themselves, and frequently have respiratory problems. Unless well protected from fire and smoke by a door or horizontal evacuation, death in a fire is almost certain.

(3) It is necessary to find means of identifying the room of a fire and informing those in the vicinity that (a) the room is the fire room; (b) it has been emptied of people: (c) that its door should remain closed until fire-fighters arrive. A non-flammable sign with this information might be attached to such a room door.

Since the fire, the mayor of the city in question has ordered all nursing homes to install a sprinkler system throughout the building and to have solid doors on all rooms.

Case 2 'near catastrophe'—a penal institution

Staff and inmate makeup

In the men's section there were usually 180 to 190 inmates, and two officers per 50 to 60 inmates. In the women's section there were usually one to ten inmates and eight guards (Figure 9.2).

This is a pretrial institution. However, some come who are sentenced because the jails are overcrowded; in fact 30 to 40 per cent of the inmates are sentenced because the state prisons are so overcrowded that some prisoners (who would normally go to State prisons) are sent to county jails (see Wardens, 1976–78). Some are only in for a short time and then go on to a halfway house.

The time of the fire was 10.45 am, May 1976.

Summary

An inmate, (the occupant of Cell B), who had threatened to set a fire to revenge herself on her next-door neighbour whom she thought was getting better treatment, set fire to her mattress at about 10.45 am.

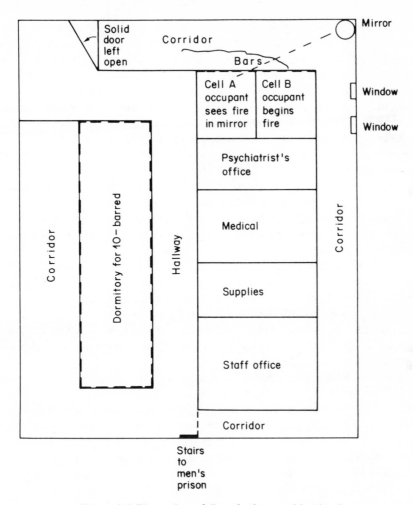

Figure 9.2 Floor plan of Case 2, the penal institution

Her next-door neighbour, against whom the fire was directed, saw the occupant of Cell B setting fire to her own mattress, in a mirror placed in the corner above Cell B, set there so that the matron could see into the corner cell. The inmate of Cell A screamed out, 'She's setting fire'.

The matron came running to Cells A and B, looked into Cell B, then went to her office and got wet towels, handing them to the ten women in the neighbouring iron-barred dormitory who were screaming as smoke poured into their dormitory. She soothed them and told them to get down on the floor; then she obtained keys to the cells. About two minutes elapsed. The matron had not yet called for help.

The matron instructed another women officer to press the emergency button of the intercom which signalled to the men downstairs that there was an emergency and help was needed. The officer on duty downstairs flipped the switch on the intercom and asked 'what's the problem?' The women officer said 'Fire'. Since small insignificant fires were frequent, and emergency calls from the women's section were usually in connection with a fight or illness among the women, only two men were sent up. They brought no fire equipment, nor did they know the layout of the women's section. Since the Joan Little trial, where a woman prisoner killed a male guard because she said he was trying to rape her, there has been a ruling forbidding male guards in prisons and jails to enter the women's area, which is usually upstairs or in a separate small wing of the men's prison area. One or two men, however, had been to this women's section before the ruling and knew the area.

Men rushed up with keys to open the doors, but without firefighting equipment; they were not expecting a large fire. The officer released the inmate from Cell B, placing her far from the fire in another cell on the women's floor. The matron went back to her office and got the arsonist a wet towel which was rejected. The matron tried to get down the extinguisher which was placed too high for her to reach and might fall on her anyway. The smoke was getting dense. Corporal H turned on the exhaust fans which were brought up by Sergeant K.

The men went downstairs for masks, and returned with masks and an extinguisher. The smoke was now 0.7 to 0.9 m (2½ to 3 feet) off the ground. Captain B put out the fire with the extinguisher, while the other men evacuated all the women to a downstairs office and locked them in with the women guards.

The fire and rescue squad arrived, extinguished the fire and hospitalized several people for smoke inhalation; they gave out Aspirin and Darvon for headaches and discomfort.

The women were moved from the office room to an isolated part of the men's floor (the ground floor), and locked in there.

Most of the staff felt shaken up, especially the matron who experienced severe conflict between wanting to let the prisoners out of the fire area without letting them escape. There was only a 1.8 m (6 foot) boundary fence around the prison, and the prison faced directly on a highway. Had any prisoner been able to get as far as the fence he or she could have conceivably hitched a ride, particularly as prisoners wore coveralls, not distinctive prison garb.

Changes in regulations introduced for fire safety after the fire

(1) Regulation fire-hoses (like the one in the corridor) to be installed in the main corridor—at female officer height.

(2) The 30 m (100 ft.) garden hose to be used until the regulation fire-hose is installed.

(3) One more water fire-extinguisher to be installed in the workroom.

(4) All three fire-extinguishers to be lowered to height of average female officer.

(5) Two pairs of asbestos gloves to be purchased and kept in women's section.

(6) One MSA air mask to be kept in the women's section.

(7) One small, 110 volt smoke exhaust ejector fan to be purchased and kept in the women's section.

(8) Complete set of women's section keys to be made and kept in control for emergency use.

(9) Hooks to be installed in ceiling to hold large smoke exhaust fan.

(10) One 220 volt receptacle to be installed to plug large smoke exhaust fan.

(11) Lock to be replaced to door to dormitory across from the cell area.

(12) Non-toxic mattresses already on order to replace those currently in use—proposed delivery date July 1.

(13) Cluster of small keys for the women's section to be divided into two clusters, one for crucial keys, and one for operational keys.

(14) Mattresses and pillows to be changed to non-flammable, non-toxic bedding.

Some possible design implications

(1) Communications between men and women could include a special alert for fire, not a general emergency alert.

(2) The lack of keys and/or their failure to operate under fire conditions pervades all prisons, for example, the Danbury fire in 1977 where five prisoners died while guards searched for the keys.

(3) Direct communication with the fire department might be available. This was recommended in the Danbury fire report.

(4) An evacuation plan should be initiated and practised.

(5) Staff need to be taught the importance of closing doors on a fire so smoke does not escape to injure others.

(6) Perhaps in prisons where there are open-barred cells automatic shutters could come down, except of course, in the room of the fire. This would protect inmates from smoke inhalation in non-fire cells or dormitories. Although officials fear that prisoners may damage sprinklers, they could perhaps be recessed into ceilings.

(7) In view of the large number of fires in American prisons described by Wardens (1976–78), and the difficulties in evacuating prisoners quickly it is important to have either improved surveillance of prisoners through closed circuit TV, and /or smoke detectors or sprinklers in cells and dormitories recessed to avoid vandalism.

(8) Prisoners can be informed of the consequences of fire, and some can possibly be trained in firefighting on each floor.

Case 3 'Success'—a Jewish home for the aged and retired

This case was selected as an example of well-organized staff functioning and fire-drill training together with good construction and fire protection. These factors served well in the face of a totally unexpected fire emergency. The only area where the institution needed, and has since obtained, better functioning was in the area of communication between staff as to when there was a fire, and between staff and the fire station as to the existence of a fire. It should be noted also that there was little panic.

Staff and resident makeup

There were about 267 residents, the ratio of women being three to one (men), and the average 80 (the Home only takes those from 65 upwards). There were 700 staff of all levels from directors to cleaning staff; this staff is unusual in having very low turnover. Most of those interviewed had worked there between seven and 16 years. The Home is five storeys high and shaped as shown in Figure 9.3 (not drawn to scale).

The floors were occupied as follows:

Basement	Maintenance, supplies, laundry.
Ground (or first) floor	Reception desk, switchboard, office.
Second floor	Skilled care (that is, hospital care) and senile patients.
Third floor	Intermediary, that is people who are not sick enough for the second, nor well enough for the fourth.
Fourth floor	North wing, skilled care; west wing, self-care.
Fifth floor	North wing, self-care.

There was a kitchen and dining room, and TV room, on each floor; smoking is only allowed in the TV room. At the entrance to the home, which has a glass front, there was a reception desk and a switchboard. This fact is important for understanding the progress of the fire. The building was fireproof, with smoke detectors in the hallways which cause fire doors to shut automatically and the fire alarm to go off upon activation.

The fire occurred at about 9.10 am in April 1973. This is the only fire studied by the author outside the 1976–77 period; but it has such interesting features that it has been included as one of the cases in this chapter.

Summary

A former, disgruntled employee arrived at the Home about 3.15 am in his car, and waited until the coast was clear for him to begin firesetting. Having worked at the Home (but been released as unsuitable for the job of orderly), he knew that the guard went off duty at 3.30 am, that the daily newspapers were delivered by 4.0 am, and thereafter the front area of the home was usually unguarded until the morning shift came on. He had backed his car into the entrance of the home, and left the boot open where he had stored about twenty molotov cocktails. He also had a further twenty in a milk case container which he carried with him. He picked up a heavy metal chair left outside the home for sitting in the sun, and smashed the glass front of the home with it. As soon as he entered the building, by a door next to the switchboard, he pulled out all the telephone connections disabling the lines. When the nurses on the second floor picked up the 'outside' phone they found the lines dead.

At the sound of the glass smashing nurse S, who was doing paperwork in the board of directors room on the ground floor, ran to the front of the building (the north end) and found a tall man dressed in white like an orderly. Facing the intruder she said, 'Get out of here'. She had, like most people, never seen a fire bomb or a molotov cocktail. It was nurse S's opinion that the bomber had not expected anyone to be on the ground floor as there were only offices there and the guard had

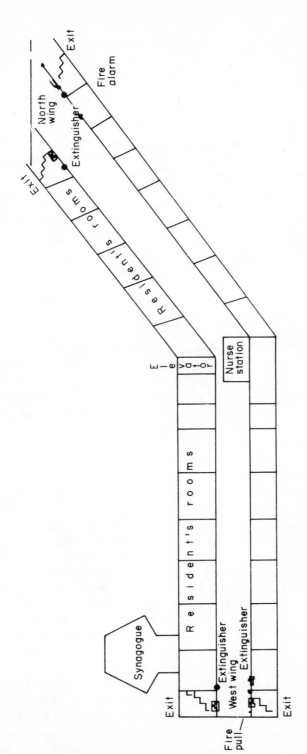

Figure 9.3 Floor plan of Case 3, the home for the aged and retired

left. The intruder said, 'I am from the FBI and I am going to burn this place up'. He threw a molotov cocktail in front of nurse S, in the foyer, then he threw several bombs into offices. Nurse S ran behind him shutting doors. He then ran to the middle elevator and threw a bomb in it which activitated the fire alarm. Nurse S decided that she was dealing with a 'nutcase' and ran in the direction of the boardroom, intending to run upstairs to the second floor where there was an outside telephone to phone for help. She thought the man was behind her; in fact he ran in the opposite direction, up to the fourth or fifth floor (the ambulatory floors) where he began throwing bombs into patients' rooms. Residents remained fairly calm while the alarm was going off, and nurses were running behind the bomber either smothering the fires with pillows, or pulling burning mattresses out of the rooms, or pulling sleeping people out of the rooms, taking the whole bed (on wheels) out of the ambulatory, but always closing the doors behind them on the fire.

Meanwhile nurse S had gone to the second floor, pulled the firebox alarm, which is on each floor, and telephoned the fire department, but the line was dead. She was unaware, as were all the staff, that there were outside lines from some of the administrators' telephones on the ground floor, and forgot the fact that some patients or residents had private lines which could have been used to dial outside. She encountered nurse H on the second floor and said 'we have a problem here. Some man is throwing bombs.' In the presumed absence of any outside line nurse S borrowed money from the medical aide (an aide somewhat more trained than a nurse's aide) and called the operator on the public telephone. According to the nurse the operator delayed the call by asking a half a dozen questions; impatiently nurse S hung up, and directed David, a college boy serving as orderly, to go out and find a policeman. David said that he 'snuck out' the side as if he were in Vietnam, 'not knowing where the bomber was, and later feeling shaky because [he realized] the bomber could have shot me'. He saw the bomber's car still standing in the front driveway with the boot open and full of molotov cocktails. At the same time, the engineer, alerted from the basement by the fire alarm, ran up to the front of the ground floor with the fire-extinguisher, saw the molotov cocktails in the car's open boot, and doused them with the extinguisher, Meanwhile nurse H also borrowed money and reached on the pay telephone to the fire department. David, the orderly, simultaneously found a policeman in a car at the end of the road who called the fire station, and who returned with him to try and catch the bomber.

The bomber was thought by some to have raced up the stairs and gone from the fifth to the fourth floor, the latter being where he used to work. He threw bombs into the rooms of people with whom he had particularly worked. On the fourth floor the arsonist was recognized by a former classmate (female) in an orderly's training class. Saying to her, 'Get out of my way,' he threw more bombs into rooms, some falling where there were patients rather than residents (those able to care for themselves). Nurses hurried after him to either smother the fires with pillows, or to wheel the patients out of the room if they were disabled. They always shut the doors so that the smoke did not spread after evacuation, or elimination of fire.

All the smoke sensors had activated the automatic fire doors so that there was maximum smoke containment, and these doors activated the fire bells which clanged deafeningly. Even so, the fifth and fourth floors were described by staff as very smoky, and people were coughing badly. Nurses began evacuating patients to the third and second floors, which fortunately was not attacked by the bomber as the latter was occupied by disabled and wheelchair patients. Presumably he heard the arrival of the fire trucks and police as he worked his way bombing through the fifth and fourth floors, and then fled. Police and a nurse stood at the entrance to the front door to identify everyone coming in and going out, but the bomber managed to escape, it was thought, by a basement door. As nurse S said ruefully, 'There are too many doors here. Although people cannot get in, there are many ways for them to get out.' The same man was suspected of having begun another fires two hours earlier in another nursing home. It is interesting to note that he made for his customary floor and customary patients.

Another nurse, nurse V, upon hearing the glass smash and the fire alarm, turned off the oxygen even though this meant disconnecting some patients. Fortunately no one suffered through this action. Nurse V was in charge of drugs and put her keys to the drug cabinet in her inside pocket, as she presumed someone was breaking in to steal drugs.

(At the time of the interviewing, between 11.00 pm and 4.00 am, one could almost hear a pin drop. All the residents were asleep, and the staff is minimal at night and speaks quietly. This accounts for the fact that the sound of smashing glass and exploding bombs carried so clearly.)

Nurse H, an RN who has worked at the Home for nearly twenty years in both its locations, was on the fourth floor nurses' station waiting for a nurse's aide to come back on to the station. RNs are usually at the nurses' station doing paperwork and medicine allotments, while other personnel, such as nurse's aides, work on the floor itself.

The nurse's aide had gone to answer a light and helped the resident into the bathroom. (Each resident has a switch by the bed which when pressed lights up outside his door and at the nurses' station signalling that aid is needed.) Nurse H heard the fire bells ring and saw the fire door swing shut. At the same time the man with the molotov cocktails appeared on the fourth floor and began throwing them into patients' bedrooms. He threw one into the bedroom where the nurse's aide was in the bathroom with the patient. The mattress caught fire, and the aide dragged it out of the room.

She went to get a fire-extinguisher but did not need it as the fire went out by itself. Nurse S told the nurse's aide to see that the doors were shut and that any patients in burning rooms were evacuated. Nurse H ran down the stairs to the second floor to telephone the fire department, as the only known outside line was on the second floor; but the line was dead. Nurse H went down the corridor to the pay phone as had Nurse S previously. The nurse's aide 'fed her dimes'. The two women, nurse S and nurse H, met on the second floor. Nurse H finally got through to the fire department. Since the second floor was quiet they ran back to the fifth floor where they could hear bombs going off (the fourth floor was being held down by the two nurse's aides that nurse H had left in charge). On the fifth floor they closed doors where there was a fire in a room without a resident, and took the residents of rooms with fire in them to the TV room, a corner room used for evacuation in drills. Soon the fifth floor became heavy with smoke, partially, firemen said, because of the frequent opening of the doors between the fourth and fifth floor, apart from the actual fires lit on those floors. Smoke from fire bombs on the fourth floor rose to the fifth floor each time doors were opened by the bomber, the nurses running between floors, and finally by the firemen who arrived within about ten minutes of the bomber's initial entry. Nurse H estimated the time between her awareness of the fire and her call to the fire department to be about two minutes. The large number of events described in this time period gives some idea of the speed with which people were moving. Nurse S said of nurse H, 'She moved like a streak of lightning'. Fire bells were clanging constantly.

The firemen who arrived began to have trouble breathing on the fourth and fifth floors, so they put on masks. By now most of the fourth and fifth floor patients and residents had been evacuated to the second and third floors which had not been attacked. The firemen directed the nurses to use one elevator for wheelchair patients. The nurses had told patients and residents that this was a fire drill, but most of them, particularly the self-sufficient ones on the fifth floor, laughed and said they knew it was a real fire. A number of those on the fifth floor just walked themselves down to the second floor when asked. The nurses got everyone down to the second floor, and kept people calm, and quickly provided drinks for the residents, many of whom were coughing from the smoke. Most of the residents were quite calm. They had had a number of fire drills, as had the nursing staff, and an evacuation plan was well formulated.

It should be noted that while the resident number remains the same at night, the staff is much smaller, thus there were only about fifteen people available to deal with 267 patients and residents.

It was hardest to evacuate people from the fourth and fifth floors, partially because the north wing of the fourth floor held some wheelchair cases, and also due to the heavy smoke on these two floors.

Nobody panicked; Quarantelli and Dynes (1972) observe that the idea that people always panic is overrated. As Canter noted in correspondence (1978), in the third case mainly women staff were involved. They acted as a team, perhaps better than when men were involved. In some cases untrained men may have hampered the resolution of the problem by trying to attack the fire themselves and delaying a call to the fire department.

Changes in fire protection since the night of the fire

(1) There has been a guard on duty all night from 10.30 pm until 7.30 am. Before that there was one only until 3.30 am.

(2) Telephones connected to the outside have been installed on every floor.

CONCLUSIONS

These three cases hopefully throw some light on the awareness and activities of people involved in fire in three total institutions—a nursing home, a penal institution, and a home for the aged and retired. In cases 2 and 3, fire was begun by deliberate firesetting, in other words, arson. In case 1, arson was *believed* to be the cause. This coincides with the finding of Haber (1976) that six out of seven fires in health care institutions were begun by humans, five of them deliberately. In the earliest paper on fires in health care institutions Haber noted that four out of six fires set by humans had begun in corner rooms; in fact fires by inmates were largely begun in corner rooms. In the 1976 paper it was proposed that extra attention be paid to people occupying corner rooms and or those far from the nurses' or guards' station. Of these three cases, the one fire which was begun by an inmate—the prison fire— was begun in a corner room (or cell). Those in both the home for the aged and in the nursing home were neither begun in a corner room, nor by residents or inmates.

It is important to note that each of the three cases cited represent about ten minutes only, from first awareness to the fire's extinction. The time frame of a fire is very very short; it is believed by some that a fire begins and becomes dangerous within *two minutes*. This points up the need for fast action and communication. We cannot overemphasize the need for fast action, such as an immediate call to the fire department, fast evacuation where necessary, and closing the door of the fire and all other doors. From case 1, we see clearly that lack of a door killed 21 people, whereas patients right next to the fire with the door shut survived.

From case 2 we see the need for specific communication, the need for fast evacuation without escape for inmates, and the need to protect those in open barred cells or dormitories.

Case 3 points up a highly effective team working in a surprise attack, yet who were unaware of all the means of communicating with the fire department for help.

In all three cases staff were taken by surprise at the speed of the fire, the density of the smoke and its overpowering effects.

In no case could the staff, residents or inmates be said to have panicked, except perhaps the women in the barred dormitory of the prison. However, since there was no possibility of them escaping their barred dormitory they cannot be said to have panicked, since panic allows the (reduced) possibility of escape. Likewise in case 1, the aged who died had no options; they could not move nor be moved fast enough, so they died within five to ten minutes.

ACKNOWLEDGEMENTS

I wish to express my appreciation to all the city, county and state fire departments in the United States, Canada, and the Ministries of the Interior of France, Italy and Egypt, who helped in gaining access to buildings and for giving information on fires; to institutions permitting access to their buildings, administrators and supervisors, staff, patients, residents and inmates; to the National Bureau of Standards (NBS), Long Term Health Care, HEW; to Professor John Bryan, Professor and Chairman of Fire Protection Engineering, University of Maryland, and to Doctor David Canter, Department of Psychology, University of Surrey, Guildford, England, for thoughtful and constructive comments on an earlier draft of this chapter.

The author is responsible for all statements made in this paper.

REFERENCES

Bader, J. E., Maxwell, R. E. and Watson, H. W. (1972). 'Doomed status and the ecology of institutions', Philadelphia Geriatrics Center, 25th Annual Meeting of the Gerontological Society, San Juan, December 1972.

Bryan, J. (1977). *People Project,* University of Maryland.

Canter, David and Matthews, Rowan B. (1976). *The Behaviour of People in Fire: The Possibilities for Research.* Building Research Establishment, Fire Research Station, Borehamwood, CP11/76.

Canter, D. (1978). Correspondence on an earlier draft of this paper.

Federal Bureau of Prisons (1977). *Board of Inquiry into the Danbury (Prison) Fire, 7 July 1977.*

Goffman, E. (1959) *Asylums.* (New York: Anchor Books, Doubleday and Co, Inc.).

Haber, Gilda. (1976). *Human Response to Fire as Environment,* National Bureau of Standards.

Haber, Gilda M. (In preparation). *Human Response to Fire.*

McKinnon, G. P. and Tower, K. (1976) *Fire Protection Handbook,* National Fire Protection Association, January.

Merton, R. K., Fiske, M. and Kendall, P. L. (1956). *The Focussed Interview,* (New York: Free Press).

Quarantelli, E. L. and Dynes, R. (1972). 'When disaster strikes', *Psychology Today,* **5 February,** 66–71.

Quarantelli, E. L. (1977). 'Panic behaviour: some empirical observations'. In Donald J. Conway (ed.) *Human Response to Tall Buildings,* Community Development Series, (Stroudburg, Penn.: Dowden, Hutchinson, and Ross), Chapter 28.

Report No. 70.51. (1977). *Federal Prisoners Confined on 9-30-1977,* US Bureau of Prisons, October.

Selltiz, Claire, Jahoda Marie, Deutsch Morton, (1959). *Research Methods in Social Relations,* Revised One-Volume Edition (New York, Chicago, San Francisco: Holt, Rinehart and Winston).

Slater, James A. (1976). *Synthetic Polymer Fire Accident Case Study,* National Bureau of Standards.

Turner, Ralph H., and Killian, Lewis M. (1972). *Collective Behavior, Second Edition.* (Englewood Cliffs, NJ: Prentice-Hall).

Wardens. (1976). Discussions with 24 prison wardens and safety personnel in prisons.

Wood, Peter G. (1972). *The Behaviour of People in Fires,* Fire Research Note, Fire Research Station, Building Research Establishment, Borehamwood.

Fires and Human Behaviour
Edited by D. Canter
© 1980 John Wiley & Sons Ltd.

CHAPTER 10

Fires in Nursing Facilities

LARS LERUP, DAVID CRONRATH AND JOHN K. C. LIU
University of California

INTRODUCTION

This chapter describes a case study technique for analysing the fire situation that includes and utilizes behavioural data for the purpose of improving the design of nursing facilities. The technique involves the use of maps and diagrams to depict both human and fire behaviour. The findings touch on the nature of emergency behaviour, directions for improvement of the physical setting and the importance of a management component, as well as the utility of case study techniques.

The purpose of the work is to address the improvement of the physical setting in order to obtain a higher order of life safety for its inhabitants. However, it has become clear that physical improvements have to be coupled with the improvement of both the inhabitants' knowledge and management of the physical situation. Thus the findings have both a physical and a behavioural component.

Work by the scientific community has resulted in the clear possibility that we will be able to predict the course of development of fires. This skill will allow us to simulate fires, along the lines of the scenarios described by Clark and Ottoson (1976) and may subsequently develop some lifesaving strategies in accordance with the fire development simulation. We also are coming to a better understanding of the tendencies of humans in relation to fire, and this general understanding of behaviour coupled with the specific understanding of fire will allow us to train people to react and act correctly in fire situations. We may thus circumvent the vagueness of our understanding of behaviour and still arrive at reasonable levels of life safety in buildings. And in the light of people's documented creativity in dealing with emergency situations, such an approach is likely to be the most successful.

Nonetheless, we feel strongly that the utility of current findings is severely limited by the following pragmatic circumstances: (1) There are at present no official strategies for life safety that can stipulate a correct sequence of actions to follow;

The original research was prepared for *HEW/NBS Fire/Life Safety Program* Grant number 6-9013 under the sponsorship of Robert Blake, Director of the Office of Facilities Engineering and Property Management of HEW, John W. Lyons, Irwin A. Benjamin and Harold E. Nelson all of the Center for Fire Research of NBS.

(2) the physical environment cannot be designed to literally reflect a specified sequence of actions; (3) no two fires are alike; and (4) fires can start in many different parts of a building, demanding different lifesaving strategies and actions.

This said, we still argue that a combination of increased knowledge, new lifesaving technology applied to the built setting, and improved management is the correct way to attack the fire problem.

DATA SOURCES

The main data source for this project is ten case studies of fires in nursing facilities which were developed in an earlier project by ALSG (Architecture Life Safety Group) presented in Lerup *et al.,* (1975). Supporting data are drawn from the FIDO (Fire Incident Data Organization) file maintained by NFPA. The nature and utilization of these two sources are discussed below.

The ten case studies were drawn from accounts of the actual fires, information ranging from newspaper clippings to interviews with survivors and reports by experts in the fire field. Because the cases were reconstructed from secondary accounts, various deficiences are inevitable. Most prominent is accuracy and similarity to the actual situation in the fire, though the problem here is not so much the actual events and types of behaviour, but rather their exact sequence. We therefore relied on discrete events and behaviours as our basis for determining the implications.

The material underlying the ten case studies was selected by Harold Nelson, the grantor and Chief of Design Concepts at the National Bureau of Standards, from the NBS file (largely coinciding with the FIDO file). The cases were selected for a variety of reasons, none entirely systematic, simply because the data do not allow it. A range of cases was sought, recording fires in small as well as large buildings, in resthome and hospital-like facilities, and in many parts of the country. All the fires had incidence of death, but as fires go these ten emergencies were easy to control once the fire department arrived and, typically, smoke rather than flame was the cause of death.

We saw early in our investigation that these fires were much like residential fires. This led us to compare certain circumstances with what Clark and Ottoson (1976) have referred to as 'residential death scenarios'. We ourselves refer to these circumstances and their links as *emergency chains*—that is, a series of circumstances that are linked together in a dependent manner such that they produce an emergency. Such a chain is produced, for example, by bedroom → bed → cigarette. Clark and Ottoson (1976) describe these circumstantial links more elaborately as:

type of loss → type of occupancy → time → ignition → source → item
ignited → direct cause of loss.

We modified their list to serve our purpose and established the following array of circumstances (or emergency chain):

building type → people → activity → ignition → source
→ agent is spread.

The slight discrepancy between Clark and Ottoson's death scenarios and ours still allows us to make a direct comparison. This is done in Table 10.1.

The striking feature about our scenarios is that all the fires occurred at night or in early morning when most patients were in bed. Three cases (5, 7, and 10) co-incide with the highest ranking death scenario defined by Clark and Ottoson, two of the ten coincide with their second ranking scenario, and one with the fourth ranking. This correlation is significant, as showing that nursing home fires of the type we have investigated are much like any residential fire, especially in terms of ignition source and ignited agent. Again, smoking is the most prominent ignition source. The ten scenarios are portrayed in Table 10.1.

The physical facilities in which the ten fires occurred are all quite different. This contributes to considerable variation in fire development and human behaviour, although, as the death scenarios suggests, the original circumstances surrounding the outbreak of the fires may have been similar. The facilities range from refur-bished single family homes to eleven-storey high-rise buildings.

Most of the buildings studied were one-storey frame houses, defying the myth that low-rise buildings are easy to escape from. In Case 1, for example, the building lies in a garden, each room with windows facing it; yet 31 of the 46 patients died of smoke inhalation inside the building, just a few metres from safety. There was considerable variation in terms of fire-directed hardware, such as sprinklers, smoke detectors and smoke doors, etc.

We had a factual basis in the form of building remains, but behavioural data were scant. This poverty of information made it necessary to manipulate data several steps removed from the investigator. A further complication was that fire reports tend primarily to address the physical circumstances, and note behavioural aspects only incidentally. We therefore devised several structuring devices by which to order the data available for the ten cases.

These structuring devices or models are discussed in the next section. They com-pel the investigator to lay out the whole sequence of physical and behavioural events so that missing elements become obvious and can be accounted for. Poverty of in-formation is a fact in fire reporting that will plague us for a long time to come. It is unlikely that data will radically improve even if data collection practices do change for the better. For example, even when the fire reporter arrives at the scene of the fire in time, only traces are left of the fire incident. Further, the investigator must interview the survivors immediately before they rationalize their behaviour; the most important witnesses may be dead, others are traumatized, all of which litters the reconstruction effort with pitfalls. Yet we hope that the models will improve both the organization of current data and suggest directions for gathering of new data.

THE MODELS

A working model for describing human behaviour in fire situations was originated by our research group in 1975 (Lerup *et al.*, 1975). We had assumed that an invest-igation of the fire situation would disclose some aspects of the interaction between people and the burning built environment, and that we could then form a picture

Table 10.1 Death scenarios

The ten scenarios CASE	1	2	3	4	5	6	7	8	9	10
BUILDING storeys	I	I	I	I + II	I	II	II	II	II + III	XI
construction	frame	frame	frame	frame	brick	frame	frame	frame	brick	concrete
PEOPLE patients'	46	15	29	7	21	16	12	17	96	105
deaths	31	15	2	4	6	10	10	6	15	10
staff	6	1	1	1	6	2	–	1	30	4
deaths	–	–	–	–	–	–	–	–	–	1
Total deaths	31	15	2	4	6	10	10	6	15	10
ACTIVITY type	sleeping	sleeping	sleeping	sleeping	sleeping	sleeping	sleeping	sleeping	breakfast	sleeping
time	10 pm	8 pm	1 am	4 am	12 midnight	12 midnight	11 pm	12 midnight	9 am	2 am
SOURCE	smoking	electrical dryer	electrical heater	smoking	smoking	arson	smoking	arson	smoking	smoking
AGENT	waste-basket	interior finish	apparel	apparel	bedsheet	linen	chair	bed	apparel	carpet
Clark and Ottoson's death scenarios										
COMPARATIVE RANKING 1–14	14	11	4	12	1	2	1	2	12	1
PERCENTAGE OF US FIRES	2	2	4	2	27	5	27	5	2	27

of the emergecny either directly or indirectly through reports. This assumption proved to be correct. By working closely with Harold E. Nelson, Chief of Design Concepts at the National Bureau of Standards, it was possible to develop the basic assumption rapidly and focus on a conceptual basis.

As one part of the model we depicted fire development through a series of illustrations of specific moments or time-spans, not unlike widely spaced frames in a filmstrip, and added an analytical dimension by separating certain instances from the rest of the fire development. These we call *critical events,* and they mark changes in the course of the fire, such as 'ignition', 'flashover', or 'fire spreading into corridor'. Between critical events, there are other bits of fire development, and these, called *states* or *realms* by Nelson and other researchers in the field like Robert Fitzgerald of Worchester Polytechnic, are characterized by a certain internal consistency, marked at the beginning and end of events. Using states and critical events, the development of a fire can be described in a simple coordinate system, by plotting time on one axis and heat energy dissipation on the other.

In the other part of the model, at the same time as the events of the fire were described, the behaviour of the occupants during the fire was analysed and conceptualized. Scrutiny of an occupant's behaviour stream revealed distinguishable 'bits' of behaviour, investigation or rescue, for example. Such bits are characterized by an internal logic and consistency, and a sense of purpose, stretching from the situation to an action. We called them behavioural *episodes,* using Roger Barker's vocabulary (Barker and Wright, 1955). Each episode is marked by a more or less distinguishable event at either end. We called these *decision points;* that is, we deduced that, for example, the nurse's aide made certain decisions to pursue a particular line of action. The similarity between states and episodes and critical events and decision points is advantageous, as long as there is no confusion about the absolute difference between the combustion process and the behavioural process. Thus, marrying the two conceptualizations produced the first working model (Figure 10.1).

The most intriguing and crucial aspect of the model was its revelation of the time link between behaviour and fire. Both researchers and professionals in the field of fire research commonly assume that people *respond* to the burning environment. Our evidence shows that people *manage* their actions rather than respond to the stimuli of the fire in a deterministic fashion. This is demonstrated by the relative absence of a significant set of recurrent responses to fire. If fire were experienced by all occupants in exactly the same way, we would get clear recurrencies; we would find *attack* when the fire is discovered at the time of ignition, *alarm* when smell is noted, *rescue* when the fire is small and not really dangerous, and *escape* once the fire is large and dangerous.

Instead, the ten fires we studied revealed a variety of actions in the face of what seemed fairly similar severity levels. This has led us to hypothesize that people interpret the fire situation and interact with it in quite personal, sometimes unique ways. Fortunately, the uniqueness can still be associated with certain general principles of emergency behaviour, sufficient to make research conclusions possible, even if not resulting in prediction of human behaviour. These general principles we have transformed into a set of *lifesaving actions,* which are made up of a family of behavioural episodes.

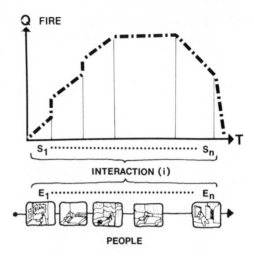

Figure 10.1 Fire—interaction—people

Each behavioural episode (E) may occur for a variety of purposes. For example, sounding an alarm may serve to: alert staff, arouse patients, or call the fire department. We refer to these varied purposes as *objectives* (O). Furthermore, these actions may occur in a variety of ways, or *mechanisms* (M), which include pulling the alarm, phoning the fire department, seeking help from neighbours, shouting at other occupants, etc. This array of episodes, objectives and mechanisms is presented in Figure 10.2.

The uniqueness of human behaviour is thus confined within the limits of these lifesaving actions. Nevertheless, there is still enough internal inconsistency to prohibit immediate hopes for predictability. Even if many of the ten cases display the same behavioural episodes, the sequence in which they occur is rarely the same. Furthermore, the state of the fire at which the same behavioural episodes occur is rarely the same. This does not imply that human behaviour occurs independently of the physical environment and the fire, but it does point out how differently any situation can be interpreted.

For this project, these findings forced us to rely on the general principles of human behaviour as manifested in the vocabulary of episodes, to steer around the lack of recurrency, and to accept that any one situation may have several interpretations. This led us to a general principle for evaluating the limitations produced by any one built setting; one should be able to pursue any behavioural episode as long as possible or until the environment becomes uninhabitable (due to flame or smoke). The logic behind this principle is that we do not yet know which is the most appropriate sequence of episodes to pursue in any one fire (appropriate in terms of the limitations set by the individual or group and also in terms of the correct way to act to minimize life safety).

In the further development of the working model to a second version the relative independence of the fire and the behaviour become more evident. Thus each specific instance of interaction between people and the burning environment can be des-

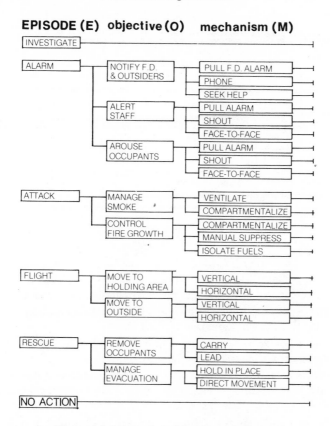

Figure 10.2 Episode—objective—mechanism

cribed and sometimes fully understood, and in fact, some recurrent patterns may be discerned by a generous investigator (Figure 10.3). But there is simply not enough recurrency to predict behaviour in the light of any given state of the fire. Nonetheless, our mission, to arrive at design implications through the understanding of behaviour, is still possible. The fire situation is thus structured around both a description of the fire itself and human behaviour. The description of the fire divides into *critical events* and *states*, while behaviour divides into *decision points* and *episodes*.

The fire and behaviour information is presented by means of an illustration technique which we have called mapping (Lerup, 1977). Behavioural mapping, mapping dealing directly with occupant behaviour in the fire situation, will be discussed in detail in the following section.

BEHAVIOURAL MAPPING

One purpose of this project is to explore the nature of emergency behaviour, but such an exploration must be inconclusive, because of the limited amount and type of data that are currently available. Nevertheless, certain aspects of behaviour are useful to discuss at this stage.

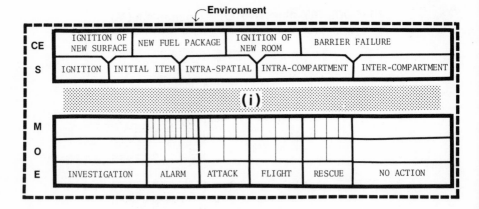

Figure 10.3 Environment

A concept of *recurrent* behaviours is essential to anyone seeking an understanding of behaviour, since it allows a certain level of predictability. At least two categories of recurrency may be defined: sequences of actions and types of actions. Both these categories are important for a complete understanding of emergency behaviour in relation to lifesaving actions.

Within our purpose of developing a safer physical environment, the ideal performance criterion for an emergency in a nursing home would be to maintain the essential environment for as long as possible, thereby allowing time for the occupants to take any necessary sequence and type of action. Thus, we should design the environment in such a way that these actions can be completed.

We found that there are few if any recurrent *sequences* of such actions taken by either staff or occupants. There is indeed a clear tendency to start most sequences with 'investigation' or 'alarm', but the following sets of actions do not follow a clear pattern. Thus, we saw investigation sometimes followed by attack, sometimes followed by rescue, etc. Our findings suggest that there are no consistent sequences of actions, and it is felt that further research will support this initial finding (Canter, Breaux and Sime, Chapter eight of this volume).

Twenty-three cases were utilized for these calculations, ten of the original cases and thirteen from NFPA's FIDO file. There is some recurrency in terms of *types* of actions, among the staff (Table 10.2) and among occupants (Table 10.3).

It is clear that our assumptions about the staff's willingness to involve themselves in decisive action are confirmed, and likewise that the relative indecisiveness of patients is also confirmed. More reliable answers to recurrencies are needed but this will come only with a concerted effort to obtain more and better data. Human behaviour must therefore be dealt with in more case-specific terms, with less emphasis on recurrencies within cases. This more general behavioural information is displayed on what we call the *behaviour maps*. The behaviour maps are a subsidiary part of the theoretical model, and provide a description of the behaviour component while also taking into account the interaction between behaviour and the burning physical environment. The two maps shown in this section (Figures 10.4

Table 10.2 Frequency of staff actions

Actions		Observations	Percentage
Investigate	(I)	23	24.5
Alarm	(A)	39	41.5
Attack	(Ak)	10	10.6
Flight	(F)	6	6.4
Rescue	(R)	14	14.9
No action	(N)	2	2.1

Table 10.3 Frequency of patient actions

Actions		Observations	Percentage
Investigate	(I)	16	25.8
Alarm	(A)	17	27.4
Attack	(Ak)	4	6.5
Flight	(F)	17	27.4
Rescue	(R)	1	1.6
No action	(N)	7	11.3

and 10.5) are the complete description of the human behaviour in one of the cases under study. They represent the base data against which all further investigation of that case can be compared. The first map (Figure 10.4) is the composite of the entire emergency, showing both fire development and, briefly, human behaviour. The second map (Figure 5.10) is a behaviour map only, and shows occupants' behaviour in detail. On this map critical events of the fire are noted as vertical lines along the horizontal time axis. The stream is also divided into a series of behaviour episodes, following working model II (Figure 10.3), as they were acted out by individuals or groups. Occupants are divided into three categories: patients, staff and others.

The division into time segments reflects our research focus. Our investigation of the case studies convinced us that the period between detection of the fire and the arrival of the fire department is the most *crucial lifesaving period* in terms of the 'first compartment' (that is, the area in direct contact with the room of origin and the fire). Thus, lifesaving efforts in this crucial period must be performed by human and technological devices without the benefit of the fire department. (This is not an indictment of the role of the fire department but simply a statement of an undeniable fact: it takes time for the fire department to get to the scene, regardless of improved detection technologies.) The average response time in Britain is in the order of five minutes. For an excellent argument for the importance of time see Rexford Wilson (1962). Wilson defines here an average 'reflex time' of nine minutes (not necessarily directly comparable with the response time mentioned above). This reality has been largely neglected, and therefore attempts to improve fire protection have till now been entirely focused on either the fire department or non-behaviour-

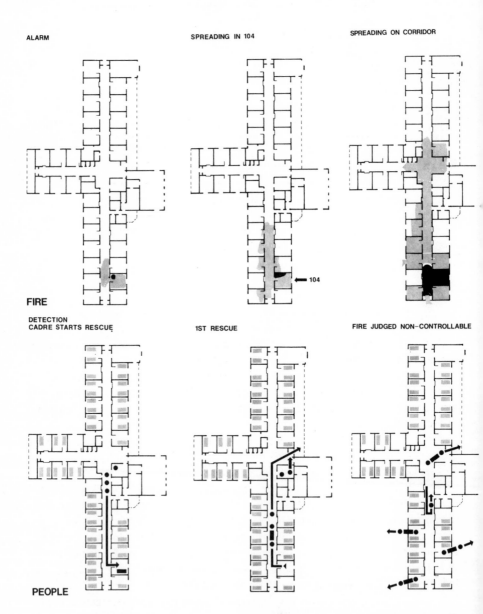

Figure 10.4 Building plan showing development of the fire

CORRIDOR FULLY INVOLVED

MAX. FIRE SPREAD

RESCUE: CADRE AND NEIGHBORS

RESCUE AND FIRE FIGHTING: F.D

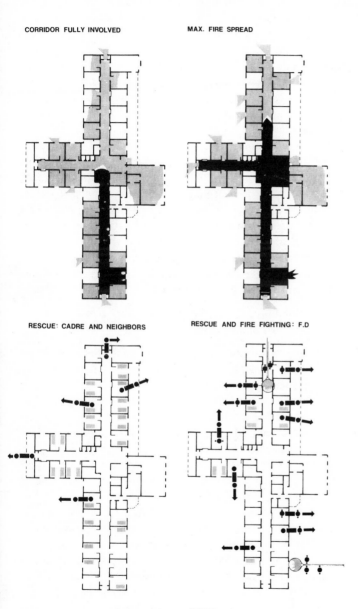

DEATHS: 31

oriented technologies. Clearly, an attempt must be made not only to shorten the time between detection and arrival but also to extend the crucial lifesaving period by maintaining a liveable environment for as long as is needed to rescue all occupants. The *behaviour* maps were a direct result of our sense of the crucialness of the lifesaving period.

Figure 10.5 also shows a death scenario. The crucial lifesaving period was included for each case when its duration was known in minutes. Since the underlying data for the time estimations were tentative, however, we were compelled not to draw any conclusions about the actual importance of time—the difference between such periods is three to 42 minutes.

The behaviour episodes are represented by cartoons and text, and those we discovered during the analyses of the case studies. Any one episode may involve more than one cartoon, or may end in one cartoon. The cartoons use simple representational techniques to illustrate most complex situations, and the sometimes diffuse definition of the boundaries of an episode reflects the available data. Nurses are shown wearing cone-shaped hats, to distinguish them from patients; firemen are shown in helmets; other persons are defined by notations in the text.

When other aspects of the investigation are brought into the model (see section on design maps below), the composite and behaviour maps serve as the frame of reference for design applications and data manipulations.

DESIGN MAPS

The practical application of the theoretical model is represented by the *design maps,* based on the data ordered by the model and the behaviour maps. They can be used to test earlier conclusions by running the situations backwards while including new physical components (such as the door-closer mentioned in the discussion below). An example of such a map, using one of the cases under study, is shown in Figure 10.6. The design maps are the centre of the project; in them we bring our analysis of the environment of fire to a conclusion. They function as a *design-generating system,* by which a selected number of behavioural episodes can be analyzed and the relevant parts of the physical setting evaluated. Where appropriate, they lead to suggestions for the modification of the physical setting.

Each design suggestion is drawn from a single episode that has been studied, and therefore the implication of each suggestion is isolated from its larger context. Thus, one case may show five design implications, which may overlap in the sense of solving the same problem, because the design map is a generating system *only, not* an evaluation of the overall appropriateness of all the design implications taken together. Such an evaluation is made later in the section on design implications (page 173).

The design maps include three elements: environmental conditions; lifesaving actions; and environmental implications. The first two horizontal sections of each map are *environmental conditions* and *lifesaving actions,* and these are drawn directly from the case study in question. Cartoons and texts are restructured to demonstrate the critical links between episodes of human behaviour in accordance with the lifesaving action tree.

Environmental conditions

This has two aspects: the physical setting and the behavioural situation. The physical setting is shown as a diagrammatic plan, in which certain important ingredients —such as the room of origin, the nursing station, and fire doors—are shown. The development of the fire is shown across the top of this section, according to the categories defined in the working model II (see Figure 10.3).

The behaviour situation is described by dividing the horizontal section into discrete areas, each covering a particular behavioural episode. Conditions prevailing at the outset and during each episode are noted.

Lifesaving actions

This consists of two parts. One shows the tree of interrelated episodes, objectives and mechanisms; each pertinent link between *episode* (E), *objective* (O), and *mechanism* (M) is indicated. The second part shows the episodes with cartoons and associated text. Here, the relevant machanism(s) are indicated by numbers that refer to the tree. Some of the episodes are linked with arrows: solid arrows indicate that the same person(s) moved from one episode to another; broken arrows show that a piece of information was transmitted to the ensuing episode.

Environmental implications

This consists of several integrated parts. The word 'environmental' here means both the social and the physical environments, and a design implication must be seen as only one element of an environmental implication. This understanding is central as well to a major message of this project: the success of a fire-related technology is only possible if it is correctly used, and this requires that the occupants have the *knowledge and information* necessary to *manage* the physical setting. An environmental implication therefore consists of: knowledge and information; management and behaviour; and design implication.

All environmental implications and the discussion of them pertain to the case under study only. For example, if in one case there is no automatic alarm system, such a system will not be discussed (but this will be done later, in a general evaluation). Beginning at the left side, the 'death scenario' is shown with a plan drawing that includes the implied design changes from the analysis. Below the appropriate episodes are the environmental implications, with major design implications shown in heavy type. Associated with each such implication is a short discussion.

Design implications

Seven design implications were generated from the ten case studies. These are summarized below in Table 10.4, followed by an outline of the evaluation system which was used to give each design solution its proper value.

The environmental implications of the design we have generated speak to the relation between the occupants and the physical setting of the institution. We have

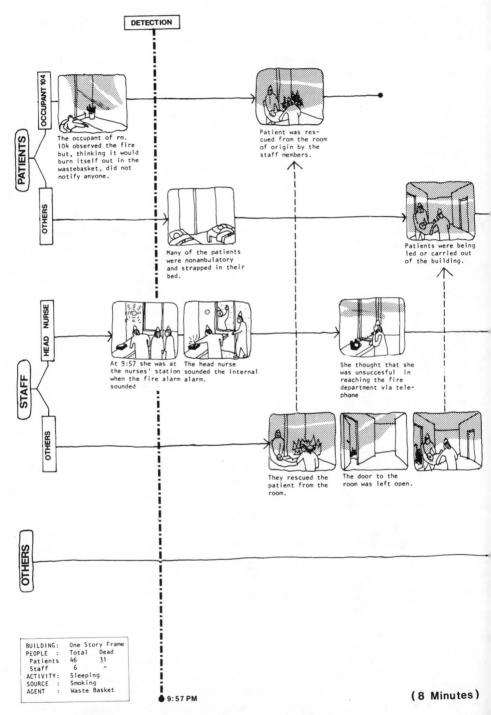

Figure 10.5 Actions in fire

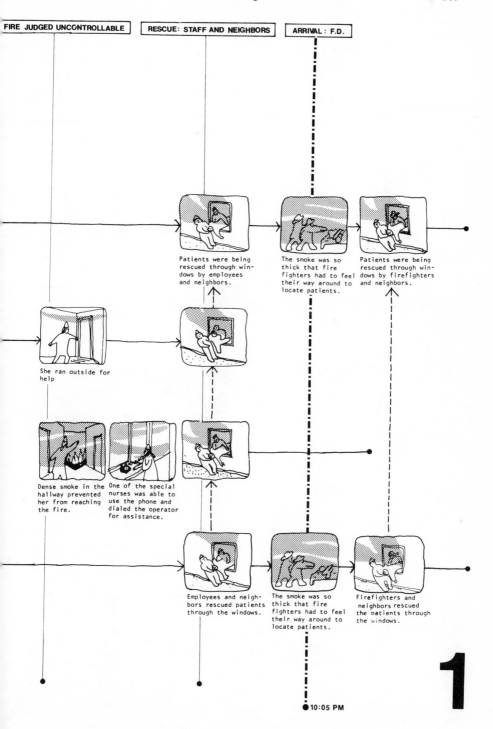

FIRE JUDGED UNCONTROLLABLE

RESCUE: STAFF AND NEIGHBORS

ARRIVAL: F.D.

Patients were being rescued through windows by employees and neighbors.

The smoke was so thick that fire fighters had to feel their way around to locate patients.

Patients were being rescued through windows by firefighters and neighbors.

She ran outside for help

Dense smoke in the hallway prevented her from reaching the fire.

One of the special nurses was able to use the phone and dialed the operator for assistance.

Employees and neighbors rescued patients through the windows.

The smoke was so thick that fire fighters had to feel their way around to locate patients.

Firefighters and neighbors rescued the patients through the windows.

10:05 PM

1

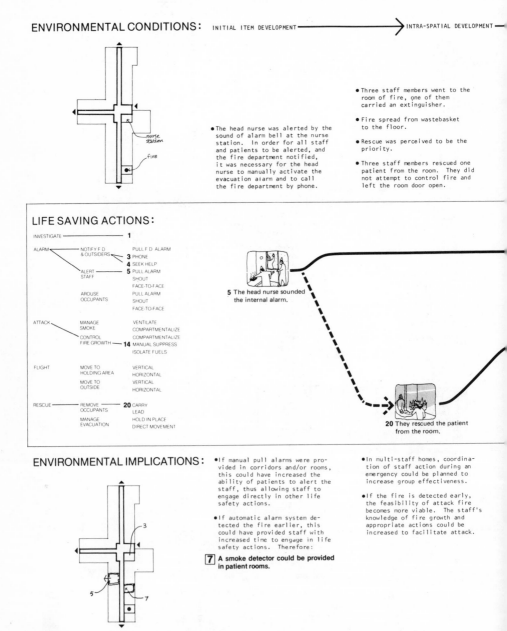

ENVIRONMENTAL CONDITIONS: INITIAL ITEM DEVELOPMENT ⟶ INTRA-SPATIAL DEVELOPMENT —

nurse station

fire

• The head nurse was alerted by the sound of alarm bell at the nurse station. In order for all staff and patients to be alerted, and the fire department notified, it was necessary for the head nurse to manually activate the evacuation alarm and to call the fire department by phone.

• Three staff members went to the room of fire, one of them carried an extinguisher.

• Fire spread from wastebasket to the floor.

• Rescue was perceived to be the priority.

• Three staff members rescued one patient from the room. They did not attempt to control fire and left the room door open.

LIFE SAVING ACTIONS:

INVESTIGATE ——————— 1

ALARM — NOTIFY F.D. & OUTSIDERS — PULL F.D. ALARM — **3** PHONE — **4** SEEK HELP

ALERT STAFF — **5** PULL ALARM — SHOUT — FACE-TO-FACE

AROUSE OCCUPANTS — PULL ALARM — SHOUT — FACE-TO-FACE

ATTACK — MANAGE SMOKE — VENTILATE — COMPARTMENTALIZE

CONTROL FIRE GROWTH — COMPARTMENTALIZE — **14** MANUAL SUPPRESS — ISOLATE FUELS

FLIGHT — MOVE TO HOLDING AREA — VERTICAL — HORIZONTAL

MOVE TO OUTSIDE — VERTICAL — HORIZONTAL

RESCUE — REMOVE OCCUPANTS — **20** CARRY — LEAD

MANAGE EVACUATION — HOLD IN PLACE — DIRECT MOVEMENT

5 The head nurse sounded the internal alarm.

20 They rescued the patient from the room.

ENVIRONMENTAL IMPLICATIONS:

• If manual pull alarms were provided in corridors and/or rooms, this could have increased the ability of patients to alert the staff, thus allowing staff to engage directly in other life safety actions.

• If automatic alarm system detected the fire earlier, this could have provided staff with increased time to engage in life safety actions. Therefore:

7 A smoke detector could be provided in patient rooms.

• In multi-staff homes, coordination of staff action during an emergency could be planned to increase group effectiveness.

• If the fire is detected early, the feasibility of attack fire becomes more viable. The staff's knowledge of fire growth and appropriate actions could be increased to facilitate attack.

Figure 10.6 Analysis of Figure 10.5

INTER-COMPARTMENT DEVELOPMENT

• Smoke spread in corridor and nurse station prevented the head nurse from reaching the fire department by phone.

• Attempt to attack fire delayed by rescue action.

• Fire spread rapidly as a result of delayed attack.

• Smoke spread in nurse station prevented the head nurse from continuing attempts to call the fire department.

• The head nurse judged the action priority to be the notification of the fire department. She sought help from outsiders and did not engage in other actions.

• Patients at other parts of the building were rescued by staff.

• Fire and smoke spread judged uncontrollable at this point.

• Conditions at the nurse station was not totally unbearable since a special nurse was able to use the phone to reach the fire department.

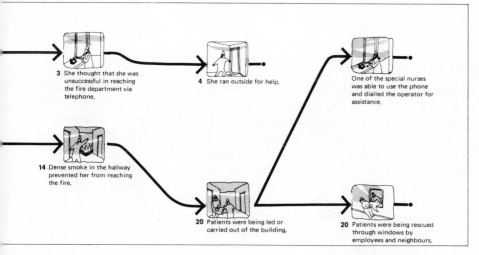

3 She thought that she was unsuccessful in reaching the fire department via telephone.

4 She ran outside for help.

One of the special nurses was able to use the phone and dialled the operator for assistance.

14 .Dense smoke in the hallway prevented her from reaching the fire.

20 Patients were being led or carried out of the building.

20 Patients were being rescued through windows by employees and neighbours.

• If there was automatic alarm connection to the fire department and staff is aware of its notification, this could have allowed staff to engage in other life safety actions. Therefore:

3 A feedback signal from fire department to staff could be provided.

• In order to facilitate communications from the nurse station without interference from smoke penetration, the nurse station could be protected from smoke.

• If the head nurse was aware of fire department notification, she would not have gone outside to seek help, thus allowing her to engage in other actions.

• If the nurse station was protected from smoke, this could serve as holding area for some patients who are unable to escape to the outside.

• When smoke fills corridor, thus reducing or prohibiting its viability as an egress route, windows from patient rooms could be used as a means of egress to the outside. Therefore:

5 An alternative means of egress from certain rooms could be provided.

BUILDING:	1 Story Frame	
PEOPLE:	Total	Dead
Patients	46	31
Staff	6	-
ACTIVITY:	Sleeping	
SOURCE:	Smoking	
AGENT:	Waste Basket	

1

Table 10.4 The seven design implications

UNCONDITIONAL
(3)   A feedback signal from fire department
       to staff could be provided.
(4)   A nurse station could be located cen-
       trally in each major compartment.
PREFERRED
(2)   A holding and releasing system could be
       provided on certain doors.
(6)   'Defend in place' could be developed as
       an alternative to 'flight'.
(7)   A smoke detector could be provided in
       patient rooms.
CONDITIONAL
(1)   The automatic closing of fire doors
       could be connected to the nearest smoke
       detector only.
(5)   An alternative means of egress from cer-
       tain rooms could be provided.

---

suggested that this relationship must become much more intimate than it generally is at present, and that any fire-related aspect of any setting must be linked with an appropriate set of management strategies, based on knowledge and information.

The matrix in Table 10.5 represents an environmental approach to fire protection. It is not within the scope of this project to develop in full the knowledge and information base, or the field of management and appropriate emergency behaviour. But we are aware that our slight references to these categories do not cover all the aspects that result in their actual importance; for example, it is inadequate to install a door-closer that is activated by a smoke detector and claimed to be failsafe, unless the staff is aware that a piece of furniture blocking it will render the installation useless when it is needed.

The entire field of knowledge and management implicated in our findings has been untouched until now. Recently, however, NFPCA's education section headed by Richard Strother has made some promising advances that suggest new attention to the field. We ourselves feel that knowledge and management must be fashioned around specific fire-related aspects and ways of assembling the physical plant. Furthermore, fire protection should be oriented towards specific types of emergencies that are most prevalent (as in Clark and Ottoson's (1976) death scenarios). This focus does not tackle the issue of 'correct emergency behaviour', which for some time must remain unresolved, nor can we claim that the matrix in Table 10.5 is a comprehensive fire protection strategy. However, we do feel that it points in the right direction.

The actual circumstances in which fire protection strategies must be worked out are usually far from ideal. Most of our emergencies occur during the night shift, when staff–patient ratios are at their lowest (one staff person per eighteen patients). Patients are also in their most vulnerable state, often drugged and strapped in their beds. The staff itself knows that least about the sociology of the situation at night, *since they usually encounter the patients sleeping and therefore know very little about the actual dynamics of the social situation, peculiarities of behaviour and habits of the individual patients, that may have relevance to the management of fire safety.* All these factors demand extreme caution and ingenuity on the part of the administrators of facilities, as well as on the part of the policy-making bodies at state and national levels.

Evaluation

Now we may legitimately raise the issue, to what extent can the seven design implications be generalized? Just how valid are they when seen in a more general context of fire protection in this type of facility? This is a question of immense complexity and we can only begin to tackle it here. We therefore developed an evaluation system in which we manipulate the model to play back and test each design solution in new cases. Here we are still working within our ten case studies, but the technique is designed to be used when the data base is more extensive.

The evaluation map, shown in Figure 10.7, is a scenario in which the fire incident is repeated after the installation of a selected fire protection device. In the case shown, an automatic fire door-closer is introduced. The fire incident is re-enacted, and arguments for and against the device are made. This could be done by a panel of experts for a more extensive and reliable evaluation (*à la* Delphi) (Middendorf, 1973).

The evaluation of the seven design implications organized them into their order of importance as shown in Table 10.4. For example, a feedback signal, telling the staff that the fire department is notified and on their way as well as a centrally located nursing station, could be provided under all conditions. Preferred design solutions, such as an area of refuge, here under the rubric 'defend in place', could be provided for patients with limited mobility. Finally, only under certain specific conditions would alternative egress from patients' rooms be provided.

CONCLUSIONS

We have used behavioural data in order to improve the physical environment, but have in this process come to understand that any physical change has to be accompanied simultaneously by behavioural changes affecting both the knowledge of the occupants and the management of the facility.

In the evaluation we simulated a fire situation in which a given set of occupants had to contend with an emerging yet unknown fire. We realized then the magnitude of the ambiguities facing the occupants in a fire emergency. From this it became clear that we must magnify the parts of the physical environment that are relevant to the occupants' lifesaving actions. It also became clear that the origins of an emergency event must be traced back into the everyday life of the institution. From here, we began to see how knowledge and management could be improved through education about the physical setting.

Table 10.5 An environmental approach to fire protection–design implications

	Knowledge and information base	Management and behaviour during fire	Design implications	
Investigate	Knowledge about environmental cues of fire: audio, visual, olfactory	Locating and verifying fire without delay: individual and group actions in investigation	4	A nurse station could be centrally located in each major compartment depending on building typology to facilitate the early awareness of emergency cues
	Reliance on patients as source of information about fire: face-to-face alert, manual pull alarm			Manual initiating alarm boxes could be located in corridors and rooms to facilitate initiating actions of patients
	Knowledge about vulnerability of smoking patients		7	Smoke detectors could be located in patient rooms to shorten the detection time
Alarm	Knowledge about the function and purpose of alarm system: initiating device, internal alarm, fire department alarm	Alertness to fire department notification	4	A nurse station could be centrally located in each major compartment depending on building typology to facilitate internal alarm and notification of fire department by phone
		Alertness to initiating signal and internal alarm	3	An automatic feedback signal from the fire department to the nurse station could be provided to eliminate duplicating alarm actions
				Manual initiating alarm boxes could be located in corridors and rooms to facilitate alarm actions of patients and staff

Attack	Knowledge about the capability and limits of manual suppression: information about fire growth and smoke spread, training for staff to control fire	Manual supression appropriate to stage fire	7	Smoke detectors could be located in patient rooms to shorten detection time and thus increase alarm effectiveness
			4	A nurse station could be located in each major compartment so that staff has direct and quick access to origin of fire
	Knowledge about the distribution and location of fuel packages	Isolate fuel packages by moving fuel items	2	Hold-open, self-closing system on room doors could be provided to keep smoke and fire in room of origin and away from other rooms
		Ensure closure of doors	1	The closing of fire doors could be connected to only the nearest smoke detector to in order to allow critical staff action within compartment before intercompartment spread
Flight	Knowledge about alternative egress routes: physical delineation of exits, egress drills for occupants	Maintain route completeness: reduce obstructions in egress routes	5	An alternative means of egress from each room could be provided so that in the event of smoke spread in corridor, there is another viable egress route from each room
	Knowledge about location of holding areas	Maintain essential environment: ensure closure of doors	6	Each room could be compartmentalized to serve as holding areas when egress routes are no longer viable
			2	Hold-open, self-closing systems on doors could be provided for room doors and stairway doors to insure the viability of egress routes

[Continued on p. 176]

Table 10.5 An environmental approach to fire protection–design implications (*Continued*)

	Knowledge and information base	Management and behaviour during fire	Design implications
			Physical and graphic delineation of egress routes could be provided to facilitate flight
Rescue	Knowledge about capability and limits of rescue action: information about non-ambulatory patients, information about staff capacity	Hold-in place for non-ambulatory patients	4 A nurse station could be located in each major compartment to facilitate removal of patients and the management of rescue
	Knowledge about alternative routes of rescue: training and drills of staff action in rescue	Maintain route completeness: ensure access to patients from outside	1 The closing of fire doors could be connected to only the nearest smoke detector to allow more time for the staff to engage in rescue action within the compartment of fire
	Knowledge about compartments: information about methods of smoke control	Maintain essential environment: ensure closure of doors	5 An alternative means of egress from each room could be provided to facilitate rescue of patients from the outside
			6 Defend in-place could be developed as an alternative to flight and removal of patients

The concept of the emergency chain—from its origin in the everyday life situation until its conclusion at the end of the emergency—must be included in any holistic environmental perspective on fire protection. An occupant's normal behaviour during routine activities can be conceived of as the potential beginning of an emergency chain. A completely formed emergency chain includes both physical and human links. Our task is to break the chain. Although the chain can be broken at any point, it is obviously better to break it at its earliest links; there is always a point at which intervention is extremely simple, such as pulling a plug, pouring a pitcher of water, or putting out a cigarette. But it is preferable to think about breaking the chain even *before* it starts, that is, before ignition. For this reason, any investigation of an emergency needs to refer to the normal, day-to-day patterns of the facility. Similar concerns with the importance of the normal situations has been expressed by Breaux, Canter, and Sime (1979). Part of the process of magnifying the fire-related aspects of the environment is to look at the emergency potential present in a given physical setting. Emergency potential can be thought of in terms of 'areas of vulnerability', that is, points in the environment that a keen observer can link together in considering likely emergency chain scenarios. We can think of the dimensions of vulnerability as signposts that help indicate the possible onset of an emergency; for example, a wastebasket that a patient uses as an ashtray is one such signpost.

Nurses are actually particularly well trained to observe these signposts in regard to the patients or residents. An important aspect of nursing is to recognize changes in the ill that signal the onset of a crisis or that indicate a specific diagnosis. We can think of what a nurse does as a pattern of intervention in the illness chain of the patient, and it is reasonable to think that this type of activity could be extended to the fire problem in the facility.

This focus on the importance of the relationship between purposeful behaviour and the physical setting indicates, as was said above, that there is no guarantee that any of the seven design solutions will improve life safety if they are applied without consideration for the behaviour of occupants. To claim the contrary goes against the reality of the nursing home situation, in which the physical environment—the building and its equipment and furnishings— is but one element of a more complex whole.

Consequently, we claim that any solution to a fire problem must be weighed in light of the whole system. A device like a hold-open automatic door-closer has to be considered in terms of how the residents and staff of the facility use the rooms and corridors. At another level, such a device has to be considered in terms of resident and staff behaviour patterns; we need to know whether a closed door would signal the abandonment of an area of the facility in the minds of its occupants, thus inhibiting efforts at rescue. The cases we have studied indicate that such problems, as well as the more common physical problems of doors jammed open because they interfere with the normal workings of the facility, continually undermine the effectiveness of such devices in a fire situation.

In a very real sense, each facility presents a unique set of human and physical attributes. The physical solutions posed to deal with a given facility's problem have

The automatic closing of fire doors could be connected to the nearest smoke detector only.

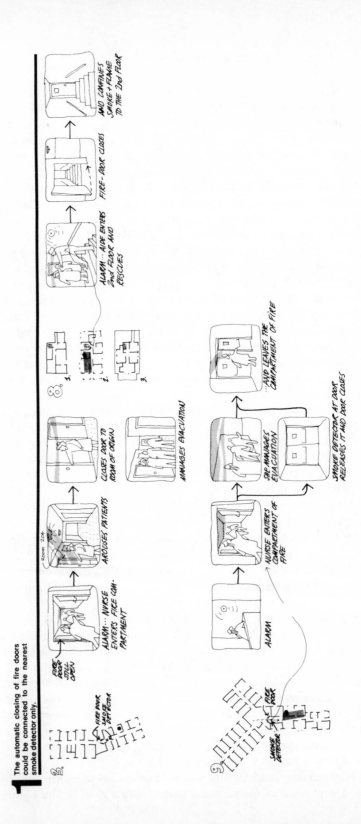

ARGUMENT FOR:

1. The fire door is a psycho-physical barrier to staff outside the burning compartment. Although this barrier is necessary and most effective in containing smoke spread, it can limit staff in assisting patients when closed prematurely. Therefore, the fire door should be activated to close only by the nearest smoke detector. If this is the case the staff will not be confronted by an unnecessary barrier and the door will respond only when smoke threatens adjacent compartments.

2. To facilitate normal operations the fire doors are frequently wedged or propped open making them useless in stopping smoke spread. A hold open, self-closing system connected to the nearest smoke detector would make the doors more reliable.

ARGUMENT AGAINST:

1. It is better to insure that each patient room is compartmented to contain the fire in the room of origin. This curtails the potential side effects of fire doors in the corridor.

2. It is better to locate a nurse station in each area defined by fire doors. In this way the closing of the fire doors do not affect the staffs actions since they are in the area of the fire anyway.

to fit into this context of uniqueness, and part of the designer's task is to gain a sufficient understanding of this context to evaluate his or her proposed solutions in light of it. A number of design tools, including participatory planning, the use of modelling and simulation techniques, and user-elicitation surveying, have been developed in the past decade that involve the occupants and staff in the design process, and this involvement greatly increases the probability that such solutions will fit within the total picture.

The selection of the appropriate level of fire protection for a population group and a building is not primarily a technical problem, but a profoundly human one with ethical, educational, economic, and technical dimensions. It cannot be decided in three research laboratories, nor among professionals, it must be brought to the community. Now, with the the help of scenarios and mapping techniques, everyone can begin to understand the complexity of the fire protection issue and their role in it.

Our conclusions can be summarized thus: fire protection must begin before, and prevail throughout, any emergency. The emergency chain is always breakable. Knowledge about fires and human behaviour in fires must be coupled with management strategies for normal as well as emergency situations. The physical setting with its assortment of fire-related elements must be appropriately designed to respond to the abilities of the occupants. Simply put, fire protection must be approached from an environmental perspective. For a more extensive treatment of this material see: Lerup, Cronrath and Liu (1978).*

REFERENCES

Barker, R. G. and Wright, H. F. (1955). *Midwest and its Children.* (New York: Harper and Row) pp. 225–273.

Breaux, J., Canter, D., and Sime, J. (1979). 'Psychological aspects of behaviour of people in fire situations', University of Surrey, Fire Research Unit.

Clarke, F. B. and Ottoson, J. (1976). 'Fire death scenarios and fire safety planning', *Fire Journal,* **May,** 20.

Lerup, L. *et al.,* (1975). *Mapping Behavior: The Case of Fire.* (Berkeley: University of California).

Lerup, L., Cronrath, Ᵽ , and Liu, J. (1978). *Human Behavior in Institutional Fires and its Design Implications* (Washington, DC: Center for Fire Research, National Bureau of Standards).

Middendorf, W. (1973). 'A modified Delphi technique of solving business problems', *IEEE Transactions of Engineering Management* **EM-20, No 4.**

Wilson, R. (1962). 'Time: the yardstick of fire control', *NFPA Firemen, November–December,* 3–8.

*This report and a number of other reports on the subject of life safety can be obtained from Harold E. Nelson, Chief, Program of Design Concepts, Center for Fire Research, National Bureau of Standards, US Department of Commerce, Washington DC.

Fires and Human Behaviour
Edited by D. Canter
© 1980 John Wiley & Sons Ltd.

CHAPTER 11

A Model of Behaviour in Fires
Applied to a Nursing Home Fire

PERRY EDELMAN, ELICIA HERZ, AND LEONARD BICKMAN
Fire and Human Behavior Research Center,
Loyola University of Chicago

The authors investigated the behaviour of staff and residents in a nursing home fire. The description of that fire, our research methods and our findings—as they relate to a proposed general model—comprise this chapter.

OVERVIEW OF THE FIRE*

The fire occurred in a nursing home with an approximate resident capacity of 250. The building had four storeys with a fifth-floor penthouse. Floors two to four of the building consisted of three wings arranged in a T-formation (Figure 11.1), the north and south wings being the same size and forming the hat of the T. The east wing was perpendicular to these two wings and somewhat shorter. Each wing had smoke doors approximately halfway down the wing which closed automatically when either the smoke detectors or pull alarms were activated.

The fire began in a clothing closet of a foyer leading to a patient's room on the fourth floor—room 411. The room was in the south wing just south of the smoke doors. Ninety-one patients resided on the fourth floor and three members of the nursing staff were on the floor at the time of the fire. The fire was first detected on the fourth floor at 8.27 pm by a smoke detector and was extinguished at approximately 8.43 pm—eight minutes after the firefighters reached the fourth floor. Fire spread through room 411 and onto adjoining areas in the hallway. The entire south

This report is a product of a joint effort of the Department of Health, Education and Welfare (HEW) and the National Bureau of Standards (NBS) Center for Fire Research (Grant No. 6–9015). The programme is a five-year activity initiated in 1975. It consists of projects in the areas of: decision analysis, fire and smoke detection, smoke movement and control, automatic extinguishment, and behaviour in institutional populations in fire situations.

*A report from B. T. Lee at the National Bureau of Standards has provided some of the background information in this section.

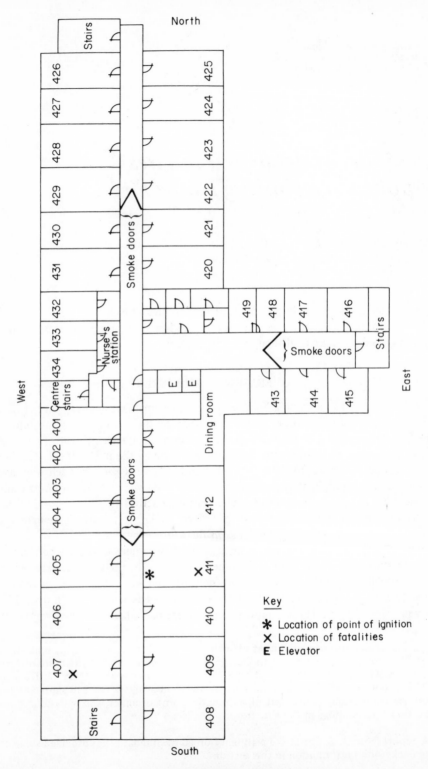

Figure 11.1 Map of fire floor

wing—rooms and hallway—and the area near the nurses' desk where the three wings met suffered smoke damage. Five members of the nursing staff assisted in the evacuation process. The loss in lives included one male resident in room 411 and one female resident in room 407; one firefighter and a number of residents were treated at a hospital and subsequently released.

APPROACH TO THE STUDY

The problems involved with studying human behaviour in fires have been previously described. Such a situation is not practically or ethically recreatable in the laboratory. It also is highly unlikely that researchers can study behaviour during a real fire. A most valuable technique, therefore, is to use *post facto* descriptions of behaviour.

In the past, these descriptions have been second and third-hand reports by fire departments and the news media (Bryan, 1977; Lerup, 1976). The inaccuracies and losses of information which accompanies such data is not surprising (Selltiz, Wrightsman, and Cook, 1976). Traditionally, the people or organizations who have collected these data have not been concerned with the interaction between human behaviour and the fire environment. Accordingly, they have not asked the kinds of questions which would lead to a better understanding of how human behaviour and the environment can be altered to help save lives and property during a fire. These and other problems are common to all archival data (Webb *et al.*, 1966).

The researchers at the Fire and Human Behaviour Research Centre, Loyola University of Chicago have developed and tested the Primary Source Interview (PSI) which has been used with the direct participants (patients, staff and firefighters) of fires in health care facilities. The interview was developed from the researchers' model of human behaviour in a fire emergency.

The researchers believe that the PSI is an excellent technique for collecting information concerning people's behaviour in fires, its two major strengths being directness and flexibility.

Directness provides for the most accurate, least distorted information possible—those people who were involved with the fire environment were interviewed directly by the researchers themselves; flexibility allowed for the collection of the kind of information needed. Rather than depend on data that is available and has been collected by other agencies not usually interested in asking the kinds of questions which researchers need to know, researchers can ask their own questions and obtain the kind of information necessary. This problem of the inadequacy of fire records has been discussed by Bickman (1976). In addition, Lerup (1976), in describing his mapping project (see Chapter 10), mentions the problem of using second and third-hand sources of information.

Also in terms of flexibility, though most of the questions comprising the interview apply to all fire situations, the interview can be modified to assess more accurately any situation which might arise. This allows for a more complete, fuller study of the man—environment interaction during a fire.

One of the major problems of the PSI is the difficulty in interpreting the results due to the retrospective nature of the technique. Respondents are asked to recall their thoughts and perceptions *after* their behaviour occurred. It is possible that this recall could have been influenced by their previous behaviour—that is, their recall of their perceptions could artifically have been made consistent with their behaviour. In this sense, their behaviour could have caused their perceptions. This is a significant limitation of the PSI and should be balanced with the use of other methodologies in this area of research.

Based on the authors' model of human behaviour in a fire emergency, the interview was constructed to discover the actions taken by respondents, as well as the factors which may have influenced those actions. Therefore, after determining what behaviours occurred, respondents were asked where they were, what they saw, heard, smelled or felt and who they were with before performing the behaviour they mentioned.

Use of the PSI has allowed the researchers to specify some of the factors which help determine human behaviour during a fire:

(1) diseases and handicaps of patients
(2) personality traits such as risk-taking behaviours
(3) leadership qualities
(4) fear of ridicule or embarrassment
(5) friendship patterns
(6) social hierarchies
(7) size and location of fire
(8) extent of smoke and flames
(9) the physical design of the floor and building in which the fire occurred (including number and location of stairways and type of communication system)
(10) location of fire-fighting equipment and fire safety equipment and fire safety features such as smoke/fire doors, fire escapes and alarm systems.

THE PRIMARY SOURCE INTERVIEW (PSI)

The format of the interview consisted of five sections of closed and open-ended questions. The interview was conducted by two of the authors. The first section describes the pre-firesetting by asking where the respondent was, what he was doing and who else was present just before the respondent became aware of the fire. The cues used to detect and define the fire were then determined.

Respondents were asked if at any time during the fire they had pulled a fire alarm, handled a fire extinguisher or opened or closed any doors or windows. In addition, the respondent's awareness of the location and use of exits and knowledge of the purpose of the smoke doors was determined. Finally respondents were asked to state in their own words what they did from the time they became aware of the fire, until they were moved to another floor. In conjunction with this, the second section of the interview—mapping the respondent's movement on a diagram of the floor—was completed at that point. Behaviour maps are not included in this report.

Using the respondent's self-description, the interviewer divided his behaviour into major events (which were used in connection with section three of the interview). These questions concerned what the respondent perceived in the environment—smoke, heat, sounds and other people. As the respondent moved through the environment, his perceptions of the environment were expected to change. Perceptions of the environment could change with the mere passage of time as well. For these reasons, it was necessary to ask the questions pertaining to the respondent's perceptions of the environment for each period of time in which it seemed likely that changes in perception might occur. Normally this meant that there would be two or three major events for each respondent. In section one, questions concerning the respondent's initial environment (usually the bedroom) were already asked, so the first major event usually involved the respondent moving down the hallway to an exit. On the south (fire) wing, the environment south of the smoke doors (fire area) was expected to differ significantly from the environment north of the smoke doors, therefore movement on the south wing was usually broken up into two major events. The final event concerned movement down a stairway to another floor.

The fourth section of the interview gathered background information on the respondent and provided the demographic data for the study. Questions concerning a number of other areas of interest were asked in this section including the following: previous practice fire drill training, previous experience being interviewed concerning the fire, reactivity of researchers' interview ('Have other respondents talked about their interviews to you?'), concern among residents about fire, effect of fire upon attitudes towards likelihood of future fires, evaluation of staff's and firefighters' behaviour and medication taken before the fire.

The fifth and final section of the interview involved an evaluation of the respondent by the interviewer. After the respondent left the room, the interviewer rated the respondent with regard to mobility, psychological status, hearing, vision and noticeable handicaps. As a validity and reliability check these ratings were compared with a social worker's evaluations.

Use of the PSI

Before the start of each interview, the interviewer introduced him or herself and the purpose of the project. Respondents were guaranteed complete anonymity and were asked to sign a consent form developed by the researchers for the nursing home administrator. The researchers had good cooperation from the residents as only two people refused to be interviewed.

Each interview lasted between 30 minutes to one hour. Answers to questions were written on the interview forms and were also tape-recorded—the tapes were later transcribed for ease of access to data. The tape-recorder, which was small and unobtrusive, did not seem to bother any of the respondents, and the questions did not provoke any overt, emotional reactions.

FINDINGS FROM THE NURSING HOME FIRE

Though the study involved only 22 respondents, thus limiting quantitative analysis, the taped interviews were transcribed allowing for detailed review. A number of meaningful findings were discovered which help to elucidate some of the components of the model. A description of these findings will be included in the subsections below: description of respondents, detection of initial cues, defining the situation as a fire, coping behaviour, summary of respondents' behaviour (initial cue detection, cognitions, behaviour), summary of determinants, and overall summary. It should be made clear that the findings are restricted to the particular circumstances of *one* fire incident. Future studies will determine ability to generalize our results.

Description of respondents

Ten male and twelve female residents of the nursing home were interviewed approximately two to four weeks after the fire occurred. Two of the residents were black, and the others were caucasian. The age of the respondents varied considerably: two were under 50 years of age, seven were 50 to 59, seven were 60 to 69 and six were over 70 years of age. The majority of the respondents had been residents of the nursing home for a substantial length of time—only five had lived there seven months or less; the rest had lived there from seventeen months to ten years and the median length of residence for all respondents was two years. Although nine of the respondents had mentioned that they took some form of medication the night of the fire—six mentioned sleeping pills or psychotrophic drugs—due to the small sample size, analysis did not demonstrate any meaningful relationships involving the use of medication by respondents. Most of the respondents were in good physical condition in terms of mobility. Only three had noticeable handicaps—one respondent used a cane, one used a wheelchair, and one had slight paralysis. This observation was verified by the interviewers' mobility ratings as sixteen respondents were rated good, five were rated fair and only one was rated poor. Psychological status was also rated by the interviewers and eleven were rated good, nine were rated fair and one was rated poor. Seven of the respondents were on the fire side of the smoke doors on the south wing (major fire area); eighteen respondents mentioned that they were with one or two other residents and only four stated that they were asleep when the fire began.

Detection of initial cues

During a fire, time is obviously a critical factor. The amount of time which has elapsed since ignition of the fire will greatly determine the size and extent of the fire and smoke. The ability of staff and firefighters to extinguish the fire is thus directly affected. During evacuation, the size of the fire and particularly the amount of smoke produced will have life and death consequences for residents. Basic to any fire emergency, then, is early discovery of the fire. The early detection of one or more cues to the existence of a fire by staff and residents of nursing homes is of central importance to successfully coping with the fire emergency situation.

The alarm (horn)

In the fire studied, the initial cues included properties of the fire, the fire alarm and indications from others. Properties of the fire such as flames and smoke were not as significant at this point as they were later in the respondents' behaviour since most respondents were in their rooms when the fire occurred and were not close to the origin of the fire. Thus, most respondents did not initially have access to these cues. The alarm was mentioned as the first indication of trouble by seven respondents and three more respondents mentioned the alarm in combination with smelling smoke or seeing flames. It seems plausible that more respondents should have described the alarm as their first cue since most respondents were not able to perceive flames or smoke. However, two findings may explain why the alarm was not mentioned more frequently as an initial cue. First, as will soon be described, indications from others served as the first cue for a number of people. Second, the alarm may not have been considered a cue by many people even if they had heard it. Staff, residents and firefighters mentioned that there had been numerous false alarms in the nursing home in the past. Many respondents upon hearing the alarm probably did not define the situation as a fire, and thus some respondents may have failed to mention the alarm. In any case, it is clear that the alarm did not have the desired effect upon the residents. Of the seven residents who mentioned only the alarm as the first indication of trouble, six stated that they were not at all certain that there was a real fire at that point (one was 'somewhat certain'). This is substantiated by their behaviour, as four of the six respondents did nothing at all and one actually closed the door to deaden the sound of the alarm (Figure 11.2).

The reasons given by respondents for this first action provide further evidence of the failure of the alarm system to motivate respondents. Of the five respondents who did nothing or closed the room door after hearing the alarm, four said the reason was that they thought it was a false alarm (the fifth respondent said he was confused). When all 22 respondents were asked if they thought the fire alarm was a false alarm, eight said 'yes', one said 'no', and the others gave no response or did *not* remember hearing the alarm. Again, this finding (of numerous false alarms) appears to be a common problem of institutional settings. The consequence of a high frequency of false alarms is a lessening of the effectiveness of fire alarms and the development of indifferent attitudes of residents of nursing homes.

Figure 11.2 is intended to demonstrate the effect of various cues as *initial* cues. Thus, first actions which immediately followed the initial cue are provided with respondents' reasons for these actions. Clearly, additional cues were perceived by these respondents and additional behaviours followed; however, this goes beyond the effect of the *initial* cue and is shown in Figure 11.3 below.

Indications from others

A final set of initial cues involves indications from others. Two respondents stated they were warned by the staff (Figure 11.2). One of these respondents said he was 'very certain' that a fire existed and left the room without delay because he was

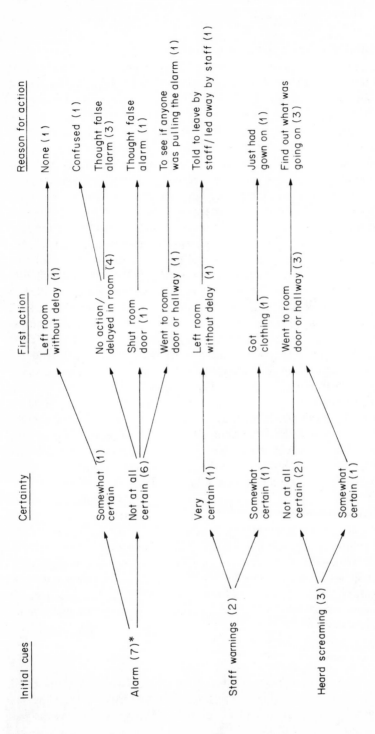

Figure 11.2 The effect of the alarm, staff warnings and screaming as initial cues

* Numbers in parentheses indicate number of respondents to which this response refers

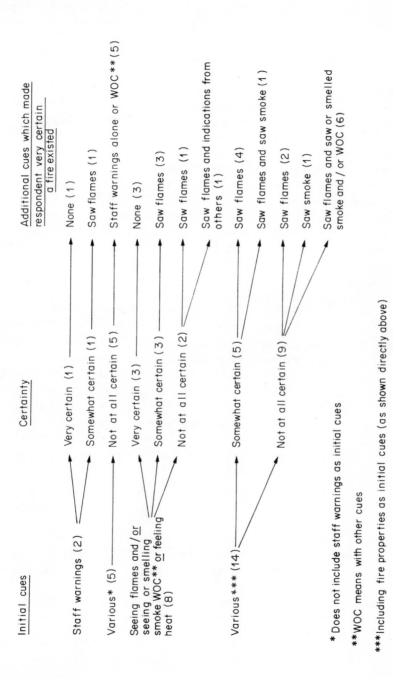

Figure 11.3 Overall effect of staff warnings and fire properties as cues

either told to leave or was lead away by a staff member. The other respondent was only 'somewhat certain' about the existence of the fire and managed to collect an article of clothing before leaving.

Three respondents stated they heard screaming and all three went to the room door or hallway 'to find out what was going on'. These responses were to the interview question, 'What first made you aware of the fire?'. However, it was one of the interviewer's impressions that more interesting information was imbedded in the respondents' full interview. Since the complete interviews had been transcribed, it was a straightforward task to review the transcripts and find quantitative support for the interviewer's belief. What was discovered was that the screaming or yelling of 'fire', either directly or indirectly, caused a number of residents to be convinced of the existence of the fire and motivated their escape behaviour.

This leads to the definition stage of the model. Whereas the previously mentioned cue of the alarm did not immediately motivate respondents to seek further information concerning the fire, hearing screams of 'fire' did motivate behaviour which caused the situation to be defined as a fire. What may be even more significant is that, contrary to most fire emergency plans and guidelines which caution against yelling 'fire' for fear of causing panic, only one instance of panic-type behaviour was recorded for the twelve respondents who had mentioned having heard someone yelling 'fire'. However, this respondent also indicated that the screams motivated her to escape. Moreover, it should be noted that this particular resident who 'panicked' was also medically classified as mentally retarded. The transcripts of five of the twelve respondents who mentioned having heard someone yell 'fire' provided indication that the effect of this cue was beneficial in that it helped the respondent define the situation as a real fire which led to an appropriate coping behaviour. This cue had no effect on six respondents (according to the transcripts) and it caused the one previously mentioned individual to report that she panicked.

Excerpts from the transcripts of four of the five respondents who benefited from having heard others yell 'fire' should clarify the effects that these perceptions had on the respondents.

Respondent 6

> Interviewer: How did you first know there was a fire?
> Respondent: Well, a woman kind of screamed, and I heard—I went out to look and I saw the fire.

Later the respondent mentioned that after seeing the fire he went downstairs. In addition, the screams stimulated immediate action as the respondent mentioned that the time between hearing the screams and seeing the fire was very short.

Respondent 11

When asked what first made him aware of the fire the respondent answered, 'Screaming and [unclear on tape] in the hallway'. After hearing this the respondent reacted as follows: 'I just went out the doorway. Then I looked across the hall and sees that smoke coming out of the doorway there. And it was smoking. Then it started to burst into flames.'

The respondent said he attempted to get some water to put on the fire but was not able to get any, so he then escaped down the hallway.

Respondent 13

'I heard a commotion and screaming out in the hall. I think that was the time when I just grabbed my sweater and went out the door.' Later she said: 'And I think I heard someone screaming. I think that must have been what happened to really make me [unclear on tape], make me go.'

Respondent 17

The interviewer did not tape this respondent's interview. He tried to keep it informal since she was still upset about the incident. However, a second interviewer asked the original interviewer the interview questions, and these answers (as recalled immediately following his discussion with the respondent) were recorded ('S1' refers to a staff member, and 'R17' refers to respondent 17 in these excerpts).

> S1 was yelling from the hallway: 'R17, R17, fire, fire! Get up!' Something like that, and at the same time that she got up, she saw flames out her doorway. So she was very certain at this point [that there was a fire]. After she heard the staff member yelling and saw the flames she went out her door and down the hallway.

These examples cast some doubt on the basis of the well-known rule in fire emergencies: 'Don't shout fire. It may cause panic.' It may be true for instance, as our data suggest, that in many situations the real problem is motivating residents—not preventing panic! Considering such factors as the residents' alertness, frequency of false alarms and lack of training in nursing homes this notion would seem to have some validity. At the very least, some modification in nursing home fire emergency plans would be useful. The age-old warning against shouting 'fire' might be changed to the following: 'Though you should try to maintain calm and order, you may have to shout fire to get residents moving during an evacuation'.

Defining the situation as a fire

Figure 11.3 involves the stage of the model at which time the situation is or is not defined as a fire. This stage is crucial in determining the individual's actions which follow since coping behaviours (described below) will only occur after the situation is defined as a fire. The overall effect of cues during this fire—initially and at a later stage—is shown in Figure 11.3.

Though screams of 'fire' were important as an initial cue, two types of cues contributed most towards convincing respondents that a fire really existed—staff warnings and properties of the fire. Warnings from the staff were expected to be particularly effective cues. From the patients' perspective, staff members are in an authoritative and credible role and there is little reason to doubt their sincerity concerning a situation as serious as a fire. Two respondents mentioned staff warnings alone as a first cue and one respondent stated that he was 'very certain' at that point. The other respondent became 'very certain' after seeing flames (Figure 11.3). Considering all 22 respondents, of the ten respondents who were not at all certain after some initial cue, five became 'very certain' after being warned by a staff member, or seeing flames or smelling smoke (Figure 11.3). Overall, seven respondents stated that staff warnings—alone or in combination with other cues—made them certain that there was a fire.

The 'immediately verifiable crisis cues' of smoke and flames were even more convincing to respondents. As previously mentioned, since most patients did not have access to these cues initially (most respondents were in their rooms away from the fire room), flames and smoke were more important *after* some initial cue was detected. Even so three respondents were 'very certain' there was a fire when the initial cue was seeing or smelling smoke or seeing flames in combination with other cues. Five other respondents who were only 'somewhat certain' or 'not at all certain' after initially detecting flames or smoke with other cues, became 'very certain' after 'seeing flames' and/or receiving 'indications from others'.

Fourteen respondents are shown in Figure 11.3 as having initially detected various cues (including properties of the fire) which did *not* convince them that a fire existed. Seeing flames made six respondents 'very certain' that a fire existed, one was 'very certain' after seeing flames and smoke, one respondent was 'very certain' after seeing smoke, and six were 'very certain' after perceiving some combination of flames, smoke and other cues.

In determining the overall effect of properties of the fire as effective cues, a total of seventeen respondents mentioned these cues alone or in combination with other cues as having made them 'very certain' that there really was a fire. (This number was derived by adding the fourteen respondents who used various initial cues and later became 'very certain' after perceiving fire properties (this includes respondents who used fire properties as initial cues) to the three respondents who were 'very certain' immediately after detecting fire properties with other cues.)

Coping behaviour

The coping behaviours which are displayed during a fire are dependent upon a number of factors. The nursing home environment—due to the physical and psychological condition of residents, and social structure of staff-resident interactions —restricts the number of different coping behaviours that residents could be expected to demonstrate. Though there were some examples of attempts to suppress the fire, information-seeking and even one panic-like behaviour, most coping strategies were classified in a few categories. The logic of these coping behaviours becomes clearer when studying them with respect to previous actions taken by the respondent and others, and with respect to the respondents' certainty of the existence of a fire. A brief description of general behavioural sequences will be presented, followed by a more detailed examination of one particularly significant finding.

General findings

As previously mentioned, most respondents were originally in their bedrooms when they first became aware of the fire. Four respondents were certain there was a fire, after they perceived initial cues, so they immediately defined the situation as a fire. Thus, their coping behaviour consisted of leaving the room and heading towards an exit without delay in three circumstances ('escape'), and going to the window in the room to wait ('self-protection') in the fourth instance. Four respondents stated that they were somewhat certain, but required further information before coming to a conclusive definition of the situation. With respect to this, three of these respondents went to their room door or the hallway after which their coping behaviours consisted of, respectively, returning to the room and leaving the area, collecting an article of clothing and then leaving the area, and immediately leaving the area. The fourth respondent, who was somewhat certain, did not go to the room door for additional information but immediately collected an article of clothing and then left the room.

The majority of the respondents ($n = 10$) were not at all sure that there was a fire after receiving initial cues, so their first behaviours do not represent fire-coping behaviours. Four of these respondents took no action for a period of time before leaving their rooms. Four others went to the door, then either returned to their rooms before leaving or obtaining further information and then left the area. Of the two remaining respondents who were not at all sure there was a fire, initially, one closed the room door then got an article of clothing before leaving the room. Finally, one respondent who was never certain of the fire's existence, first took no action then left the room.

In summary, the actions taken, for the most part, followed logically from the respondents' perception of the environment. Three of four respondents who were initially 'very certain' of the fire's existence immediately left the room. Three out of four of those respondents who were only 'somewhat certain' went to the door or hallway, probably to gain further information. There were eleven respondents who were not at all or never certain of the fire's existence. Of these, five took no action at first and one closed the room door to deaden the sound of the alarm as her first action. Table 11.1 provides a summary of the coping behaviours performed by respondents.

Use of emergency exits

Perhaps the most significant finding of the study involved the use of stairways during evacuation. Of the 22 respondents, all but six, who were assisted by firefighters or used the elevator, left the fire floor using the centre stairway behind the nurses' desk (see Figure 11.1). In fact, after questioning nursing home staff and fire-

Table 11.1 Coping behaviours reported*

Suppress/contain fire	2
Warn/rescue others	4
Activate alarm system	0
Self-protection	4
Remove property from fire environment	0
Information seeking	1
Preparation for further action	5
Panic	1
Escape	21
No action	1

*All possible behaviours are listed.

fighters, no indication was given that *any* exit aside from the centre stairway was used by *any* residents of the fourth floor (except again those few who used the elevator or were assisted by firefighters). This means that as many as 85 residents were evacuated down one of four possible stairways!

The danger in such actions are many. For one, the crowding of one stairway could slow down the evacuation procedure and expose the last of the evacuated residents to deadly smoke and fumes. It would also seem likely that an overcrowded stairway could cause accidents due to tripping or pushing which would both injure residents and hinder evacuation. The crowded conditions might also increase the chances of causing panic. Finally, the exclusive use of one stairway indicates that people were not being moved with respect to the flames and smoke. In fire emergencies, the goal is to move people *away* from the fire to a safe place. During the fire incident described, only a minority of residents were actually moved away from the fire. By evacuating them via the centre stairway, most of the residents were moved towards the fire. Moreover, a number of residents who had safer exits available passed within a few metres of the flames and smoke while heading towards the centre stairway.

In retrospect, it is quite surprising that the evacuation proceeded as smoothly as it did under the circumstances. The only problem that the evacuation caused was the obstruction of firefighters' efforts. Firefighters also used this central stairway to bring hose lines and equipment to the fire floor. The overloading of this stairway with residents definitely hindered the firefighters' actions, according to interviews with the firefighters.

Under other circumstances the result of the evacuation procedure could have been much worse. This fire incident demonstrates that inappropriate behaviour did occur and should be changed. But what were the determinants of the behaviour?

One possible reason for the exclusive use of the centre stairway could have been lack of knowledge of the existence of other stairways. It is particularly useful to determine *post hoc* what exits respondents should have used and then to study their knowledge of these exits. Looking at Figure 11.1 (the floor plan of the fire floor), with respect to the location of the fire room—411, just south of the smoke doors in the south wing—it would be reasonable to evacuate residents in rooms 401, 402, 403, 404, 412, 432, 433 and 434 through the centre stairway behind the nurses' desk. All other residents should probably have been evacuated via one of three stairways at the end of each wing in order to move residents away from the fire and to maintain the protection provided by the smoke doors. The most critical area was that most affected by smoke and flames—south of the smoke doors in the south wing. Of six respondents who were in that area at the time of the fire, and responded to the questions, five correctly named the south wing exit when asked where the closest exit was located. (In fact, two respondents who used the centre stairway were in room 407—within a few metres of the south wing stairway!) Six respondents were in the other wings when they first became aware of the fire and five of them were aware of the exits on their wings.

It is also informative to look at how many respondents were aware of the centre stairway. Eighteen out of twenty respondents who answered this question, correctly stated the location of the centre stairway.

According to these findings, it does not seem that there was any lack of awareness of the exits that should have been used during the fire. Most respondents were aware of the location of exits at the end of their own hallways. However, it is interesting that nearly all of the respondents identified the centre stairway. The reason for this is that the centre stairway was the only one used by both staff and patients during a normal day, under non-emergency circumstances. Since the flow of traffic on and off the floor had to be monitored, this stairway behind the nurses' desk was ideal for use. All other stairways were only to be used during emergencies and use of these stairways would activate an alarm. This alarm had been set off often in the past by residents who attempted to leave the floor through these exits. Thus, negative consequences were associated with the use of these emergency doors, as some respondents mentioned that residents who were caught using these doors were scolded. The association of the alarm with the use of these exits was probably reinforced by the labelling of these doors: 'Stairs. Fire Exit Door. Keep Closed. Emergency Exit Only. Alarm Will Sound.'

The data tends to support the above conclusions; 93 per cent (thirteen out of fourteen) of the respondents stated that they had not used any of the emergency exits at any time while they had resided in the nursing home. Sixteen of the 22 respondents, or 73 per cent, implied in their statements that these exits were not supposed to be used under normal circumstances. Half of the respondents, eight out of sixteen, stated that an alarm goes off when the emergency exits are used, If nothing else, these data indicate that the respondents had no practice using the emergency exits, and this finding should be considered a factor contributing to the respondents' failure to use the emergency exits.

Probably the most important determinant of the respondents' behaviour has yet to be mentioned—the behaviour of staff members. This represents a significant aspect of the social environment of the residents. The social structure of nursing homes, as in other total institutions, places authority, power and responsibility in the hands of the staff. Staff members represent information and direction and often assist residents in the simplest activities of daily living. It is logical that residents would follow the advice of staff, particularly during an emergency.

The simplest way to find out why those respondents who moved down the hallway towards the nurses' desk and the centre stairway did so, was to ask them. When this was done the number of respondents who provided a reason was small, but the results are meaningful. Nine out of thirteen (69 per cent) of these respondents stated that their reason for moving towards the centre stairway was that they were 'told to leave by the staff.' (This calculation does not include one respondent who said he was led away by firefighters.) Unfortunately these responses do not indicate whether or not these staff members actually directed the respondents to the centre exit. It is, in fact, more likely that the respondents were just told to leave in a general manner. However, the staff are supposed to be well trained, and during an emergency should provide clear, appropriate directions. So they should have directed respondents towards the safest exit, not just provided a general warning. Thus, poor training of staff seems to be another determinant of staff's and residents' evacuation behaviour during this fire.

Given that most residents were not told specifically which exit to use by the staff, what other factors may have influenced their behaviour? A direct influence, not mentioned yet, would be a staff member leading the respondent or other residents towards the centre stairway. In addition, modelling behaviour could occur in which the respondent saw other people moving in a certain direction and they simply followed their lead. It does, in fact, seem that both of these factors contributed to the residents' behaviour.

When asked about the behaviour of staff members in the hallway, thirteen respondents stated the staff members were leading them (the respondent) or other residents to an exit. Five other respondents stated that staff members were doing various other things including giving instructions to them (respondent) or others to leave the area, physically helping the respondent and/or other residents, and moving towards the fire. In all, staff behaviour could be implicated as either a model for or direct determinant of respondents' movement towards the centre stairway for 72 per cent of the eighteen respondents who answered this question.

The behaviour of other residents is also included in the social environment, and of the eighteen respondents who answered this question, eleven (or 61 per cent) said that they saw one or more residents moving down the hallway towards the area of the centre stairway.

To summarize briefly, there seems to be a number of factors that contributed to the exclusive use of the centre stairways. The negative consequences associated with setting off the alarm when the emergency exits are used may have deterred residents from using those stairways. Since the centre stairway was the only one *not* connected to the alarm, it was the only stairway used during the day. This meant that both staff and residents had limited, if any, experience using the emergency exits. Finally, the behaviour of staff and residents, directly through physical assistance in evacuating and indirectly by presenting a model for behaviour, was very likely the most significant determinant of the use of the centre stairway by perhaps 85 residents or 93 per cent of the resident population of the floor.

Summary of respondents' behaviour: cues detected, cognitions, and coping behaviour

The behaviour of respondents—residents of the nursing home—can be best illustrated using the researchers' model. Behaviour in this model includes cognitive behaviour such as perceptions of stimuli (detection of cues) and processing of these stimuli (defining the situation), as well as overt behaviour such as information-seeking (cue-seeking) or any of a number of coping behaviours.

Figure 11.4 describes the behavioural sequences which were reported by respondents. The sequences are broken down into three sections, one for each type of initial cue which made respondents aware of the fire: the alarm properties of the fire and indications from others. The sections are not mutually exclusive, that is, a respondent who has perceived both the alarm and fire properties appears in both sections. Fire properties include feeling heat, seeing flames and seeing or smelling smoke. The alarm is self-explanatory and indications from others included hearing commotion, screaming or warnings from staff, residents and firefighters. When one of these initial cues were used in combination with one or more other cues, 'WOC' (with other cues) follows the numbers on the flow chart in Figure 11.4 which indicates how many respondents this refers to.

'WDIC' (with different initial cues) is used at the last segment of the flow diagram before defining the situation as a fire. This indicates that some other cue besides the particular one referred to on the flow chart was detected initially, and the detection of the flow chart cue convinced the respondent that a fire existed. For example, for the flow chart in which fire properties are the initial cue, nine respondents defined the situation as a fire upon detecting properties of the fire *after* the alarm and/or indicators from others had been the cues detected.

Coping behaviours were considerably limited. 'Preparation for futher action' and escape behaviour accounted for 26 of 38, or 68 per cent of all reported coping behaviours. Due to the small number of respondents no significant relationships could be discovered between cue detection/situation definition sequences and coping behaviours. Coping behaviours are therefore reported in summary form only.

The fact that more respondents were *not* involved in different coping behaviours is of interest itself. As discussed previously, the nursing home environment in which most residents are physically or psychologically handicapped, limits the residents' behavioural opportunities. In addition, staff members maintain the authority and power over residents and are also expected to care for residents. Therefore, from the residents' perspective, behaviours such as fighting the fire and rescuing others are seen as the staff's responsibility.

Two things should be noted about the coping behaviours reported. First, any individual may have engaged in more than one coping behaviour during the fire incident. Secondly, although all respondents obviously escaped the fire, only 21 respondents reported escape behaviours. One respondent who was found sitting near a window at the end of the hall could not recall how she escaped. Her coping behaviour was listed as 'no action'. Figure 11.4 shows that the alarm alone served as the initial cue for seven respondents, and these respondents were not initially certain that there was a fire but became certain after detecting other cues. Three other

Coping behaviour

Suppress / contain fire (2)

Warn / rescue others (4)

Activate alarm system (O)

Self- protection (4)

Remove property from
fire environment (O)

Information seeking (1)

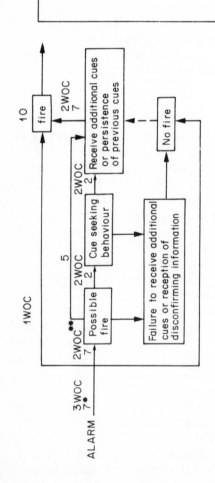

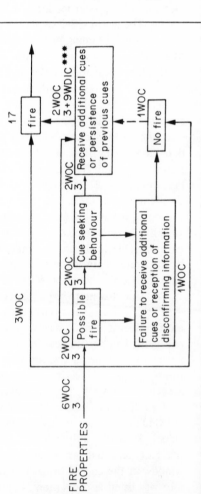

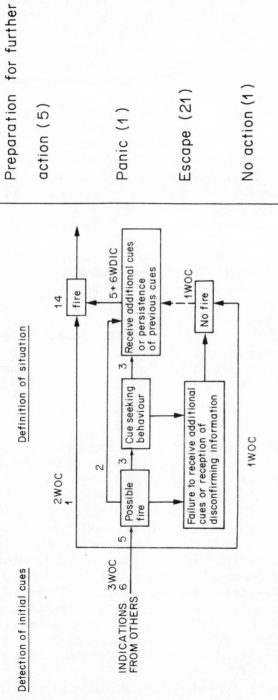

Figure 11.4 Summary of respondents' behaviour: cues detected, cognitions, coping behaviour

* Numbers indicate number of respondents that this response refers to.
* WOC means with other cues.
*** WDIC means with different initial cues and refers to those situations in which the particular cue for the flow chart (alarm, fire properties, indications from others) was detected *after* other cues, and convinced the individual that a fire existed.

respondents first became aware of the fire by detecting the alarm with other cues; one of these immediately defined the situation as a fire, the other two defined the situation as possibly involving a fire. Of the ten previously mentioned respondents (seven alarm only and three alarm WOC), nine defined the situation as possibly being a fire. Four of these individuals then sought out further cues and upon detecting these cues realized that a fire really existed. Five other respondents, though they did not actively seek cues, did detect additional cues (or previous cues persisted long enough) and thus determined that a fire did exist.

Properties of the fire were the most frequently perceived cues of respondents though in all but one instance they were detected with other cues. Three respondents immediately decided that the fire was real, while one respondent was never certain that a fire existed. Five other persons determined that a fire might possibly exist. After seeking and receiving further cues, they were certain that the fire existed. The nine WDIC indicate that after detecting some other cues, detecting properties of the fire convinced these nine patients that a fire existed.

Indications from others served as the only cue, initially, for six respondents, and five of them originally described the situation as a possible fire. After receiving additional cues (three of the respondents actively sought them out) all five individuals decided that the fire really existed. The other respondent who used indications from others as the *only* cue, immediately decided that a fire existed. Six respondents were convinced a fire existed when they received indicators from others, after having previously detected some other cues. Two persons immediately defined the situation as a fire (WOC) and one (WOC) was never quite certain of the fire. The large number of respondents who mentioned indicators from others as a cue—with or without other cues—clearly indicates the effectiveness of this cue during the nursing home fire.

Summary of determinants of behaviour

Most fire emergencies represent complicated situations involving numerous factors. The social scientist studying the behaviour of people in fires is often faced with an array of these factors influencing different behaviours at different points in time. In addition, the environment—physical and social—of the fire emergency situation is constantly changing, further complicating matters. This chapter has presented data from a nursing home fire using the previously described model of human behaviour. As an overall summary of the fire and the use of the model, the major findings of the nursing home fire are presented in Table 11.2, which describes the interactions between determinants of respondents' behaviour and stages of the model.

Determinants of detection of cues can be found in Table 11.2. The physiological/ physical determinant was not found to influence this stage or the other stages of the model, largely due to the good physical and psychological condition of the residents. Though one respondent who was mentally retarded did display panic-like behaviour, for the most part the authors' data did not demonstrate any particular significance of this factor.

The alarm was discussed as a cue which was used as an initial cue more than

Table 11.2 Summary of determinants of behaviour

| Determinants | Alarm | Cues | | | Definition of situation as involving a fire | Fire coping behaviours escape via centre stairway |
		Fire characteristics	Staff warnings	Screams of fire		
Physiological/ physical						
Intrapersonal*	–Past experience of false alarms				–Past false alarms	+Negative associations with use of emergency exits
Educational/ preparation	–Lack of fire emergency training				–Lack of emergency training	+Lack of fire emergency training +Lack of use or observed use of any stairway but centre
Social			+Staff authority		+Staff authority	+Staff assistance and directions; modelling of staff and resident behaviour
Fire characteristics				+Location in south wing	+Seeing flames, seeing or smelling smoke	
Physical environment	+Location in room	–Location in room +Location in south wing or dining room +Location by door or in hallway				+Signs on emergency exit doors, which strengthened negative associations of use of emergency exits

*The negative (–) and positive (+) signs indicate that this determinant decreased or increased detection of the cue, definition of the fire or the use of the centre stairway

others, mostly because it could be heard in respondents' rooms where most respondents were originally located during the fire. 'Location in room' appears as a factor of the physical environment which increased detection of the alarm. The alarm, however, was less effective than expected and past experience with false alarms and lack of resident training for fire emergencies is probably at fault.

'Characteristics of the fire' were difficult to perceive from the respondents' bedroom, but respondents were more likely to detect these cues if they were in the south wing or in the dining room (a location from which the south wing could be easily observed). This is substantiated by the finding that all nine respondents who mentioned fire characteristics alone or in combination with other cues as initial cues, were located in the south wing (eight respondents) or the dining room (one respondent). In addition, going to the bedroom door or hallway brought respondents in contact with fire characteristics. Screams of 'fire' were also used as cues by some people, and location in the south wing was a determining factor since the resident who was screaming was located in that area. (The three respondents who mentioned screams as an initial cue were located in the south wing.)

Defining the situation as a fire most frequently occurred after perceiving flames or smoke (this was shown in Figure 11.3.) This finding seemed to substantiate labelling such cues as 'immediately verifiable crisis cues' (Breaux, Canter, and Sime, 1976). Staff warnings were a social factor that also facilitated the definition stage.

Previous false alarms in this nursing home is a factor which was detrimental to defining the situation as a fire. The frequency of false alarms tended to cause respondents to define fire alarms in general as false; this explains the lack of immediate action, in many instances, after hearing the alarm. In addition, both residents and staff should be trained to react to every alarm as if it meant that a real fire existed. In essence, staff and residents should immediately define the situation as involving a fire when hearing an alarm. Since both the frequency of false alarms, and the lack of fire emergency training are responsible for the negative effect of the alarm in terms of defining the situation, they have both been listed in Table 11.2, as intrapersonal and educational/preparational factors, respectively.

The major findings of the nursing home fire study in terms of coping behaviours was the overuse of one stairway as a means of escape. Lack of training—an educational/preparational factor—in use of other exits, as well as the daily use of *only* the centre stairway by both residents and staff, provides a partial explanation. In addition, negative associations, which were probably reinforced by signs on the exit doors, had been established with the use of emergency exits. These associations are considered an intrapersonal determinant in the model. Finally, indicated as a social determinant, staff members assisted or instructed respondents to use the centre stairway, and both staff's and residents' movement towards the centre stairway served as models of behaviour which some respondents may have followed.

SUMMARY

The researchers' model of human behaviour in a fire emergency provided a useful framework for describing the nursing home fire investigated. The alarm and 'indications from others' served as initial cues detected by most respondents, however, only 'indications from others' were effective in motivating escape behaviour. The ineffectiveness of the alarm was apparently due to lack of fire emergency training and frequency of false alarms in the past. 'Characteristics of the fire' were usually mentioned as cues which convinced respondents that a fire really existed. A number of different coping behaviours occurred during the fire, but escape behaviour proved to be the most prevalent, and the most disturbing. Escape behaviour was not conducted in the safest way, as most residents were evacuated down the same stairway. Interviews with residents indicated that the possible determinants of such behaviour include: lack of practice using fire emergency exits, negative associations with the use of emergency exits, and signs on the doors of emergency exits which strengthened these negative associations, as well as inappropriate instructions provided by staff during the fire emergency.

These findings, placed in the context of the model, indicate that increased and improved fire emergency training for both staff and residents is needed. Such training would decrease the time involved in detecting cues and defining the situation as a fire by teaching staff and residents to react to the alarm as if a real fire existed, and would improve coping behaviour by teaching staff and residents to use all fire exits.

REFERENCES

Bickman, L. (1976). *Methodological Considerations in the Social Psychological Study of Human Behavior in Fires.* Paper delivered at the Meeting of the Society of Experimental Social Psychology, Los Angeles, 28–30 October.

Breaux, J., Canter, D., and Sime, J. (1976). *Psychological Aspects of Behavior of People in Fire Situations. Paper* presented at Karlsruhe, Germany, September.

Bryan, J. L. (1977). *Smoke as Determinant of Human Behavior in Fire Situations (Project People).* US Department of Commerce, National Bureau of Standards, Center for Fire Research, June.

Lerup, L. (1976). *Mapping of Recurrent Behavior Patterns in Institutional Building under Fire: Ten Case Studies of Nursing Facilities.* US Department of Commerce, National Bureau of Standards, May.

Selltiz, C., Wrightsman, L. S., and Cook, S. W. (1976). *Research Methods in Social Relations.* (New York: Holt, Rinehart and Winston).

Webb, E. J., Campbell, D. T., Schwartz, R. D., and Sechrest, L. (1966). *Unobtrusive Measures.* (Chicago: Rand McNally and Co.).

Fires and Human Behaviour
Edited by D. Canter
© 1980 John Wiley & Sons Ltd.

CHAPTER 12

Patient Evacuation in Hospitals

JANET HALL
34a, Kings Parade, Blackwater, Surrey, U.K.

A fire in a health building can present staff with a problem rarely encountered in fire situations elsewhere. The main function, particularly of nursing staff, is the care of patients. If a fire occurs they must take the best course of action to preserve the safety of those patients, which may be contrary to their own initial safety. They must decide whether it is best to try and control or fight the fire until the fire brigade arrives, whether to leave the fire and contain the patients in the safety of a fire compartment (an area structurally protected from fire with alternative exits usually recognised by fire doors), whether to evacuate the patients to an area of greater safety and perhaps risk their own and the patients' lives in the process. If they decide to evacuate the patients, they must know when, where and how to do so otherwise greater harm can result from the evacuation than from the possible effects of fire. The dilemma of which decisions should be made is still not properly resolved although work is in progress and opinions have been voiced which have clarified the choices and provided a basis for reasonable fire precautions training for hospital staff (Hall, 1977, c). The study described below assumes that the decision to evacuate has been made and evaluates evacuation methods currently being taught. It is hoped that the result will enable fire officers to teach only the optimum methods and help staff to decide whether it is advisable or even feasible to evacuate if and when they ever have to make the choice.

Three carefully constructed evacuation exercises were conducted at Hackney Hospital by the DHSS and the Fire Research Station in 1975 (internal reports on these exercises are available on request to the above bodies (Appleton and Quiggin, in press and Hall, 1975, a). One reason for the exercises was to study the speed and operational difficulties of the patient evacuation methods chosen by the staff. The results of this study together with a body of opinion built up from ongoing studies concerning fire precautions training in hospitals and the effects of various floor coverings on commonly used evacuation methods (Hall, 1976, b) has shown that there are many difficulties with the methods currently being taught to hospital staff. There are also several methods being taught which have not been tested in an actual fire situation. These differences, plus general lack of information on evacuation methods, have led to disagreement as to which method(s) are the most appropriate and to an increase in the diversity of techniques advocated.

Attempts are being made to standardize fire precautions training (Fire Precautions Act, 1971, unpublished). Consequently in order to lessen the confusion of choice and to decrease the instruction time involved in teaching so many methods, this study was initiated to evaluate the most commonly taught methods and to recommend those methods which are the most appropriate for given environmental or fire situations.

A survey determined that the methods illustrated in Figure 12.1 were those most commonly taught. Their effectiveness was evaluated on the following test criteria, determined in consultation with fire and nursing officers, architects, engineers, hospital administrators, ergonomists and patients.

(1) Speed
(2) Minimum physical effort
(3) Minimum number of operational staff
(4) Logical, easily learned procedure
(5) Safe for patients and staff
(6) Minimum risk of malfunction.

The experiments were conducted in four stages:

(1) Measurement of the pulling strength and heart rate at rest of each nurse.

(2) Recording for criteria (1)-(6) above for each nurse while evacuating a patient of average male weight (75 kg, 165 lb) (Kirk, 1971) by the various methods, from a lying position on a bed in a far corner of an unfurnished Nightingale ward to either (a) the middle of a furnished ward opposite (horizontal route), or (b) to the top of the central stairway and down two flights of stairs negotiating one landing (vertical route). (see Figure 12.2 for evacuation directions and distances.)

(3) Training the nurses in each method using the minimum numbers of teachers and minimum amount of training time.

(4) Repeating (2) above, noting any changes of time, heart rate, etc. from the first trial run. A furnished ward was used to obtain information on the possible problems caused by furniture obstructions.

The staff sample was carefully chosen and grouped to obtain the maximum amount of information from the minimum number of nurses. There were seven pairs of nurses whose collective average height, weight and pulling strength were the same as that of an average woman. However, the range of their individual characteristics were representative of those found in the British population as a whole, for example 1.52–1.78 m (5–6 ft) tall, 44–80.5 kg (97–177 lb) in weight (Kirk, 1971). It is reasonable to say, therefore, that the average figures achieved for each method are applicable to a women of average stature.

Evacuation is typically defined as having three main stages. Stage one is movement of patients horizontally, generally from their present position to an adjoining or closely situated compartment. Stage two is the vertical movement of patients, generally to an adjacent floor or occasionally away from the particular block from which they are escaping. Stage three is seen as the complete evacuation of a hospital block or in the worse case the whole hospital.

The evacuation process for the purpose of this study consisted of four sections.

(1) Preparation: preparation of the patient, timed from the start of the evacuation to the moment before horizontal movement.

(2) Horizontal: movement of the patient, timed from the start of horizontal travel to:

(a) the finish point in the ward opposite, or

(b) to the start of vertical movement (see Figure 12.2)

(3) Vertical: movement of the patient, timed from the first step on the stairs to the finish point down two flights and around two bends.

(4) Waiting: pauses in movement due to requirement for a rest period.

A survey was recently conducted in 42 health districts (Hall, undated, c) to discover the level of training standards in the country and estimate the cost of aligning the present standards with those recommended in the Draft Fire Precautions Code (Fire Precautions Act, 1971). Costs were to be made on the most efficient (in terms of numbers of staff receiving 1 hours training per year) (Draft Fire Precautions Code) training systems available or those considered viable considering present limitation. Part of this study showed that an average example of the time, number of staff and patients present for a horizontal evacuation, using information from fire drills and actual evacuations, is 25 mixed (non-ambulant semi-ambulant and ambulant) patients and eleven staff evacuating them in 10.5 minutes, that is, two staff appear on average to be able to evacuate five patients of mixed capabilities horizontally in two minutes (range: 40 patients, mostly ambulant but mentally disturbed, eight staff in three minutes, to twenty patients, twenty staff in twenty minutes). The average example is comparable to a ward with an average number of patients (Hall, undated, a, and Webber and Treasure 1974) being evacuated during the late evening period when staff on the ward being evacuated have help from staff from two adjacent wards.

The Home Office advise that any fire compartment should allow for complete evacuation in 2.5 minutes. There are still no time regulations for hospital evacuations.

Little information is available for a stage two type evacuation. Stairs limit the choice of possible evacuation methods. Cursory studies of evacuation drills, plus, more scientific study using the Hackney exercises, together with information from evacuations conducted in real fire situations, would suggest that approximately a quarter to half an hour would be necessary to evacuate an average ward of 25 patients of mixed dependency with approximately ten staff down four flights of stairs. This figure would be dramatically influenced by several factors. For instance a narrow exit could seriously delay an evacuation particularly when using methods like the mattress, bed or wheelchair. One exercise at Hackney showed that with a combination of a narrow 900 mm (36 in.) exit, 28 patients (twelve non-ambulant), the use of the mattress method and predominantly weak staff it was likely that no matter how many staff were present it would be impossible to evacuate all those patients before the estimated time of smoke logging for the ward (that is, nine minutes for the ward used at Hackney).

The time of day will affect the time to evacuate. At night only two staff are likely

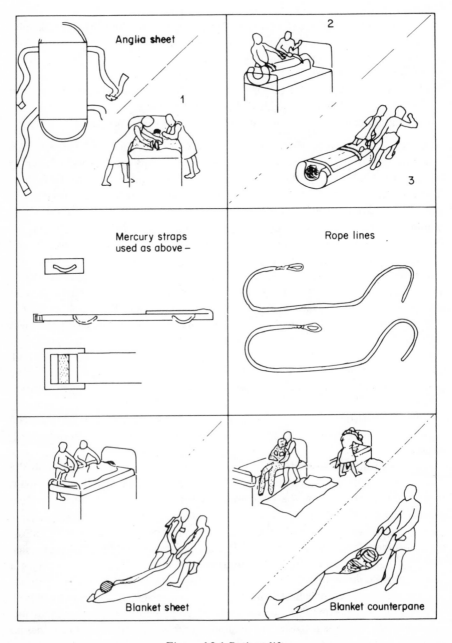

Figure 12.1 Patient lifts

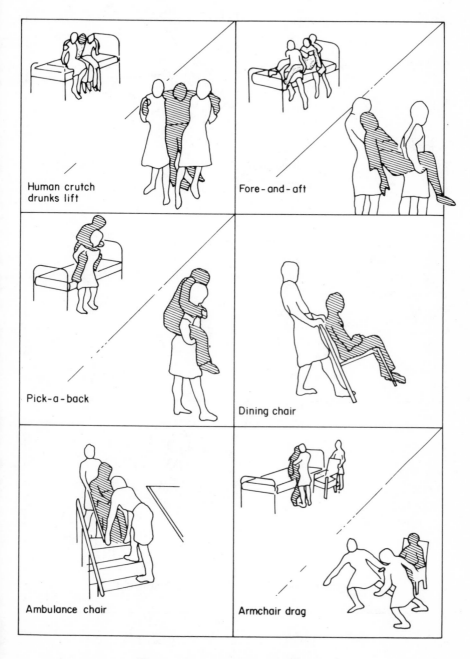

Figure 12.1 Patient lifts (*continued*)

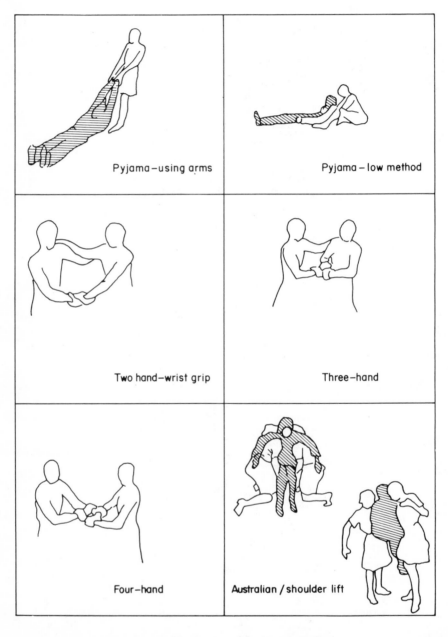

Figure 12.1 Patient lifts (*continued*)

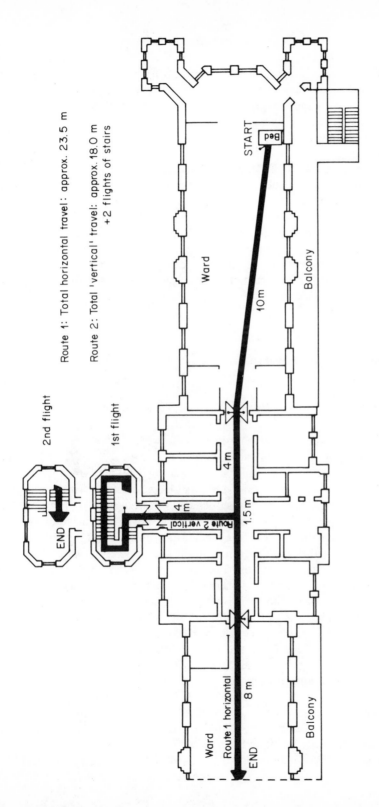

Route 1: Total horizontal travel: approx. 23.5 m

Route 2: Total 'vertical' travel: approx. 18.0 m +2 flights of stairs

Figure 12.2 Evacuation route: Layout and dimension

to be on duty, supervising on average 25 to 30 patients (Webber and Treasure, 1974). Generally one nurse will be young and probably inexperienced, the other will be a much older, more experienced, but probably less strong person (Hall, undated, b). Also, fewer staff are available in the vicinity to give rapid assistance, and most (approximately 72 per cent on average) of the patients will be sedated (Hall, undated, a).

The design layout and maintenance of the hospital will affect the evacuation time. Older buildings invariably consist of numerous narrow and winding corridors with unexpected stairways. Indiscriminate storage often block escape routes, and cramped wards and too much furniture hinder movement. At Hackney the continual relocation of obstructing furniture took 2.5 minutes of the evacuation time, the time recommended by the Home Office for the complete evacuation of a fire compartment. These and other such factors were carefully considered when compiling recommended methods for evacuation. The effects of some of these factors are discussed in more detail below.

SUMMARY OF RESULTS

Post-training figures were used in order to obtain realistic comparison values between methods.

Speed

Tables 12.1 and Figures 12.3 and 12.4 clearly show that all the mattress methods take considerably longer overall and in each of the timed sections (that is, preparation, horizontal and vertical movement and waiting time) than any other method. If a spring interior mattress is used times will increase further. Excessive waiting time was caused by the need for rest periods and to rearrange the equipment.

Most fire officers agree that vertical evacuation should be avoided or used as a last resort. Consequently, some of the methods are not designed for vertical movement. As a result of this, and time limits for the experiments, not all methods were tested vertically. Of those tested the four-hand and fore-and-aft lift were easily the quickest. However, rudimentary vertical tests for the two-hand lift and blanket drag have shown these to have comparable times. The blanket drag requires one person to clasp his hands under the patient's arm and around his chest (or over the patient's folded forearms) and then drag the patient backward downstairs, keeping the patient's posterior on the steps if he is too heavy to lift.

All the lifts, but particularly the 'drunk's lift', together with evacuation by bed were the quickest for horizontal movement and only the chair and pyjama methods were slightly quicker than those methods in the preparation section, although very quick lifting methods have drawbacks. Limited strengths and complexity of some hand-holds resulted in seven pairs of nurses being unable to utilize the three and four-hand and Australian lifts, and two pairs the two-hand lift for horizontal and vertical movement. Four pairs could not complete vertical movement with the fore-and-aft method. This was mainly due to one nurse at the back being unable to see

Table 12.1 *Average* times and heart rates for all methods *after* training

Evacuation method	Speed			Total per complete method			Heart rate				Training per 10 pupils
	Prepara-tion	Horiz-ontal sec/metre	Vertical sec/floor + bend	Waiting	Preparation/ Horizontal	Preparation/ Horizontal Vertical	Prepara-tion	Horiz-ontal	Vert-ical	Recovery time	
Anglia sheet	46 sec	2.3	21.0	4 sec (2 groups)	1 min 39 sec	2 min 21 sec	156	183	183	6 min 7 sec	21 min 0 sec
Mercury straps	64 sec	3.1	23	20 sec (1 group)	2 min 18 sec	-	155	192	-	6 min 7 sec	18 min 42 sec
Mattress lines	95 sec	5.1	25.5	8 sec (1 group)	3 min 35 sec	4 min 27 sec	158	184	184	6 min 9 sec	17 min 30 sec
Blanket	38 sec	2.5	to be tested	15 sec (1 group)	1 min 38 sec	-	150	181	-	5 min 8 sec	13 min 0 sec
Bed	13 sec	1.1	-	-	0 min 39 sec	-	129	163	-	3 min 5 sec	2 min 30 sec
Two-handed lift	12 sec	.9	8.0	-	0 min 35 sec	-	136	169	-	3 min 5 sec	6 min 20 sec
Three-hand lift	17 sec	1.3	-	-	0 min 47 sec	-	142	169	-	3 min 7 sec	5 min 30 sec
Four-hand lift	14 sec	1.0	8.0	-	0 min 37 sec	0 min 52 sec	137	172	176	-	10 min 20 sec
Australian lift	14 sec	1.2	-	-	0 min 41 sec	-	138	167	-	4 min 8 sec	15 min 30 sec
Fore-and-aft lift	21 sec	1.0	8.5	-	0 min 43 sec	1 min 01 sec	156	172	178	6 min 3 sec	8 min 40 sec
Human crutch											
Drunk's lift	9 sec	1.0	-	-	0 min 33 sec	-	135	172	-	3 min 5 sec	1 min 40 sec
Pick-a-back lift	10 sec	1.3	to be tested	-	0 min 39 sec	-	149	184	-	4 min 4 sec	7 min 30 sec
Pyjama	12 sec	2.2	to be tested	-	1 min 4 sec	-	146	188	-	6 min 11 sec	12 min 30 sec
Dining chair drag	13 sec	2.2	-	-	1 min 4 sec	-	137	180	-	4 min 5 sec	7 min 20 sec
Armchair drag	11 sec	1.8	-	-	0 min 53 sec	-	143	179	-	4 min 7 sec	2 min 30 sec
Wheelchair	10 sec	1.3	-	-	0 min 41 sec	-	133	159	-	2 min 4 sec	3 min 0 sec
Ambulance chair	19 sec	1.2	17.5 sec	3 sec (1 group)	0 min 45 sec	1 min 20 sec	134	159	177	4 min 3 sec	6 min 30 sec

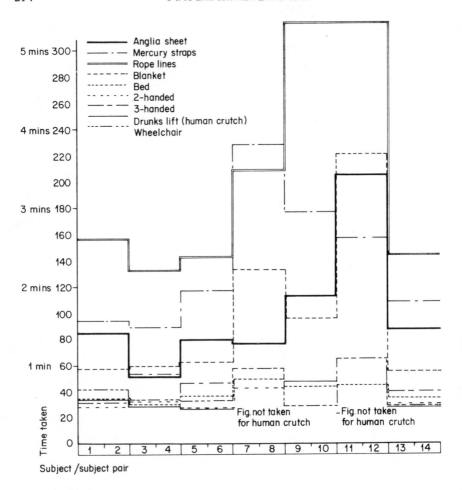

Figure 12.3 Average combined preparation and horizontal times for each method of each subject pair

where she was placing her feet and forcing a nurse in the front to descend too quickly, and overbalance as the weight of the patient shifted. A strict call of pace rhythm is necessary for the success of this method.

Correct preparation drastically decreases horizontal, vertical and waiting times, and does not presume longer preparation times. Also, methods inherently requiring long preparation times, such as the mattress method, do not result in shorter horizontal times.

Minimum physical effort

The time and physiological information obtained in the experiments were for the evacuation of one patient only. It is likely that times, etc., will be affected if a

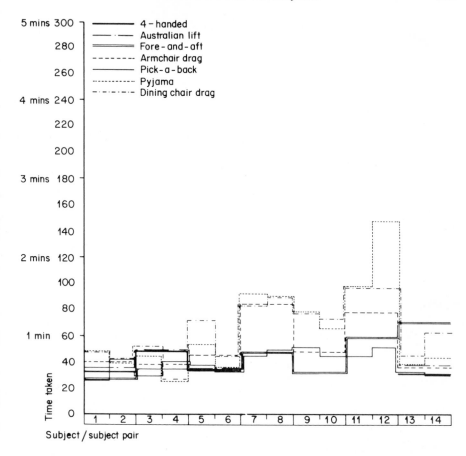

Figure 12.4 Average combined preparation and horizontal times for each method of each subject pair

succession of patients have to be moved. Indications of probable changes were gained by examining the nurses' heart rates and from subjective observations of each method.

Heart rates, recorded on portable ECG machines, and recovery times indicated that in general evacuation on wheeled equipment or by lifting methods caused the least effort for preparation and horizontal movement. The mattress, blanket, pyjama and pick-a-back caused the biggest increases in heart rates, The same applied to those methods used horizontally. Recovery times correlated with the degree of increase in heart rate; wheelchairs, for instance required the least recovery time, and the mattress, required the most. Methods requiring a backward, low pulling action most increased operation times and heart rate.

There is a clear correlation between each nurse's continuous pull strength and her capacity to complete certain methods (see Figure 12.5). If a nurse's continuous (30 second duration) pull strength is less than 30 kg (66 lb), most of the lifts would

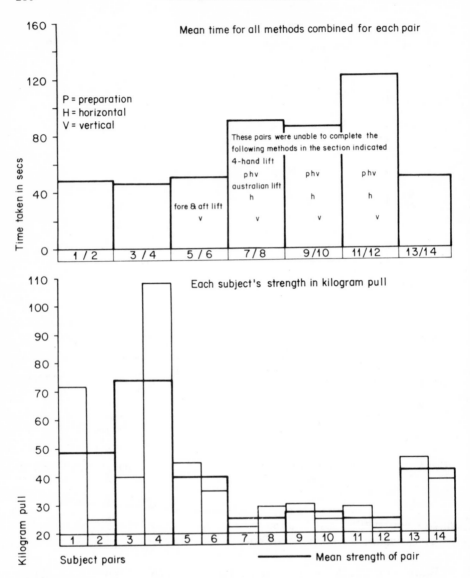

Figure 12.5 Comparison of subject strength and time taken for all methods combined

prove impossible or very strenuous when the patient is of average male weight. The drunk's lift and pick-a-back proved the only exceptions, probably because the weight of the patient is taken across the shoulders or on the hips.

Methods with complicated hand-holds, noticeably the three, four-hand and Australian lifts, particularly where the patient's weight was taken mainly on the nurse's wrists, frequently failed. Often the patient was dropped onto the floor. A difference in height between nurses will hinder some lifts, particularly the Australian lift.

Fire officers differ in recommending precise operational procedures for some methods, but the experiments strongly suggest that much more effort is necessary when some methods are executed in particular ways, and in some cases methods could become dangerous to patients and staff. Some examples of beneficial procedure are given below.

Pick-a-back method

The patient's *thighs* should be well above the nurse's hips before she attempts to carry him. This position applies no matter how heavy the patient is. The position would necessitate the patient adopting a standing or squatting position on his bed before the start of the method. If he sits on the edge of the bed the nurse is obliged to squat in front of him and small, less strong nurses frequently find it impossible to rise with the patient from this position. Generally, lifts that start from, or use a series of, static positions fail more frequently than those which use the momentum of the lift to start horizontal movement. The mattress method was shown to be quicker vertically than horizontally when its momentum of descent was used effectively.

Fore-and-aft method

The patient should be sat up in bed first; one nurse then sits close behind him and clasps her fingers over his chest or preferably slips her hands under his arm, then up, and overgrasping the patient's folded forearms. The second nurse then sits between the patient's legs with her arms around his thighs as far above the knee as she can, or his ankles if the patient is large. Even heavy patients can then be swung gently sideways off the bed. If the patient's knees are grasped, most of his weight will fall back towards the nurse behind, and his legs will often slip from the grasp of the nurse in front until he assumes a kind of sitting position. This puts excessive strain on the nurse at the rear.

Blanket and pyjama method

Little difference was seen in effectiveness between lifting the patients off the bed already wrapped in a sheet and placing him onto a counterpane previously placed on the floor (the patient slid across the floor better on the counterpane). However, effort and time was reduced when the nursed pulled from the patient's head end. With the pyjama method the nurses grasps the patient's wrists. When pulling from the other end the nurse would be lifting the weight of the legs as well as having to pull the patient along. The nurse at the head end would also be in a better position to reassure the patient.

Mattress method

No matter what equipment is used to pull the mattress it will be noticeably easier to move and manoeuvre if it is very tightly cocooned, as it presents less surface area to increase the drag coefficient (Hall, undated, b) and less bulk to negotiate obstacles. When transferring mattress from bed to floor two techniques can be employed. One nurse stands beside the bed and pulls the mattress to the edge, a hand on each strap, resting it on the front of her thighs. While a second nurse holds the bedhead and assists in pulling the head end off the bed, the first nurse drags the mattress fully onto her thighs, steps back and lets the mattress slide or even drop to the ground (the patient will remain safe) keeping her back straight all the time. Otherwise each nurse takes a strap and the mattress is pulled off the bed, feet end first and then the head end. The nurse at the head end steadies the bedhead with her free hand all the time. This prevents the bed from moving with the mattress, or the mattress catching on the bed springs. A pillow should be placed over the patient's chest when tightening the rope lines or straps with Britax buckles as these often bite into the patient's skin.

Chair method

The patient is sat on the edge of the bed and the chair seat placed beside the nurse who faces the patient at a slightly diagonal angle. The nurse grasps the patient under the arms clasping her fingers behind his back. She then places her knees against the patient's legs, leans back and lifts him off the bed, swinging him round into the waiting chair all in one continuous movement. If another nurse is present she can hold the chair steady.

Minimum number of operational staff required

Although methods requiring one nurse show competitive times for evacuating one patient, the high heart rates and increased physical exertion required indicate that the efficiency of these methods are suspect if a succession of patients have to be evacuated. Consequently methods using two nurses are normally preferable when staffing levels permit, except when wheeled equipment is used, as the study showed that the reduction in time and effort using two persons for such methods was negligible; the second person is only useful in opening doors and assisting with guiding the equipment. There is no indication that more than two nurses will improve times or minimize effort significantly for any method.

Learning different procedures

Table 12.1 shows that the mattress methods took the most time to teach (17.5 to twenty minutes). Australian lift, blanket and pyjamas took the next longest (twelve to 15.5 minutes). All other lifts plus the chair methods took the next longest (5.5 to 8.75 minutes). The quickest methods to teach were the bed and drunk's lift (2.5 and 1.25 minutes, respectively).

Training for each method consisted of verbal instruction, while the teacher demonstrated practically each stage and a period for questions. Then the pupils conducted a practical run of a particular method of all the methods as required. Where the pupils and teacher considered it necessary several practical runs were made with different pupils; after this questions were answered until all the pupils stated or demonstrated that they understood how to operate each method.

The practical demonstrations took the most time in the training process, and as a consequence those methods with long operational times tended to have longer training times. Pupils considered, however, that the practical demonstration formed the most valuable part of the training. Even pupils who had seen the training film 'Hospital Evacuation and You' (distributed by Consort Films), which describes in detail the techniques of most of the methods tested in the study, said the practical demonstrations were very useful.

A pupil group of ten to fifteen persons was found, after trials with groups of two to 30 pupils, to be the best number for one instructor to manage in one session, that is, it allowed for the maximum number of people to be taught to minimum requirements, with the least number of instructors and pieces of equipment in the shortest time.

Training decreased operation times for nearly every method, dramatically so for the mattress and chair methods. Figure 12.6 illustrates by 23 per cent the improvement of overall method time and shows where that time-saving occurred. Heart rates also decreased for most methods after training, but the mattress and pyjama methods showed increased heart rates. This may be an additional cause for the dramatic decrease in operational time for these methods. The nurses no longer required the waiting periods to adjust equipment as in the first trial runs. As these waiting periods also provided time to recoup from the effort of these methods, heart rates were lower. It is likely that those methods which benefited most from training (mattress and chair methods) would suffer the most with untrained staff, in terms of decreased efficiency. Training noticeably reduced the number of incidences of potential or actual accidents and discomfort and also reduced the risk of malfunction for each method. Lack of training actively increased the likelihood of malfunction.

Safety for patients and staff

Clearly a method which may be safe for one patient could be harmful to another. Given this, safety was assessed, assuming that the method chosen was suitable for the particular patient.

The ambulance chair had the most recorded dangerous incidences, which occurred mainly before staff were trained and were a direct result of the narrow-gauge track of the equipment. When the patient was pushed or pulled fairly fast the chair became uncontrollable and both it and patient frequently tipped onto the floor (Training rectified this by steadying the speed of horizontal movement). Also, when weaker nurses used it on the stair they were unable to lift the patient and bounced the chair down each step. Several times the nurse at the front trapped her fingers, trapped or dropped the front bar. Unable to restrain the chair adequately the remaining nurse permitted the chair to tip backwards. Then patient and chair collapsed onto the stairs, so hard that on one occasion the seat-retaining bracket

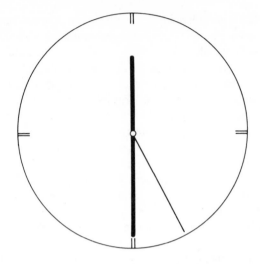

Total time taken for all methods combined before training = 30 min 26 sec

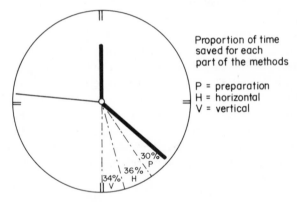

Proportion of time
saved for each
part of the methods

P = preparation
H = horizontal
V = vertical

Total time taken for all methods combined after training = 22 min 46 sec

Overall reduction in total method time = 23 %

Figure 12.6 Reduction in total method time

buckled. However, stronger nurses considered the method easy and efficient as they could lift the patient and chair easily.

The staff found the white pulling tabs on the mercury straps impracticable as when out of use they slipped around the mattress. The belts and lines bit into their hands making them sore. Rope lines would make it virtually impossible for patients independently to escape from the cocoon if the fire/smoke came too close. The low backward pulling movement in methods like the blanket drag caused considerable strain and discomfort in the nurses' lumbar region and thighs.

The backward tilt of the dining chair and sideways overbalance of the mattress when straps slipped proved disconcerting for the patient. With the former, the patient instinctively leaned forward decreasing the efficiency of the method. The patient was acutely aware also of the possibility of a splinter from the floor penetrating spine or buttocks with the blanket and pyjama method. Also the fire doors often closed on the upper part of the patient's body when he was dragged through feet first. The drunk's lift caused discomfort to the patient's feet when he was not wearing shoes.

When going through fire doors with the bed method the leading nurse was 'crushed' into the door. The wheelchair, like the bed, was frequently difficult to steer. Also, unless someone opened the door first the patient was prone to bang his feet and knees, except if the chair were completely reversed, and stopping to open doors reduced speed.

With a patient of average weight any type of lifting is potentially difficult. Consequently, any untrained staff are at considerable risk. The correct positioning, use of dynamic lifting, and respect for the degree and distribution of weight to be moved will help minimize the danger. There is still much to be learnt about lifting techniques. Research was recently conducted (Steventon, 1978) to determine what strain is imposed on staff using the various evacuation lifting methods. The study was partly based on information gained from pills swallowed by the staff which are sensitive to changes in abdominal pressure and transmit this sensitivity in the form of radio waves. Results using this more accurate equipment correlated well with the findings of this study.

Overall, the bed, dining, wheelchair and armchair, and drunk's lift were the most comfortable and least accident-prone methods. The mattress method was reasonably safe and comfortable for the patient, particularly for vertical movement, although it caused some discomfort for the staff.

Minimum risk of malfunction

Generally, as the complexity of the equipment or body configurations increased so did the likelihood of equipment failure and manual faults. All the mattress methods were prone to malfunction. The Anglia sheet caught in the bed springs, the mercury straps slipped, hindering proper cocooning, or were lost under the patient in horizontal movement, forcing the mattress on its side. The straps twisted and became awkward to thread and tighten. The rope lines and mercury straps often slipped off the pulling end of the mattress, releasing the cocoon and causing several *minutes* delay. The latter problem, however, improved after training when cocoons were made much tighter.

The problems of the ambulance chair have been discussed above. The bed and wheelchair methods were jeopardized when the wheels were stiff to lower or steer. The problem of bed attachments can be realized too late and cause serious delays in evacuation; this is discussed later below.

Research was conducted recently by the author on behalf of the DHSS into the effect of various floor coverings on the efficiency of certain evacuation methods

(Hall, undated, c). Results showed that those methods which necessitate pulling the patient along the floor are seriously affected according to the type of floor covering being used; the pull is easier on lino than wood. Soft floor covering could impede progress dramatically; some coverings may prevent the method being used at all.

The methods least prone to malfunction were those using no equipment and simple 'instinctive' body holds, for example, two-hand lifts, pick-a-back, drunk's lift and fore-and-aft methods. The more complicated three, four-hand and Australian lifts frequently deteriorated into a rudimentary two-hand lift when grips failed. The blanket and pyjama methods were also minimally prone to malfunction.

The possible effects of additional variables on the efficiency and suitability of each evacuation method.

The most important variables likely to affect the efficiency and suitability of a method are:

(1) Location and extent of fire and smoke at the decision to evacuate
(2) Building layout and dimensions for escape route
(3) Patient type, physical and mental, and their degree of dependency
(4) Number and fitness of staff present

The degree of the effect of these variables will in turn depend on the type of evacuation necessary. Patient evacuation falls into three main types:

Type A: Where a fire is discovered with an immediate threat to patient safety from smoke and fire, and the fire has become impossible to extinguish or control. Evacuation must be completed as fast as possible by the quickest methods along the easiest route to the nearest area of safety.

Type B: Where there is no immediate danger to life because the spread of smoke is slower, or the fire and/or smoke are in no danger of spreading from an adjacent area. There is a little more time for evacuation but speed is still essential.

Type C: Where there is no immediate threat to life, but the fire and/or smoke is in an adjoining floor or building and *precautionary* evacuation is required. There is more time to move patients and here comfort and respect for the condition of the patients is more important than speed.

The suitability of the methods chosen depends on the evacuation situation, whether A, B or C described above. The effects of these will therefore be discussed according to the type of evacuation.

Evacuation Type A

With evacuation type A, if smoke is rapidly filling the area the first priority is to keep patients and staff as low down as possible to take advantage of the pocket of air that usually remains for some minutes near the floor, and with these conditions remove patients as fast as possible even at the expense of their comfort. Consequently, although slower and more strenuous than many methods the blanket and pyjama methods are most suitable here as both patients and staff can maintain a low-level position. Also, it is preferable that as many non-ambulant patients as possible be placed on the floor before dragging them to the nearest place of safety. If the escape requires vertical movement to a safe place the patient should be grasped around the chest and dragged backwards downstairs by one person with his legs and bottom on the stairs if he is too heavy to lift.

Evacuation Type B

Type B requires speed, and the lower concentration of smoke allows faster methods to be employed, because the necessity of keeping low on the ground is removed. Here a variety of methods which are faster and require less effort than the blanket or pyjama methods can be used, although the suitability and efficiency of these methods are affected by variables (2), (3) and (4) above.

The various lifts are shown to be quick but they require staff of adequate strength and for the three, four-hand and Australian lifts suitably trained to perform efficiently. The patient would have to want and be able to cooperate. Lifts would certainly be preferable for lighter patients, particularly the pick-a-back lift which requires only one person. The dining chair or armchair lifts would be appropriate particularly if the patient is already seated in either, but the patient must cooperate.

If the ward contains a lot of furniture it would be difficult to negotiate a route with wheelchairs or beds, particularly with difficult steering, or mattresses. In the exercises at Hackney staff spent 2.5 minutes just shifting furniture, which had frequently been moved 'out of the way'; also, if the patient is placed on the floor it is possible that he may be overlooked. However, where they can be utilized the bed and wheelchair are the quickest, most comfortable and least strenuous methods.

The bed method is probably the best to use in evacuation type B, provided that the beds have wheels, are reasonable to steer, and that the exit and remainder of the escape route to the nearest area of safety (it is possible to take patients from their beds if further evacuation is required) have an effective width of no less than 1300 mm (52 in.); although 1600 mm (64 in.) would be preferable as this would allow for a person to go past the bed, whether back to help in, or away from, the ward. The area of safety must be large enough to take all the evacuated beds. (This also applies to all the other methods and research is being conducted to determine the necessary dimensions of such areas of safety when using a particular method or number of methods to evacuate varying numbers and types of patients.) It is feasible that several patients can be taken on one bed, therefore involving the minimum of staff and 'stacking' space.

Inadequate widths would affect many of the lifts and mattress methods alike. The minimum width for both would, for comfort and efficiency, be similar to the bed, although narrower widths of 1100 mm (44 in.) could just be tolerated but would almost certainly cause congestion and prevent people getting back into the ward to evacuate the next patient. These dimensions apply particularly on the stairs. If stairs are part of the escape route to the nearest area of safety, with evacuation type B the minimal time available would probably exclude mattress methods. Lifts or ambulance chair (dining chairs are more difficult) would be the most suitable if staff are adequately strong. If they are not, then dragging the patient downstairs in a blanket, as described above, is better.

Evacuation Type C

With evacuation type C all the above methods still apply, but as more time is available there is less likelihood of wheeled equipment becoming entangled in furniture. Provided that sufficient equipment is available, these methods are the most appropriate. The wheelchair is the most suitable as it is more manoeuvrable and less space-consuming than the bed. However, if it is dangerous to transfer the patient the bed method could be used if there are no constraints, such as narrow passages and low doors when bed attachments are in use, etc.

With evacuation type C, if vertical descent is necessary lifting methods are preferable provided that staff are adequately strong; otherwise the mattress cocoon would provide an alternative method.

Table 12.2 gives recommendations for the best evacuation methods, in a descending order of priority, for the three types of evacuation situations, taking into account the limitations of variables (2), (3) and (4) above.

REFERENCES

Appleton, I, and Quiggin, P. *Hackney Fire Precautions Project: An Evacuation Model*, Fire Research Station Internal Report. (In press).

Fire Precautions Act, 1971—Draft Code of Guidance to Fire Precautions in Hospitals 1975, unpublished.

Hall, J. (1975, a). *Hackney Evacuation Exercise: A Study of the Ergonomic and Physiological Effects of Patient Evacuation*. DHSS Report, (In press).

Hall, J. (1976, b). *Report on the Findings of Experiments Conducted to Determine the Maximum Drag Coefficient for Floor Coverings that Could be Recommended for Bed Areas in Hospitals, Together with the Effects of Various Floor Coverings on the Ease of Evacuation*. DHSS Report. (In press).

Hall, J. (1977, c). *Fire Precautions Training for Hospital Staff: Part 1, Estimated Costs of Aligning Training Standards to Those Recommended in the Draft Code of Guidance. Part 2, Training Problems. Part 3, Training Methods*. DHSS Report. (In press).

Kirk, S. (1971). 'Some anthropometric dimensions of the British population', BSI Private Circulation, Document 71/62032.

Steventon, J. (1978). 'An evaluation of lifting methods for fire evacuation in hospitals', MA Thesis, Ergonomics Dept., University College, London.

Webber, C. and Treasure, A. (1974). *Ward Fire Evacuation Study (Patient Dependency Study)*. DHSS Internal Report.

Table 12.2 Recommended methods of evacuation (in each group, methods are given in order of preference)

Other variables affecting choice of evacuation method	Variable (1): Location and extent of fire		
	Fire type A	Fire type B	Fire type C
Variable (2): Building layout and escape route			
(a) Horizontal and without constraints	Blanket Pyjama Bed*	Bed Wheelchair	Bed Wheelchair
(b) With constraints such as narrow doorways or tortuous routes, but no stairs	Blanket Pyjama	Wheelchair Drunk's lift Two-handed lift Fore-and-aft lift Pick-a-back	Wheelchair Dining/armchair Mattress† Two-handed lift Fore-and-aft lift
(c) Involving stairs	Blanket Pyjama	Two-handed lift Fore-and-aft lift Blanket	Two-handed lift Mattress† Fore-and-aft lift
Variable (3): Patient type			
(a) Non-ambulant	Blanket Pyjama Bed*	Bed Wheelchair Drunk's lift Two-hand Fore-and-aft lift Armchair Blanket	Bed Wheelchair Mattress† Two-handed Fore-and-aft lift
(b) Ambulant or semi-ambulant but mentally disturbed	Blanket Pyjama	Bed Wheelchair Drunk's lift Fore-and-aft lift Blanket	Bed Wheelchair Two-handed Mattress† Fore-and-aft lift
Variable (4): Staff present			
(a) Three or more staff, some stong	Blanket Pyjama Bed*	Bed Wheelchair Two-handed lift Drunk's lift Fore-and-aft lift Pick-a-back	Bed Wheelchair Two-handed lift Fore-and-aft lift
(b) Only two staff	Blanket Pyjama Bed*	Bed Wheelchair Pick-a-back Dining chair/armchair Two-handed Blanket	Bed Wheelchair Pick-a-back Dining chair/armchair Two-handed Fore-and-aft lift
(c) Three or more staff, all unfit or weak	Blanket Pyjama Bed*	Bed Wheelchair Drunk's lift Two-handed lift Fore-and-aft lift	Bed Wheelchair Dining chair/armchair Two-handed lift Fore-and-aft lift

* Bed method only applicable in type A if smoke above chest height.
† Mattress method here refers to the Anglia type sheet or straps, preferably with Britax-type buckles; mattress lines are not recommended.

Fires and Human Behaviour
Edited by D. Canter
© 1980 John Wiley & Sons Ltd.

CHAPTER 13

Building Evacuation: Research Methods and Case Studies

JAKE L. PAULS AND BRIAN K. JONES
National Research Council Canada
Division of Building Research, Ottawa

INTRODUCTION

This paper is concerned with what happens when several thousand office workers, without prior warning, participated in two major evacuations. By referring to these case studies of two fundamentally different types of evacuation—traditional total evacuation and selective sequential evacuation—we can better understand the limited relevance of traditional regulatory approaches covering exit geometry and the increasing importance of commication systems, operating procedures and training in the very large office buildings of today.

IMPORTANCE OF STUDIES OF DRILLS

The information from studies of drills or emergency exercises is important. Knowledge of behaviour in drills and of attitudes toward drills is needed to assess the effectiveness of drills as one means of preparing for an emergency such as a fire, a massive power failure or a serious bomb threat. Going beyond this, we believe that enough similarity can exist between important conditions found in drills and those found in genuine emergencies to make study findings from drills generally useful in understanding what happens in certain emergencies.

The evacuation studies supervised by J. L. Pauls would not have been possible without the cooperation, assistance and encouragement of many people. Many colleagues at the Division of Building Research, National Research Council of Canada, assisted with the observations. The Dominion Fire Commissioner's Office staff were most helpful, keeping us informed of planned drills and occasionally helping with the observations. To those on the fire emergency organizations of the buildings studied we express gratitude for their assistance and we also apologize for the protracted analysis of study data and the late publication of detailed findings. Finally, to the many thousands of people who unwittingly 'committed data'—to borrow a phrase from a fellow researcher—within view of our observers, we hope the quality and safety of your workplaces are enhanced by our joint efforts.

Behind this belief there are our findings on behaviour of people totally evacuating a twenty-storey office building because of a fire in the basement. Building occupants could see smoke outside the building and could smell it in some parts of the building, particularly in the lower levels of the exit stairs. We learned from interviews with evacuees and from reports of fire wardens that the behaviour of evacuees was essentially the same as we had observed in an earlier drill in this building and in drills in other buildings.

An example of a serious fire being treated too lightly by people in a building is the May 1976 fire in the Beverly Hills Supper Club in Kentucky. In the room where most of the 164 fatalities occurred people's perceptions of the risk to themselves remained remarkably low during a large portion of the evacuation even though they were told of the fire and some smoke could be seen (Best, 1977). One could therefore hypothesize that a model of evacuation developed in other situations such as drills, where people assume there is minimal risk, can be applied to many fire situations in large buildings.

In drill, as in real fires, occupants are often made aware of possible hazards by indirect means—through communication systems or by ambiguous cues such as alarms. Furthermore, in the case of large buildings, a fire in one part of the building may not produce environmental cues, such as smoke, that are immediately perceptible to occupants in other areas. In other words, evacuations are sometimes carried out where large segments of the building population are unsure whether there is a drill in progress, a minor or a major fire, or some other emergency. One example of where a drill was interpreted by many evacuees as a genuine fire emergency—largely on the basis of what they heard over a building's public address system—is described in the case study of selective evacuation.

BUILDING EVACUATION CASE STUDIES

The following case studies illustrate two basically different types of evacuation. The first type is *traditional total* evacuation in which all occupants of a building attempt, more or less at at the same time, to evacuate from a building. The main controls over what happens in this type of evacuation are exit geometry, building population and movement behaviour of evacuees. The evacuation is not centrally controlled as is the case with the second type of evacuation, *selective sequential* evacuation. In this type, evacuation starts with the building occupants closest to the fire; this is followed by evacuation of adjacent floors. This may be followed by evacuation of certain other floors, for example, the upper floors of a tall building that may receive smoke carried upwards from a lower fire area. Decisions on who should be evacuated and when they should be directed to use the exits are made at a central control point equipped with public address equipment and emergency telephones to reach all parts of a building.

Description of buildings

To illustrate the two types of evacuation we have chosen two extensively studied, medium-size, high-rise office buildings in Ottawa, Canada. They have similar occupancies—basically single-tenant, government departments—and they have similar office areas. They differ markedly in plan configuration and exit stair provision. Building A has approximately four times as much effective stair width as building B, in other words it effectively has four times as much exit width at the base of the stairs—a controlling element in total evacuations. The buildings also differ when exit widths from individual floors are compared; building A has nearly three times as much effective stair width from any one floor. This is a factor in selective evacuations where floors are cleared one at a time—at least in theory.

Building A has a cruciform plan with a total of 33 000 m^2 (350 000 ft^2) of gross rentable area on fourteen typical office floors above ground level. There is a stair at the centre of the building and one stair at the end of each of the four wings. All stairs have a measured width of 1140 mm (45 in.) and a dogleg configuration (with two 180° turns per storey).

Just prior to the evacuation drill building A had a reported occupancy of 1709 workplaces on floors 2 to 15. This occupancy was not spread equally among the four wings; the largest wing—the west one—had 616 workplaces whereas the north, east, and south wings had 281, 471, and 341 respectively.

Building B has a rectangular plan with a total of 31 000 m^2 (331 000 ft ^{2}) of gross rentable area on twenty typical office floors above the ground floor. Its two exit stairs are in the centre core of the building, have a measured width of 1040 mm (41 in.), and have a dogleg configuration.

Prior to the evacuation drill building B had a reported occupancy of 2404 workplaces on floors 2 to 21. Persons considered to be physically disabled in some way numbered 69. It was also known that 48 per cent of the workplaces were assigned to men and 52 per cent to women.

TOTAL EVACUATION DRILL IN BUILDING A

With its abundance of exit stairs building A was considered suitable for traditional evacuation and this was the procedure set out in it's fire orders (based on a standard of the Dominion Fire Commissioner, 1972). The stairs were in fact sufficiently large to allow all occupants of the building to stand in the four perimeter exits and have one tread of area each.

There had been a drill in September 1972 for which people had been forewarned. Observations similar to those to be described revealed that there were 1220 evacuees in this drill. Six weeks later, on a cooler day in late October, another drill was held, this time without prior warning. This time 1453 able-bodied persons used the four perimeter exit stairs and 73 persons—those who were disabled, along with assistants and some fire wardens—used the central stair. Disabled persons knew from the fire orders and from previous drills that the central stair was designated for their use.

The total evacuation procedure entailed having all able-bodied persons moving to exit stairs at the ends of wings, joining other evacuees on the stairs and descending as rapidly as crowd conditions permitted to the ground floor and then away from the building. There were no external controls on stair use except that in the larger west wing on several floors and in the east wing on one floor evacuees were to use predesignated exits in the relatively under populated north and south wings. This procedure had been selected by the building fire emergency organization in an attempt to balance the population loading on the four stairs.

Study procedures

Seventeen observers were given six pages of detailed instructions (outlined in Appendix A) and portable cassette recorders to record all observations during the drill. Those with prior observation experience were given more demanding tasks near ground level. Five observers were stationed at the ground floor level at each exit discharge in order to count evacuees (recording their running count into a recorder). Twelve moving observers were assigned, one to each of the four perimeter stairs, to evacuate with people from floors 5, 10, and 15. They provided data on initial behaviour when the alarm sounded, entry to the exit stairs, mixing of evacuees entering the exits with those already inside, speeds of the evacuees they accompanied down the stairs, crowd densities and crowd configurations. When they reached ground level the four observers from floor 5 reported density and speed conditions of evacuees on the last flight of the stairs.

Each observer's recording was transcribed and correct time-scales were calculated. An example of an observer's transcribed report is given in Appendix A. Here it will be seen that this observer counted 23 evacuees from the west wing of floor 15 entering the exit during the first 90 seconds of the drill.

Exit entry

Figure 13.1 shows the evacuees' entry flow, calculated at five second intervals, at floor levels 5, 10 and 15. Because floor 10 was one of those where evacuees had to move to distant predesignated exits it is not surprising that the entry flow of 82 evacuees to the south exit is extended to nearly four minutes. At least three specific factors affected the length of time required: brief confusion of evacuees as to which exit to use; a long travel distance—as much as 75 m (250 ft)—to the designated exit; and extensive queueing at times to enter the stream of slow-moving, often-stopped stream of evacuees in the heavily used south exit.

Observations in this drill and in others indicate that evacuees already in the exits usually defer to those entering from a lower floor. In cases where queueing occurs in a floor area at the entry to an exit there is likely to be a queue in the stair as well so that no further evacuees can enter until those at lower levels move downwards.

Movement down exit stairs

Immediately after recording the entry flows of evacuees into exits, and before becoming separated from those evacuees, the observers began their descent of the stairs. From position data recorded on each observer's continuously running recorder the traces of observer movement shown in Figure 13.2 can be plotted. From the chart we can quickly see where any particular observer was at any time during the drill. Such charts are useful for data reporting and for simulation purposes (Pauls, 1971, 1977a, b and 1978).

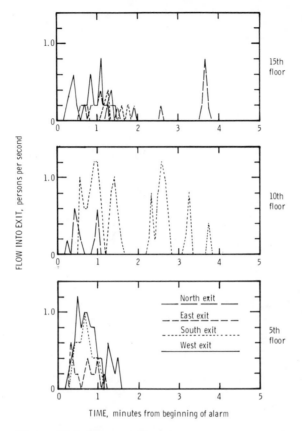

Figure 13.1 Observed flow into exits during total evacuation of building A

Figure 13.2 also shows when flow into exits began on floors 5, 10, and 15; when first evacuees reached ground level; and when main flows (not including stragglers) finished in each exit. Here it should be noted that the exits were used by different numbers of people, despite the attempt to balance usage; 291 persons used the north exit, 329 used the east, 385 used the west, and 448 used the south (that is, too many persons were diverted to the south).

The slopes of traces indicate speeds of movement down stairs. It can be clearly seen, from the generally uniform slopes of the traces, that descent speeds at the lower levels of the stairs were similar from one stair to another. Below floor 7 the evacuees and the observers moving with them (except for floor 15, north exit) had a mean speed of 0.44 m/s along the slope of the stairs. On the average they took 112 seconds to move from floor 7 to ground level. It should be noted that in this evacuation, as in others in even taller buildings, there was no indication of a fatigue factor slowing the descent of evacuees as they approached the bottom of the stairs. In fact there was evidence that speed increased with descent. Finally, with regard to Figure 13.2, it should be clear that horizontal portions of the traces indicate no movement (that is, there was queueing) in the exits; this occurred in the east, south,

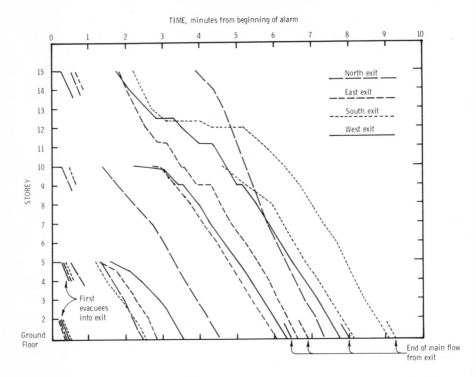

Figure 13.2 Movements of observers with evacuees in exits during total evacuation of building A

and west exits at upper floors. (There is one exception to this; the observer in the east exit from floor 15 stopped briefly at floor 9 and was passed by evacuees.)

Figure 13.3 is derived from Figure 13.2; the two traces shown highlight the movement of observers in the west exit from floors 10 and 15. The latter's log of observations is reproduced in Appendix A. In the west exit below floor 9 the two observers and the 66 evacuees between them had an average descent speed of 0.39 m/s (one storey every 23 seconds); the average density was 1.8 persons per square metre (p/m^2) or one person on every two treads of the stair; and the average flow was 0.78 person per second (p/s).

Above floor 9 their densities, speeds, and flows had not reached what might be called steady-state conditions. There were stops and queueing as the stair was temporarily overloaded. For example, around the three-minute point in the drill the west stair at upper levels had high densities of evacuees—as high as 3.0 p/m^2 (six persons on every seven treads). Figure 13.3 depicts, with the three dotted lines and arrows to the upper trace, the upward progression of stoppages and starts of movement. With higher densities—and the related slow descent speeds—the progression of stoppages can be very rapid.

The highest density at which movement occurred in the west exit was 2.6 p/m^2 (three persons on four treads)—an average figure between floors 9 and 12 at 3.5

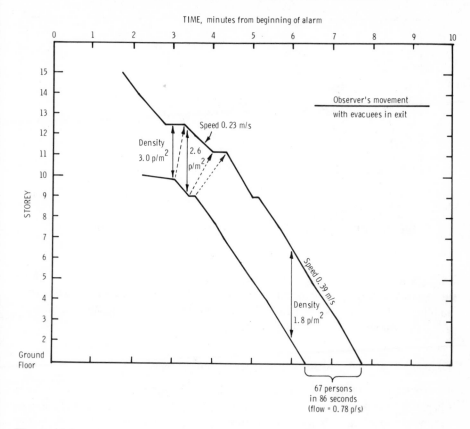

Figure 13.3 Analysis of movement of evacuees in west exit during total evacuation of building A

minutes into the drill. Data from the two observers' reports of numbers of evacuees per flight of stairs indicated that there were briefly sustained, localized densities ranging between 2.0 p/m² and 3.2 p/m². The speed of movement was 0.23 m/s (a slow shuffling pace moving one storey in 34 seconds). The flow was 0.63 p/s.

Movement conditions at exit discharge

In Figure 13.4 the discharge flows from the four exits at the building's perimeter are plotted against a time-scale. The mean flows, over the course of the evacuation, were 0.80 p/s, 0.83 p/s, 0.85 p/s, and 0.85 p/s in the north, east, south, and west exits respectively.

There were 58 samples of speed and density measured on the last flights of the four stairs. Mean speed was 0.66 m/s with a standard deviation of 0.13 m/s. Mean density was 1.38 p/m² with a standard deviation of 0.27 p/m². Stated in everyday terms, each storey of descent took fourteen seconds and the density was equivalent to having each person using between two and three treads of the 1140 mm (45 in.)

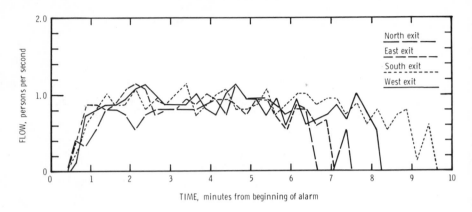

Figure 13.4 Flow from exit stairs during total evacuation of building A

wide stairs. These flow, speed, and density conditions were much like those reported at exit discharge points for all of the total evacuations studied in Ottawa office
buildings. The full extent to which this total evacuation was typical of those studied will be shown in the following section.

Comparison of predicted and actual evacuation performance

In this section extensive use is made of empirical findings reported by Pauls (in
Chapter 14) to assess how well the evacuees in this building performed in compar-
ison with expectations based on performance in all the total evacuations studied in
Ottawa.

For this performance prediction let us assume that we know a few things about
the building's physical features but, as far as occupancy is concerned, we know only
that it is a single-department, government office occupancy. Starting from a known
gross rentable area of 33 000 m² (350 000 ft²) we assume an area per actual occupant of 22 m² (240 ft²), a reasonable estimate in a fairly large government office
building. This gives an actual occupancy of 1500 persons from which 4 per cent is
deducted to account for disabled persons and some assistants who are supposed to
use the centre exit. This gives an evacuation population of 1440 persons who we
will assume are distributed equally among the four perimeter exit stairs. Each
1140 mm (45 in.) stair, with an effective width of 840 mm (33 in.), is thus assumed
to carry 360 persons.

Using an equation given in the accompanying paper, the mean evacuation flow is
predicted to be 0.89 person per second. We can attempt some modification of this
mean flow figure because we know the stair walls are slightly rough and evacuees
use outdoor clothing for this cool-weather evacuation. Based on reductions suggested in the accompanying paper we can assume the mean flow of 0.89 p/s should be
reduced by 4 per cent to account for stair wall roughness, and 3 per cent to account
for use of coats.

The final predicted mean flow—based on knowledge of building area, stair widths, stair wall texture, and evacuees' use of coats—is 0.83 p/s. This compares very well with observed mean flows of 0.80 p/s, 0.83 p/s, 0.85 p/s, and 0.85 p/s in the north, east, south, and west exits respectively. It is not surprising that observed flow is lower in the north exit; it had an actual evacuation population lower than assumed. The observed flows in the south and west exits were higher because their evacuation populations were higher in the drill (see Chapter 14 for details on the effect that population has been found to have on flow on stairs).

Again using an equation given in the accompanying paper, we can make a prediction of total evacuation time. This predicted time is 7.0 minutes, assuming equal numbers of evacuees use each stair. The times actually measured for all evacuees to reach ground level, except stragglers separated from the main flow, were 6.6, 7.0, 9.3, and 8.2 minutes in the north, east, south, and west stairs respectively. The average observed time was 7.8 minutes.

The range of times is mainly the result of non-uniformity of evacuee distribution among the four perimeter exits. The difference between the predicted 7.0 minute evacuation time and the average measured time of 7.8 minutes is due to the evacuees' use of coats, the slightly reduced flow rate due to stair wall roughness (making the true effective stair width even less than mentioned earlier) and some delays on several floors, as perhaps confused evacuees had to use a distant, designated exit instead of a close one.

It should not be surprising that there is a good fit between anticipated performance and measured performance. Two total evacuation drills in building A provided eight of the 58 cases studied to provide a basis for the preceding predictions. Good fit simply results from internal validity.

Performance in evacuation judged to be due to real emergency

Early in 1973, on a cool rainy day, building A had another total evacuation—the third in less than a year. People occupying the building were accustomed to only one drill each year. With the unusual timing, the bad weather and the absence of any observation activity that might suggest a drill, some people interpreted the drill to be an evacuation due to some emergency.

The building population was reported to be approximately 1900 persons but since this was not an actual count of evacuees it should be assumed that the figure is too high. Assuming that it is too high by the same ratio found earlier (when 1709 occupants were reported but only 1526 took part in the unannounced fire drill) the evacuation population could be assumed to be 1700 persons, of whom 1620 might be assumed to be able-bodied persons using the four perimeter exits.

Based on the equation used earlier to predict total evacuation time and considering performance in earlier drills, the predicted time for the latest evacuation ranges between 7.6 and 8.7 minutes. Two estimates of evacuation time, reported by building fire emergency staff, were 7.9 and 8.5 minutes.

The evacuations of this building—well supplied with exit stairs and occupied by people familiar with total evacuation—epitomize what can be accomplished with traditional total evacuation. Incidentally, the evacuations also show what can be accomplished by way of performance prediction for such evacuation procedures.

Costs, benefits, and problems with total evacuation

An annual total evacuation drill apparently might be adequate to keep occupants of an office building familiar with and highly proficient in the total evacuation procedure. The cost or time loss for such an annual exercise is estimated to be approximately 1000 person-hours (p/h) for building A. This includes the direct time loss for building evacuation, for movement of evacuees to areas well away from the building, and for re-entry of the building. (Incidentally, on the matter of getting evacuees to move well away from the building, this was done only with great difficulty in most of the observed drills.) The 'cost' could be somewhat higher if time loss from post-drill disruptions of building occcupants' work is added (for example, due to informal discussions about the drill).

In view of the major time loss related to the three evacuations of building A within a one-year period it is of some consolation to know that at least there was a major benefit by way of data and knowledge because two of the drills were well documented. Not counting time loss of fire emergency staff in the building, nor time taken for its evaluation as well as our research, it is estimated that the Canadian government paid for at least 3000 hours of staff time in connection with the three evacuations in this one building. The high cost of such drills simply underlines the importance of getting value for the cost; in other words, evaluate the drills to see whether they accomplish the assumed fire safety objectives.

It should be made clear that it was mainly because of the initiative of one of the authors that the Ottawa evacuation drills were documented as they were. The drills are held periodically quite independent of the research activities of the author who has no responsibility for the frequency, timing or procedure of such drills.

Against the large, recurring costs of total evacuation drills must be compared the cost of public address, emergency telephone, fire suppression and smoke control equipment as well as substantial staff training and occasional selective evacuation drills that are needed for selective evacuation procedures to work dependably.

In the following section is given an example of a selective evacuation where fire-threatened areas are given priority in evacuation—unlike the situation in total evacuations where evacuees from the immediate fire area are not necessarily given priority in exit stair use (particularly if the fire alarm is sounded throughout a building before the fire floor is evacuated). Furthermore, in selective evacuations, non-threatened areas are not cleared indiscriminately. Selective evacuations are, however, not without problems of their own. This is illustrated in the next case study in building B.

SELECTIVE, SEQUENTIAL EVACUATION DRILL IN BUILDING B

Building B, a 21-storey office tower, was one of the first in Ottawa to have a selective, sequential evacuation procedure (or phased evacuation procedure as it was termed locally) for use in case of fire emergency. In this procedure the occupants of the fire floor are evacuated immediately under local control while occupants on other floors respond to the fire alarm by going to main corridor areas near the building's core where public address loudspeakers carry directions issued from a central control station at ground level. If the fire situation warrants, the two floors immediately adjacent to the fire floor are evacuated next; then others, in sequence starting from the top of the building, are evacuated if necessary. Incidentally, one basic difference between the traditional, total procedure and the selective, sequential procedure is that in the former the building is cleared from the bottom to the top floors; in the latter, after the fire area is cleared, the building is cleared from the top downwards. (An exception to this was the first selective evacuation documented by Pauls (1971).)

Usually in the selective, sequential evacuations observed in Ottawa only several floors were evacuated. In some cases all the floors above an assumed 'fire floor' were evacuated in the sequence described above. In the case of the evacuation of building B, with the 'fire floor' chosen to be floor 3, the entire building was evacuated. The drill therefore contained all the elements of a selective, partial evacuation—the limited evacuation now often practised in drills in Ottawa.

Building B was an obvious candidate for the selective evacuation procedure. With a population of just over 2100 persons and two exit stairs only 1040 mm (41 in.) wide the building would require 20 minutes to totally evacuate in the traditional way. During much of this long time the occupants—perhaps even the ones most threatened by a fire—might be queued either on their office floors or close to their floors in the exits. With 2100 persons occupying the two exit stairs there would be only 0.16 m^2 (1.7 ft^2) of stair area per person or just over one-half tread each. In other words, the stairs would be jammed with people and no movement would take place except at the bottom of the queue.

In late 1970 building B had a bomb threat that led to a total evacuation employing the traditional procedure with everyone making a rush for the exits. This evacuation went badly according to reports by fire emergency staff, and people were upset by the experience. Aside from the problems caused by overloaded exits there was a problem of inadequate information during the evacuation. A public address system and an emergency telephone system were installed before this evacuation but they were not used. Thus there was no information from a central, authoritative source available to those waiting outside the exits at upper floors or attempting to move down the crowded exit stairs.

It should be recognized that the selective evacuation procedure was only instituted officially in building B in December 1970 and, in the months between this and the May 1971 evacuation drill, no extensive training in the new procedure was given to fire emergency staff or to other building occupants. This lack of training was an important factor in the actions of those who had to operate the communication systems during the May drill.

Study procedure

Although the May 1971 evacuation drill was held without advance warning to all but a few of the people in the building (the chief fire warden and some assistants) preparations for its study were made over a period of months by one of the authors. Thirteen staff members of the National Research Council's Division of Building Research worked as observers using the instructions outlined in Appendix A. Officials of the Dominion Fire Commissioner's Office—the fire authority for federal government-occupied buildings—as well as provincial and local fire officials were also in attendance and some assisted with the research effort directly.

Observers were positioned throughout the building. Those who were to move with evacuees were stationed on floors 3, 4, 12, 14, and 21. Four of these observers then took up fixed positions at floor 2 and ground floor levels. Altogether there were six observers engaged in various tasks at the ground floor. Two observers proceeded down the stairs, keeping just above the stair-entry activities of each successive floor group evacuating according to the sequential procedure, from the top of the building downwards. All observations plus the use of public address and emergency telephone systems were recorded using portable or fixed tape recorders.

Within hours of the drill's completion 200 copies of a lengthy questionnaire were distributed, ten per floor. Of the 178 returned by mail, 176 came from people who had taken part in the evacuation. The questionnaire, along with a record of responses to each question, is found in a separate paper (Pauls, 1979a).

Use of communication systems and effect on evacuees

The key to the evacuation lay in the use made of the communication systems and the decisions made by building fire emergency staff not only at the communication control console in the ground floor entry lobby but throughout the building where wardens were stationed at emergency telephones beside the exit doors at each floor.

The drill started like other drills with fire bells ringing throughout the building. The alarm was initiated from floor 3. People on this floor evacuated under the local direction of a floor warden, a process that was completed about 140 seconds after the alarm was set off. On other floors a few people moved into the exits but returned immediately to the office floors; many people however treated the alarm as only a malfunction of the alarm circuit (such malfunctions were after all not uncommon). Table 13.1 summarizes questionnaire responses on this matter of initial interpretation of the situation as the alarm bells rang. The rows of the table describe what people's interpretation of the situation was; the columns describe their reasons for the interpretation.

The alarm bells stopped sounding 85 seconds into the drill; thirteen seconds later the public address system operator attempted to make an announcement to all building occupants informing them to stand by for evacuation instructions. Due to an incorrectly set control the announcement was not carried by the public address system. Therefore, when there was a long silence after the alarm stopped, wardens and others on the office floors were wondering what was the reason for the alarm.

Soon the emergency telephone was busy; the first warden to call the central control console asked, 'Is there a fire?' On floor 20 the chief building fire officer—who of course knew that a drill was in progress—then used the emergency telephone to inform central control that the public address was not functioning. This comment was echoed by wardens on two other floors.

Table 13.1: How and why people initially interpreted the situation

Interpretation	Reason given	Previous incidents	Always treat alarm as emergency	No fire trucks	Other reason	Did not answer	Total
Circuit malfunction		100				3	103
Fire emergency		1	18		5	6	30
Drill		5		3	16	5	29
Equipment test		2			1	1	4
Bomb threat				1	1		2
Other		2			2		4
Did not answer						4	4
Total		110	18	4	25	19	176

It was not until two minutes and 46 seconds (2.46) into the drill that an announcement was heard on the public address system throughout the building and it went as follows:

> Ladies and gentlemen. We have to evacuate this building. The alarm has been set on the third floor. Please evacuate. Other floors stand by. *Votre attention s'il vous plaît. L'alarme a été pose au troisième plancher. Il faut évacuer. Il faut évacuer.*

(It should be noted that bilinguial messages are *de rigueur* in the office buildings studied in Ottawa.) The announcer sounded somewhat agitated after failing to have his first message carried. The English portion of his message was ambiguous, suggesting both evacuation and non-evacuation; it did not make clear that only those on floor 3 were to evacuate immediately. The French portion of his message indicated immediate evacuation for all.

Observers reported changes in occupant behaviour when the message was received. Some moved prematurely into the exits instead of standing by to hear further instructions. Here the questionnaire findings give some valuable insight. Although only 17 per cent of the respondents reported interpreting the situation initially as a genuine fire emergency, 43 per cent reported that this was their interpretation after hearing the first public address announcement. Some respondents even reported that in the announcement they thought they heard, 'a fire has been reported on the third floor'. Tables 13.2 and 13.3 summarize data on this matter of situation interpretation. Note that 32 out of 51 persons who changed their initial interpretation of the situation to belief that there was a fire emergency reported doing so because of the public address announcement.

At this stage of the drill there was considerable confusion as floor fire wardens

Table 13.2: Initial and subsequent intrepretations of the situation

Initial interpretation	Subsequent interpretation Circuit malfunction	Fire emergency	Drill	Equipment test	Bomb threat	Other	Not applicable, already evacuated	Did not answer	Total
Circuit malfunction	7	46	23		4	4	6	13	103
Fire emergency		22	6			1		1	30
Drill	1	3	24		1				29
Equipment test		2	1				1		4
Bomb threat					1			1	2
Other	1		1			1		1	4
Did not answer			1			1		2	4
Total	9	73	56		6	7	7	18	176

and other occupants wondered what to do. Some of the wardens contacted central control by emergency telephone for clarification. This led to a second announcement about five minutes into the drill (5.02) that went as follows:

> Please stand by. The third floor must evacuate and then the rest to follow. Fourth floor and second floor, please evacuate. *Le quatrième et le troisième plancher s'il vous plaît évacuez.*

This was followed by a further announcement at 5.46.

> Twenty-first floor, please evacuate. Twenty-first floor, please evacuate. *Vingt-et-unième plancher s'il vous plaît évacuez. Vingt-et-unième plancher évacuez s'il vous plaît.*

There was no further announcement made until nearly eight minutes into the drill.

Evacuee movement

Having some appreciation of what people heard during the crucial first minutes of the drill and how they interpreted the situation, we should now examine Figure 13.5 which graphically depicts the course of the sequential evacuation both predicted and actual. The prediction was drawn up just prior to the drill and was based on an evacuation population of 2336; this was the number of work-places considered occupied by able-bodied persons. It was assumed that equal numbers of evacuees would use each stair. The prediction also took into account the

Table 13.3: How and why people subsequently interpreted the situation

Subsequent interpretation	Reason given for change in interpretation		Interpretation same	Did not answer	Total
	Public address announcement	Other			
Circuit malfunction			6	3	9
Fire emergency	32	5	22	14	73
Drill	4	14	24	14	56
Equipment test					
Bomb threat	2	1	1	2	6
Other	1		1	5	7
Not applicable or did not answer				25	25
Total	39	20	54	63	176

known, non-uniform distribution of workplaces on the different floors. In the evacuation there were actually 2006 persons using the stairs and approximately 70 disabled persons were evacuated by elevator. Therefore the observed evacuation time (29.5 minutes) could be expected to be less than the predicted time of 31 minutes based on the number of workplaces. The fact that it was not closer to 26.6 minutes [that is, 2006/2336 (31) minutes] is partly explained by the unequal use of the two exits and other factors discussed below.

In Figure 13.5a are plotted traces of observers' movement down the exit stairs from floors 3, 4, 12, 14, and 21. Also shown are asterisks representing the beginnings of announcements directing specific floors to evacuate. Such public address announcements were made for all floors except floor 3 and floor 18. Floor 3 was cleared under the immediate direction of the floor warden; floor 18 was missed by the public address announcer and the order to evacuate came later by telephone. This mistake caused additional confusion during the evacuation as people from various floors simultaneously attempted to enter the exits—something the selective, sequential procedure was supposed to avoid. The heavy horizontal bars in Figure 13.5a indicate observations of substantial entry flows from particular floors into the south exit. Unfortunately, because we did not have observers at all points where such entry was occurring, Figure 13.5a only depicts a portion of this activity. Also noted in this figure are the telephone reports, from floor wardens to central control, indicating that floors had been cleared of people; these calls are marked ($\sqrt{}$).

A striking feature of the evacuation depicted in Figure 13.5a is the absence of a consistent relationship between a floor's order to evacuate and the actual evacu-

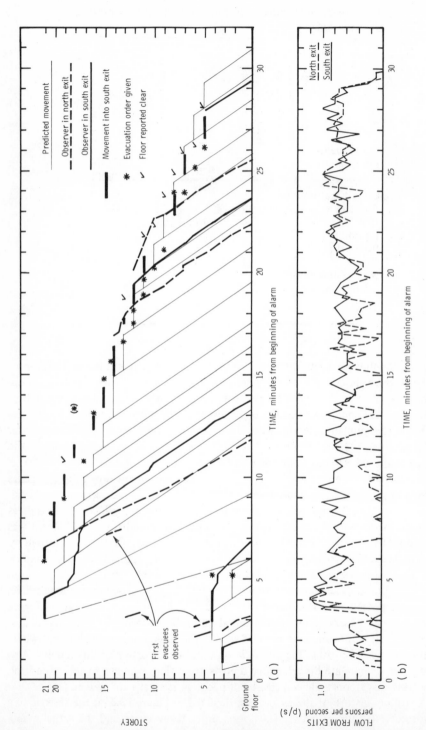

Figure 13.5 Selective, sequential evacuation of building B

ation of that floor. Similar discrepancies occurred between the wardens' telephone reports that their floors had cleared and the actual end of clearing as it was observed on many floors. This was partially due to the extensive, unnecessary use of the emergency telephone system; it was in use 36 per cent of the time during the drill. Getting the message through may have been difficult if not impossible.

Comparison of evacuation movement in each exit

Many people on many floors chose to evacuate down the south stair at about the three minute point of the drill just after the first public address announcement described earlier. As a result the south stair below floor 18 was crowded with evacuees. The observer moving with these prematurely evacuating persons experienced extensive queueing on the stairs. He took nearly four minutes to move down one storey past floor 18 and ten minutes overall to reach the ground floor. By contrast, the observer in the north exit, moving with evacuees who were following the instructions given at 5.46 for floor 21 to evacuate, left floor 21 at 6.30 and reached the ground floor at 11.50, thus taking only half the time used by the other observer. It is estimated that as many as 350 people—nearly one-fifth of the total number of evacuees—may have evacuated prematurely after hearing the first public address announcement. Most of this premature evacuation occurred in the south exit.

Figure 13.5b which shows flow from the exits over the course of the drill, indicates clearly the very high flow from the south exit during the early part of the evacuation when flow should have been minor and restricted to that of evacuees from floors 3, 4, and 2. If the drill had been intended as only a partial evacuation of those three floors alone it would have taken 7 minutes to clear the exits of people from that area plus people incorrectly evacuating from other floors.

During the last eighteen minutes of the drill (between 11.30 and 29.30), when each exit should have had a fairly regular discharge flow of evacuees leaving in sequence from upper floors, the mean flow in the south exit was 0.69 p/s with a standard deviation of 0.16 p/s while the mean flow from the north exit was 0.57 p/s with a standard deviation of 0.25 p/s. It was only during 8 of the last 18 minutes that both exits were equally used with consistently high flows (0.73 p/s with a standard deviation of 0.15 p/s). Over the entire course of the drill the south exit was used by 1175 evacuees and the north exit was used by 831 evacuees.

In passing it might be of some interest to note that there were often large single-sex groups in the streams of evacuees (partly as a result of a 'ladies first' procedure at the entries to exits). A detailed analysis of density and flow data for the lower portion of the south exit showed no significant differences in the mean values for these variables comparing samples composed entirely of females, samples with a mixture, and samples with males only. Samples composed of a mixture had slightly higher, but not significantly higher, flows and densities than samples of one sex alone. The latter showed no differences in the evacuation performance of males and females.

Reasons for the differences between the two exits

An obvious question is why was there such a difference from one side of the building to the other; after all, the building is nearly symmetrical and each exit is in the
central core.

There are two parts to this question. The first deals with the imbalance in both
population and flow during the evacuation. The second deals with the massive premature evacuation that occurred primarily down the south exit stair.

One possible reason for the difference in evacuation populations—1175 using the
south exit while only 831 used the north—is that the building was not symmetrical
in its occupancy. The north side has by far the more interesting view (of a large
river, parkland, and scenic hills), a feature that may well have been exploited by
executive and management personnel who command more office area per person.
Thus the south floor areas may have had a disproportionately large share of the
total population. The south exit, moreover, discharged into the front entrance
lobby of the building. These characteristics would lead one to expect a higher evacuation population and flow.

Explanations for the great difference in premature evacuation are more speculative. If the north side was indeed occupied more by executive and management
staff they may have more quickly resumed working after the alarm bell stopped
sounding. Many could have missed hearing the influential first announcement. On
the south side, once a small number of people moved into the exits, there may have
been a 'snowball effect' as others followed along. Evacuation is, after all, a social
event. With large windows in the doors to the exits it would be quite apparent to
people waiting on other floors on the south side that extensive evacuation was under
way; they too might begin premature evacuation because the sight of other evacuees
might tend to confirm one of the messages in the ambiguous first public address
announcement. Meanwhile, as the behaviour of those on the south side of the building's central core was not visible to those waiting on the north side, it is not surprising that the contagion did not extend to the north side. (On the left side of Figure
13.5a see the note, 'first evacuees observed'.)

Questionnaire data

Some useful findings of the questionnaire, filled out by 176 evacuees, have already
been discussed in relation to interpretation of the ambiguous situation faced by
people in this drill.

In the following discussion of the questionnaire we simply point out some of its
failings and illustrate a few findings.

An important deficiency with the questionnaire was the absence of questions
dealing with respondent's location just before the drill and movement in the floor
area prior to entering an exit. One crucial bit of information—on the specific exit
used—was missing as a result of this oversight.

Other information that should have been obtained relates to the social aspects of
the evacuation. For example, to what extent was it a group decision and action to

evacuate prematurely. Admittedly the probing required to elicit this kind of information is difficult if not impossible to do in a questionnaire. Interviews with evacuees would be a better way of probing such subtle phenomena.

The questionnaire touched on five areas:

(1) Identification of the sample population (age, sex, fire safety organization responsibilities).
(2) Fire safety awareness and attitudes; fire experience.
(3) Pre-evacuation behaviour during drill.
(4) Experiences while evacuating down exit stairs.
(5) Normal use of exit stairs and capability with stairs.

The sample population of 176 persons included 32 who had some formal responsibility in the building's fire emergency organization; 27 of these were floor wardens or their assistants. Note that the distribution of questionnaires was intended to be random with nine copies distributed by floor wardens of each floor according to a small marked plan which was different for each floor. An additional questionnaire was filled out by the person doing the distribution on each floor. Although we were not able to check on how well the intended procedure was actually followed, there was at least one indication of good sampling; respondents not on the fire emergency organization consisted of 51 per cent men, a figure that compares well with the 48 per cent of workplaces reported to be occupied by men.

Many interesting comparisons can be made between the responses of the two subsamples—people on the fire emergency organization and those who were not—but space permits mention of only two. The first is that, of those on the organization, 22 per cent reported that they always treat an alarm as a real emergency; of those not on the organization, only 6 per cent gave this response. The second comparison raises a question about the fire safety awareness and knowledge of those on the organization; their responses to the question on elevator hazards in fire were similar to those of other respondents. Many of the respondents knew very little about the hazards of elevator use in fires.

On the question about awareness of reasons behind the selective or phased evacuation procedure there was again evidence of inadequate understanding by people, particularly those on the fire emergency organization, of the reasons for the procedure. Of the 64 per cent of the respondents replying that they knew of a reason for the procedure, most mentioned the avoidance of congestion in the exits—with panic often suggested as a consequence—but only a small minority mentioned that potentially hazardous areas were given priority in the evacuation sequence.

Although responses on speed of descent and experience of congestion in the exits showed good agreement with our observations on the speeds and congestion experienced by evacuees from certain floors (such as floors 4, 11-14, 18, and 21) this was not the case with the question on crowding. The fact that responses, indicating the stairs were much too crowded, did not correlate with floor groups that actually experienced high densities, suggest that the perception of crowding may be more person-related than situation-related. Admittedly our sample was very small;

we mention these findings simply to suggest areas for further study. Incidentally, floor groups that experienced high densities for part of their evacuation had the equivalent of between one and two treads of area per person (in other words, a density of 2.3 p/m^2). At the lower levels of the south exit—the more crowded one—there was an average of two to three treads per person (that is, a density of 1.4 p/m^2—very similar to the mean density observed in total evacuations using the traditional, non-selective procedure).

About half of the respondents (51 per cent) reported using the exit stairs in building B at least once each day. The upper and lower halves of the building differed significantly with 30 per cent and 72 per cent respectively reporting this frequency of normal use. Other cross-tabulations revealed that 59 per cent of the men and 40 per cent of the women reported this extent of normal use, and 58 per cent of those under 40 and 45 per cent of those over 40 reported this frequency of use. Because of the importance of normal stair use as a factor increasing both the awareness of exits and the competence in using them in an emergency evacuation, and because of implications for physical fitness of office workers as well as possible benefits to building economics, this building and two others in Ottawa were extensively studied early in 1977 to document normal stair use. At this time normal stair use was found from observations, questionnaires, and logs to average four uses per day per office worker in building B. Much higher use was found in lower office buildings (Johnson and Pauls, 1977).

Lessons learned in the case study of building B

Much could be said about what was learned from this study, but space permits only brief comments to indicate some conclusions.

First of all it must be recognized that what happened in building B is unlikely to happen now in government-occupied buildings in Ottawa. Partly as a result of the bad experience with building B, the Dominion Fire Commissioner instituted rigorous training programmes and testing procedures. As a result, ambiguous announcements and improper use of telephone and public address systems should be less likely now than was the case during the early 1970s when the selective procedure was introduced.

It must be stressed, however, that such attention to training of fire emergency staff and to periodic drills is the exception and not the rule in most other high office buildings! What happened in building B in 1971 could easily happen in other buildings now. Therefore awareness of the 1971 experience is hardly academic or irrelevant today.

As an example of one specific suggestion on the selective procedure, we propose that, instead of having someone at the central control console determining when a floor in the sequence should begin evacuating, this can be done better under local control, with the warden of the floor where evacuation has been completed simply informing the next floor directly so that its evacuation is neither premature nor late. Supervisory personnel at central control can monitor this activity and make public address announcements to floors where the evacuation is about to begin. Our observations in selective, partial evacuations in other office buildings show clearly that up to one-half of the total time needed to partially evacuate a building using a sequential procedure controlled from a central control console is taken up with cumbersome communication and central decision-making, (see, for example, the small case studies given by Pauls, 1977a, b, and 1978).

Based on our research experience with this drill and others where communication systems are used, we recommend automatic actuation of recording devices so that all messages carried on public address and emergency telephone systems can be subsequently played back with accurate preservation of time-scales. This documentation, in conjunction with use of post-event interviews and questionnaires, will do much to improve our understanding of emergencies such as major fires in large buildings.

CONCLUDING REMARKS

In this paper we have attempted to show by way of examples that there is much that can be learned about evacuation of large buildings. Although we did not always stress it, it should be clear that the methods used for research into evacuation are readily accessible to people not engaged in full-time research. Considering what is spent to provide safety systems, train building fire emergency or security staff, and hold periodic drills, there is much to be gained by collecting some information on what actually happens when the alarm sounds. For example, in the two drills described, there was an expenditure of several tens of thousands of dollars covering working time lost simply during egress and re-entry. This was by no means the entire cost. Against such costs should be compared the fire safety benefits that might be gained. Whether sufficient benefits actually are gained from the efforts taken in fire safety programmes remains a question that will require much additional research and evaluation effort. We hope one outcome of this paper's publication will be that such additional work is achieved.

REFERENCES

Best, R. L. (1977). *Reconstruction of a Tragedy: the Beverly Hills Supper Club Fire.* (Boston: National Fire Protection Association).

Dominion Fire Commissioner (1972). 'Fire emergency organizations in federal government occupied buildings'. Technical Information Bulletin No. 22. (Ottawa: Canada Department of Public Works).

Johnson, B. M. and Pauls, J. L. (1977). 'Study of normal stair use'. In R. J. Beck *Health Impacts of the Use, Evaluation and Design of Stairways in Office Buildings.* (Ottawa: Health and Welfare Canada, Health Programs Branch).

Pauls, J. L. (1971). *Evacuation Drill Held in the B.C. Hydro Building 26 June 1969.* National Research Council of Canada, Division of Building Research, Building Research Note No. 80.

Pauls, J. L. (1977a). 'Movement of people in building evacuations'. In D. J. Conway (ed.) *Human Response to Tall Buildings.* (Stroudsburg Pa.: Dowden, Hutchinson and Ross, Inc.), pp. 281–292.

Pauls, J. L. (1977b). 'Management and movement of building occupants in emergencies'. *Proceedings of Second Conference, Designing to Survive Severe Hazards.* (Chicago: IIT Research Institute), pp. 103–130.

Pauls, J. L. (1978). 'Movement of people in building evacuations', *Canadian Architect,* **May** (also in *Buildings,* **May**).

Pauls, J. L. (1979a). *Use of a Questionnaire After a Major Evacuation Drill in a High-rise Office Building.* National Research Council of Canada, Division of Building Research, Building Research Note No. 141.

APPENDIX A

Extract from a detailed, six-page set of instructions given to observers assigned to move with evacuees in evacuation drills in tall office buildings studied in Ottawa, Canada. The checklist includes observation tasks usually carried out only in selective evacuations (shown in parentheses).

(1) Record name, floor and exit on tape cassette before drill.

(2) Go to your assigned floor about five minutes before drill.

(3) Start recording when alarm sounds, noting any delay.

(4) Record a reference time read from your watch.

(5) Keep recording continuously throughout the drill.

(6) Note what happens in the vicinity of your assigned exit.

(7) (Record a count of people descending past your assigned floor.)

(8) (Note clarity of public address announcements and evacuee responses.)

(9) (Record a count of people entering exit at your floor level.)

(10) Try to be one of the last in the main group leaving your floor.

(11) Record each floor level immediately as it is passed.

(12) Note people entering exit, open doors, etc. at other floors.

(13) Note how people are spaced on the stair.

(14) Count and report, at each storey, the number of persons per flight.

(15) Note any changes in your rate of movement, particularly stops.

(16) (Note clarity of, and reactions to, public address announcements in the exit.)

(17) Report to observer at bottom of exit, giving your origin floor.

(18) Record a reference time from your watch once outside the building.

(19) Make additional comments, particularly general impressions.

(20) Do not rewind your cassette; return it as is to J. L. Pauls (NRC).

A sample log, made from one observer's recording in an evacuation in building A, follows (Table 13.4) (See also the first case study.)

Table 13.4 Sample evacuation drill observation log (DBR/NRC/Canada)

| Bldg: Centennial | Floor: 15 | Exit: West | Observer: Staff | Date: 72/10/26 |

| Time (hr:min:sec) | | | | | Observations |
Correct	Reference	Playback	Floor	Count	Comments
0.00					(Bell alarm starts)
		0.00	15		Recorder running
	11:29:45	0.04			11.29.45.
		0.14		1	(First person into exit)
		0.15		2	
		0.20		3	
		0.25		6	
		0.40		8	
		0.45		9	
		0.50		12	
		0.55		13	
0.59	00:00:59	0.58			Bell stops briefly
		1.00		14	
		1.05		18	
		1.10		19	
		1.15		20	
		1.20		22	
		1.30		23	
1.45		1.43	15		One or two people still on floor as observer enters exit stair
				12	(12 persons on flight)
2.06		2.04	14	8	Speed very slow
2.36		2.34	13	10	
2.50		2.48	13/12		Stopped between floors 13 and 12
3.16		3.13	13/12		Moving again
3.30		3.27	12	18	Slow No one entering from floor 12
4.03		3.59	12/11		Stopped again
4.21		4.16	12/11		Moving again, quickly
4.23		4.18	11	8	
4.42		4.36	10	8	
4.48		4.42			Moving quickly
5.02		4.56	9	15	Stopped
5.14		5.08	9		Moving again
5.31		5.24	8	12	
5.50		5.42	7	10	
6.09		6.00	6	10	
6.27		6.18	5	12	Moving steadily
6.49		6.40	4	12	
7.09		6.59	3	10	
7.25		7.15	2	10	Moving steadily and quickly
7.46	7:46	7.35	1		Passed observer at exit discharge
7.54		7.43			(evacuee 372). Through doorway
		8.10			Most people across the street
8.54		8.43			Observer now across street
	11:38:45	8.53			11.38.45.
					General comment: because alarm stopped, people were confused; some tried to go back up stair
		9.20			Recording ends

Note: Recorder playback does not necessarily provide accurate time-scale. Reference times must be used to make corrections.

Fires and Human Behaviour
Edited by D. Canter
© 1980 John Wiley & Sons Ltd.

CHAPTER 14

Building Evacuation: Research Findings and Recommendations

JAKE L. PAULS
National Research Council Canada
Division of Building Research, Ottawa

INTRODUCTION

This paper presents a number of studies, carried out over the past ten years, into evacuations from high-rise office buildings in Canada. It focuses on two of the most important aspects of building evacuation. The first is the relationship between the rate of flow of people leaving a building and the width of stairs down which they walk. The second is the overall time it takes to evacuate a building.

These two aspects of high-rise evacuation form the cornerstone of building regulations which deal with means of escape. Yet prior to the research reported below there have been few empirical studies against which these regulations could be evaluated. Those studies which have been carried out have frequently been either inappropriate, or the conclusions drawn from them have been invalid. A review of the literature on building evacuation and related topics is given elsewhere (Pauls, 1979a).

CONFUSIONS IN PREVIOUS REPORTS

Two major confusions have occurred in earlier research and the recommendations derived from it. The confusion of 'experiemental' flow conditions with normal use, and the confusion of sustained mean flows with peak flow.

In the experimental, or test, situation under controlled artificial conditions, tests are run using a small group of selected individuals who temporarily disregard normal conventions of behaviour, sacrificing personal space and perhaps even normal concern for safety to achieve very high flows at very high densities accompanied by high speeds. Such tests are reported in the literature (National Bureau of Standards, 1935; Joint Committee, 1952) and have been done as part of the Canadian study of crowd movement. Figure 14.1 illustrates such a test on a 1120 mm (44 in.) stair, a typical exit stair, where a small group of office workers briefly achieved a flow of 1.5 persons per second (p/s) or 1.85 persons per second per metre of effective stair width (p/s m. eff) as discussed later, the same rate as the traditional rule suggests for mean, design flow.

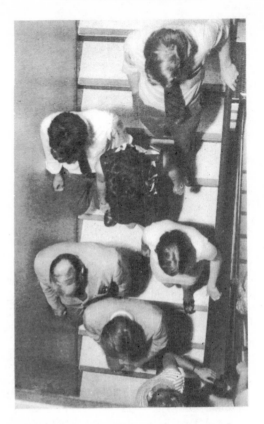

Figure 14.1 Arrangement of people on a
1120 mm (44 in.) stair during a test of maximum flow

Such test results must be treated with extreme caution as indeed must the tests
themselves. The former *caveat*—often unheeded—is not original to this paper. Note
for example the excerpt from the 1952 report of the Joint Committee on Fire
Grading of Buildings, a group that did not go far enough in discounting the test results mentioned here:

> It is obvious from the results of the French tests that the urgency
> motive is a main factor causing a variation in the rate of discharge. It
> will be noted that, for normal speeds, the results of these tests are com-
> parable with those obtained in the various other series of tests, but that
> extraordinarily high figures were obtained when the firemen carrying
> out the tests, were deliberately hurrying and pushing to achieve max-
> imum rate of discharge (Tests 21, 27 and 32). Obviously these high
> figures, while interesting for comparison purposes, cannot be applied
> to actual fire conditions, as an undisciplined crowd, discharging in con-
> ditions which might be less favourable than those of the tests, could not

be expected to attain the high discharge rates reached by a body of disciplined men. It is also worthy of note that in one of these high speed tests (No. 13) a mishap occurred due to some of the men falling on the stairs. This mishap slowed down the movement and gave a figure for discharge rate comparable with that for normal speed. The occurrence of such an incident provides a further reason for discounting the discharge figures for high speed tests.

The error of misapplying test findings to other situations such as evacuation was not confined to the National Bureau of Standards 1935 report. Other papers, referencing the 1935 report, have failed to draw a distinction between evacuation and test conditions. An additional, related error made in the literature is to confuse mean flows with peak or maximum flows, the subject discussed next.

As regard the confusion of flow rates too often in the past, flows that are actually only briefly sustained maxima have been used as a basis for design or for performance prediction. Figures for maximum flow, particularly the high figures obtainable in test situations, are harmless only if reported in the research literature with the necessary qualifications. However they should not be quoted indiscriminately, nor should they be used in codes and handbooks for design.

As an example of how badly understood is the distinction between peak or maximum flows and sustained mean flows, consider the following. Literature referencing the London Transport Board (1958) report on commuter movement in the London Underground often   quote a descent flow of 21 persons per minute per foot width of stair. It is quoted as a mean or design flow—with reference to evacuation—even though the original London Transport Board report calls it 'maximum flow', 'practical maximum flow', 'maximum working flow', 'average maximum flow', and notes that it is 'less than 21 persons per minute per foot'.

Looking at the London Transport data the figure is 20.3 persons per minute per foot width. The 20.3 figure is the average of the ten *highest* flows observed on ten stairs during *high* flow conditions.

A more direct indication of the weaknesses of existing assumptions about evacuations is provided by comparing actual flows observed in recent Canadian studies with the traditionally assumed flow. As illustrated in Figure 14.2, the average assumed flow rate was never observed.

Figure 14.2 is based upon the observations of 58 total evacuations in high-rise office buildings. A case study example of what is meant by the term total evacuation, and the details of the procedures used in these studies is given in chapter 13. Two types of empirical evidence were drawn from these observations to reach conclusions on stair width and flow. The one described first is a statistical regression approach in which mean flow data are plotted against stair width. The second is an analysis of crowd configurations on various widths of stairs. It should be understood that stairs were being used at capacity levels; in other words there was queueing at stair entry points.

Figure 14.3 shows an early analysis of raw data for flow plotted against measured stair width in 58 cases. The 'measured stair width' is the wall-to-wall clearance if a

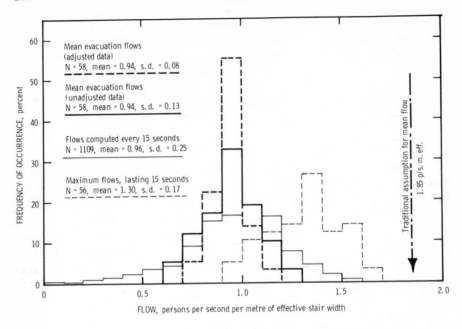

Figure 14.2 Flows in total evacuations of office buildings

flight of stairs is bounded by walls on each side. It is the wall-to-handrail clearance plus 100 mm (4 in.) for more common types of stair with a wall bounding only one side of a flight. Measured stair widths included in the 58-case sample were 910 mm, 1070mm, 1120 mm, 1140 mm, 1190 mm, 1220 mm, 1420 mm, and 1520 mm (36, 42, 44, 45, 47, 48, 56, and 60 in.). Additional widths studied subsequently in other crowd movement observations follow essentially the same pattern as is described for these widths.

The shaded areas in Figure 14.3 indicate the scatter of flow data; specifically plus and minus one standard deviation from the mean for each width is shown. A linear regression line, with flow proportional to measured stair width, is a reasonable first approximation of the relation using unadjusted, raw flow data. In fact in early reports on the evacuation drill study the relation was described as one of simple proportionality between mean evacuation flow and stair width (Henning and Pauls, 1974; Pauls, 1974, 1975a).

The traditional step-function relation used in building codes to relate exit widths and evacuation population—indirectly evacuation flow—is based on the highly unlikely step-function regression also shown in Figure 14.4. Comparing this with the proportional regressions suggested by the evacuation data it will be noted that there are marked differences in shape and location. These differences are discussed further elsewhere along with recommendations for stair width based on analyses on handrail graspability and other factors (Pauls, 1979b, 1979c).

Further examination of the data suggested that other linear regressions could be drawn. With raw data scattered somewhat a linear regression line intercepting the

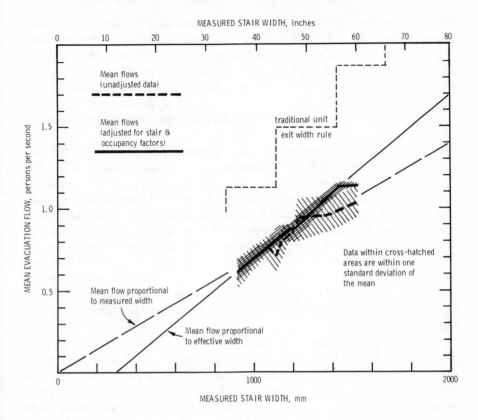

Figure 14.3 Observed relation between mean flow and stair width in 58 cases of total evacuation

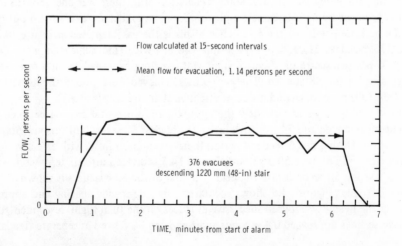

Figure 14.4 Flow from exit stair during total evacuation

width axis at about 300 mm (12 in.) also was a possibility. This in fact was the best regression once the possible influences of stair and occupancy factors were taken into account. Here was some evidence for a concept of effective stair width—the measured stair width minus a fixed width of 300 mm (12 in.). This concept evolved from the reanalysis, in early 1976, of evacuation drill data collected between 1970 and 1974. Figure 14.3 shows the results of this reanalysis along with the original analysis. The later analysis supported the concept of proportionality between flow and *effective* stair width.

Careful study of the literature subsequently revealed precedents for the 'effective-width' approach. A 1958 report of crowd movement studies by the London Transport Board concluded that for level passageways flow is proportional to width. An examination of the report's graph, with flow data plotted, suggests however that a regression line should intercept the width axis at about 300 mm (12 in.) and not at zero as drawn. In the same report a graph for ascending flow plotted against stair width has a regression with zero flow at zero width; however, it appears that the stair widths were reported in an unconventional way as the clear width between handrails—that is, at least 175 mm (7 in.) less than the usually reported measured width (for example wall to wall). In other words, the London Transport Board reported data that supported an effective width model but then failed to draw the appropriate conclusions.

Fruin (1971) recommends, 'the effective width of corridors must be reduced by 18 inches on each corridor side, to take into account the human propensity to maintain this separation from stationary objects and walls except under the most crowded conditions'. Fruin did not extend this observation to include crowd movement on stairs; however it seems reasonable that a similar behaviour would be the case on stairs, including crowded stairs. This was the hypothesis tested in subsequent studies by the author using video and film records of crowd movement on stairs observed in Canada between 1971 and 1977.

During the study of office building evacuation drills there was one opportunity, in 1971, to set up a video camera and recorder in an exit stair—as it turned out one of the best-used stairs of the 58 studies. (This is the case depicted in Figure 14.4.) The 376 evacuees descending a 1220 mm (48 in.) wide stair achieved a mean flow of 1.24 p/s per metre of effective stair width, a flow exceeded in only one other evacuation of the 58 studied. The peak flows achieved were among the highest recorded—1.60 persons/second metre of effective stair width (p/s m eff).

Figure 14.5 shows in plan view the approximate horizontal spacing between 83 evacuees (nine men and 74 women) during the busiest 60 seconds of this videotaped evacuation. During this 60-second period there was a mean flow, with effective stair width factored out, of 1.50 p/s m off. Figure 14.5 is divided into six ten-second segments starting 50 seconds into the drill. The first of the 83 evacuees is shown at the lower left side. Figure 14.5 shows accurately the side-to-side spacing and approximately the headway between successive evacuees on the final flight descended. The transverse lines are spaced 0.4 seconds apart and can be viewed as separate treads of the stair.

Even though the stair is 100 mm (4 in.) wider than a standard 1120 mm (44 in.),

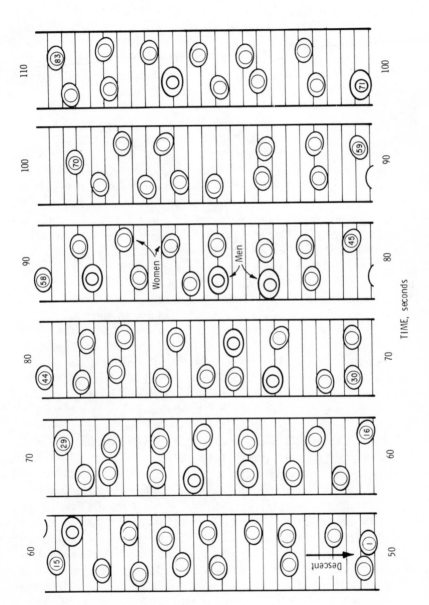

Figure 14.5 Relative locations of evacuees on 1220 mm (48 in.) stair

two-unit stair (in code terminology), there is only a small incidence of two-abreast, shoulder-to-shoulder configuration. The general pattern is better described as two staggered files or overlapping lanes, one at each side where evacuees could easily grasp handrails. About 85 per cent of the people used a handrail, a not unexpected figure in view of the stair's long flights, steepness and undersized treads.

Although almost everybody stayed at the sides of this stair, only rarely did a person's shoulder make contact with a wall. Such contact might be expected due to body sway, a phenomenon described by Fruin (1971) and discussed elsewhere by Pauls (1979c). Further analysis, focusing particularly on the distribution of people across this and other stairs, seemed useful and therefore Figure 14.6 was drawn up.

Figure 14.6 shows three pairs of distributions of people across the widths of three stairs each having extensive, capacity-level unidirectional flows. Figures 14.6(a), (b), and (c) show respectively 1220 mm (48 in.), 1680 mm (66 in.) and 2240 mm (88 in.) wide stairs all of which were filmed or videotaped during comparable conditions. Figure 14.6(a) is derived from the videotape used to produce Figure 14.5. Even though 14.6(b) and 14.6(c) are from crowd movement studies in public-assembly occupancies, that is, not evacuation drills in tall office buildings, it is appropriate to consider them here to illustrate the concept of effective stair width.

The sharply peaked distributions are obtained by plotting the location of the centre of each person's body—specifically the head—as it passes through an observation plane created by the video or film record. The lower flatter distributions, for which a separate scale is shown at the right side of the figure, are obtained by adding 250 mm (10 in.) to each side of the body-centre position. In other words a body envelope or shoulder width of 500 mm (20 in.) is assumed. (This is the shoulder width of a 50th-percentile US adult with medium-thickness outdoor clothing.) The resulting distribution shows to what extent the width of each stair is used.

The important thing to note about the sharply peaked distributions is that there is a pronounced 'lane effect' only at each side of a stair where there is a handrail attached to a stair wall. There are not three equally pronounced peaks across the 1680 mm (66 in. or three-unit) stair and there are not four equally pronounced peaks across the 2240 mm (88 in. or four-unit) stair. This provides further evidence for not using the unit–exit–width concept (with 560 mm or 22 in. units) in crediting various stair widths for carrying capacity.

From the sharply peaked distributions alone one might conclude that not all parts of a stair width are equally used. The limited extent to which this is correct is suggested by the accompanying lower distributions. They show that, due to over-lapping of shoulders, there is no major dip at the centre region of any of the stairs. There is, however, a rapid falling-off at a consistent distance from the walls of each stair. In other words, there is a boundary effect, a low incidence of a person's shoulder coming within 100–200 mm (4–8 in.) of a wall. The entire width of a stair is not equally used. Thus we have additional evidence to be used in conjunction with what can be learned from statistical regressions of data such as shown in Figure 14.3

The effective width of a stair is the measured width minus 300 mm (12 in.). For

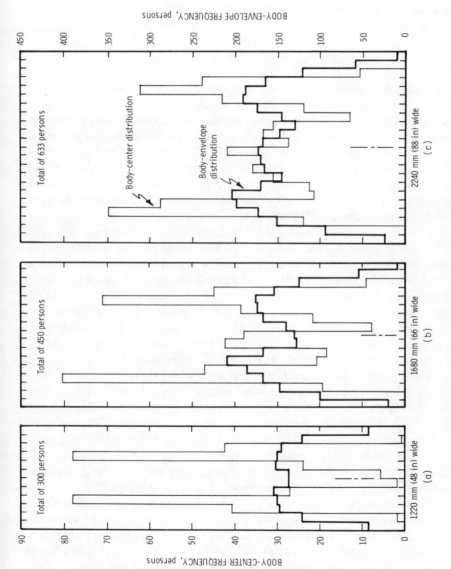

Figure 14.6 Distributions of people across intensively used stairs

example, the stair described in Figure 14.6a has a measured width (wall-to-wall) of 1220 mm (48 in.) and an effective width of 920 mm (36 in.). Further analysis will provide a better basis for determining what the reduction of measured width should be. For example, rough-textured walls enclosing a stair—a not uncommon feature of exits—or a precarious-looking open well in the middle of a stairwell might mean that a figure 50 mm (2 in.) or 100 mm (4 in.) greater than the 300 mm (12 in.) proposed should be used.

Use of the effective width model

Readers confused earlier by the somewhat complicated expression, 'per metre of effective stair width' (abbreviated /m.eff.), should now understand its derivation and utility. Evacuation flows and other variables on various widths of stairs can now be related in a realistic way. For example, flow is noted as 'persons per second per metre of effective stair width' (p/s.m.eff). The number of persons using an exit stair during an evacuation is described as 'persons per metre of effective stair width' (p/m.eff.). This population variable is considered in the next section of the paper following a slight digression to point out how the effective width model can be used to explain a finding of an earlier investigation of crowd flow.

The London Transport Board (1958) studied the descent-flow on a stair 1830 mm (72 in.) in width—between handrails—used by crowds of London Underground commuters. Flow measurements were made before and after installation of a central handrail. Their observations showed a reduction in the estimated mean flow, from 130 persons to 105 persons per minute, after installation of the handrail. A simple effective width model would have predicted this reduction quite accurately. The total effective width available on the stair was changed from 1730 mm (68 in.) down to 1430 mm (56 in.) for a ratio of 130 to 107—almost exactly the ratio of flows quoted. Note that the flow per effective width remained virtually unchanged at 1.24 p/s.m.eff.—a mean flow figure similar to what was achieved in the best-used exit stairs in office building evacuations and far below the traditionally assumed mean flow figure discussed earlier. Here it should be noted that, based on the author's own informal observations of the London Underground, the results of crowd movement measurements made there should only be applied with the greatest of care to other contexts, even public transit facilities in other places. London Underground users appear to have a high coping ability and can achieve flows and speeds that might warrant honourable mention in an athletic competition.

It should also be noted that there are some inadequacies in the London Transport report that make it difficult to really test the effective width model in any but an approximate way. For example, the data are not sufficient to quote mean flow figures with great confidence. The author's own examination of the data suggests that the before and after mean flows should be closer to 127 and 94 persons per minute with standard deviations of six and thirteen persons per minute respectively. Moreover, as will be clearer a little later in the paper, the number of persons involved, or some measure of queueing, and other factors, would have to be specified before a proper comparison could be made. The population factor could have an effect on the flows and this leads us right into the next section.

OTHER EVACUATION FACTORS POSSIBLY AFFECTING FLOW

Evacuation population and flow

In an attempt to provide possible explanations for some of the scatter or variability of flow data (Figure 14.3) a large number of possible relations between flow and other variables were explored. Figure 14.7 shows one of many graphs drawn up during the exploration. It shows flow data, in persons per second per metre of effective stair width (p/s.m.eff.), plotted against evacuation population per metre of effective stair width (p/m.eff.). An additional variable describing the evacuees' use or non-use of outdoor clothing during the evacuation is also shown and will be discussed below.

Figure 14.7 shows that population does have an effect on evacuation flow and that this effect is most pronounced for small populations. To obtain a regression curve a separate graph of the same data was drawn up using logarithmic scales. This graph had a straight line regression which translated to the curvilinear regression equation:

$$f = 0.206p^{0.27}$$

where f is the evacuation flow in persons per second per metre of effective stair width, and p is the evacuation population per metre of effective stair width.

If one wanted an equation giving the total flow on a certain width of stair the above equation would simply be multiplied by the effective stair width. The equation can

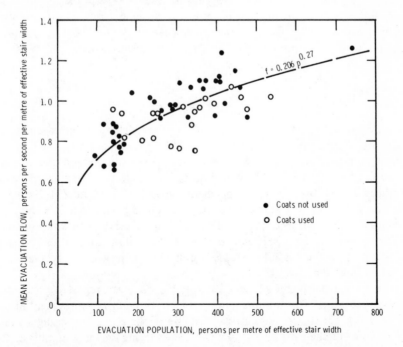

Figure 14.7 Effect of evacuation population on flow down stairs

also be changed so that the total population using the stair is considered. With these changes, and using the measured width of a stair, the equation becomes:

$$F = 0.206(W - 0.3)\left[\left(\frac{P}{W - 0.3}\right)^{0.27}\right]$$

where F is the flow in persons per second, P is the evacuation population, and W is the measured width of the stair in metres.

The equations are only proposed for populations of less than 800 persons per metre of effective stair width. In the absence of data, it is assumed for the time being that flows at higher populations will be constant at approximately 1.25 p/s.m.eff. Putting this in terms which may be more familiar to readers, a typical 1120 mm (44 in.) stair used by this high population will have a mean flow of one person per second (p/s).

The derived population-flow relation can be used to factor out the effect that population appeared to have on the evacuation flows measured. This is done by multiplying the observed mean flow by the 0.27th power of the ratio obtained by dividing a chosen base population by the observed population, with both stated in persons per metre of effective stair width. A base population used in the study was 300 p/m.eff.—a figure very close to the mean population for all 58 cases of evacuation observed.

This calculation can be illustrated as follows using the evacuation depicted in Figure 14.4. The population using this 1220 mm (48 in.) stair, with an effective width of 920 mm (36 in.), was 376 persons total or 412 p/m.eff. The observed mean flow was 1.14 p/s or 1.24 p/s.m.eff. If the population had been only 300 p/m.eff., and all other factors remained unchanged, a flow of 1.05 p/s or 1.14 p/s.m.eff. would be expected, based on the following calculation:

$$1.24\left(\frac{300}{412}\right)^{0.27} = 1.14$$

From this we can see that the apparent effect of the higher than average population (higher by 37 per cent) was to make the flow higher by 9 per cent.

Doing the calculation for all 58 cases of observed evacuations and then calculating the mean of the 58 adjusted flows had the following effects. The mean value of the 58 adjusted mean flows was 0.97 p/s.m.eff. (up from 0.94 p/s.m.eff.). The standard deviation dropped to 0.10 p/s.m.eff. (down from 0.13 p/s.m.eff.). This reduction in standard deviation begins to tell us something about the extent to which population differences alone result in scatter or variability of raw flow data. Although the work is not conclusive there is at least some basis here for improving the accuracy of prediction of flows on stairs.

It is not clear why the effect occurs. It has not been reported previously by other investigators. The author's more recent work on crowd movement in public assembly occupancies has also shown that size of the population using a stair can explain some of the variability in observed flows (using the same formula as shown above). Population has an effect that might be likened to pressure in a hydraulic model. In human terms it can be related to what some have called the urgency factor. It has something to do with the extent of queueing—the number of people waiting to use a facility. More study of the factor is warranted. Some has already been started by the author, examining in greater detail the flow as it varies over the course of a single mass evacuation whether it be in an evacuation drill or in a normal egress from a stadium.

Evacuees' use of coats and flow

This factor was one of the first isolated as having some effect on evacuation flows. Study data have always been tabulated within two classes: where some or most evacuees do wear or carry clothing for extra protection from wet or cold conditions outside a building (twenty cases), and where none of the evacuees does so (38 cases). Admittedly this factor is not well defined; it should be defined in quantitative terms describing the number of evacuees using certain types of clothing.

Mean evacuation flows in the twenty cases where coats were used were six per cent lower on average than were the flows in 38 cases where coats were not used. Also, on average, the onset of flow at the beginning of evacuation was delayed by approximately 15 seconds when coats were used.

Only one building evacuation was observed during the coldest part of a Canadian winter when boots, requiring some time to put on, or galoshes are *de rigueur* when outdoors. In this evacuation, down two stairs, flows were 6 to 10 per cent lower than expected even after consideration had been given to the apparent effects of many stair and occupancy factors including 6 per cent when coats were used. Here, as in many other cases, opposing factors may have been operating at the same time. One could have the effect of increasing flow (for example, there might have been a greater sense of urgency because an alarm during midwinter is less likely to be interpreted as being for a drill). Another factor might have the effect of reducing flow (for example, because extra bulky clothing and footwear are awkward on stairs). Generally this case illustrates the difficulty of analysis when there are many poorly defined variables operating simultaneously.

Exit discharge through main lobby and flow

There are several reasons why a stair that discharges through a main lobby at ground level might be expected to have a higher evacuation population and a higher flow. Such a stair may be more visible from normal circulation routes at each floor level. Because such stairs have a higher likelihood of being used for normal, everyday circulation their use will seem to be the natural thing to do in an evacuation. Evacuees desiring not to leave the building unless absolutely necessary can seek shelter in the main lobby at ground level once they have descended the exits. Moreover, the lobby is a good location for learning something about the reason for an evacuation because this is where building management and fire brigade personnel can often be found in an emergency.

These and other reasons might lie behind the observed differences in populations choosing to use a lobby stair, as opposed to a non-lobby stair in the observed evacuations. In a subsample of ten pairs of similar stairs in ten buildings the one lobby stair of each pair consistently had a higher evacuation population. On average the ten lobby stairs had 38 per cent higher populations than their non-lobby counterparts. As might be expected lobby stairs also had higher evacuation flows; in fact there was an average 9 per cent difference in flows between the ten pairs of stairs —exactly the difference expected in view of the population effect discussed earlier (that is, $1.38^{0.27} = 1.09$).

From the foregoing discussion, of the greater preference for and therefore greater effectiveness of lobby stairs, it should be evident that not all stairs in a building are equally useful. This fact should be kept in mind during building design and during subsequent use of a building. For example, lobby stairs should be given greater design attention, be wider and be credited with greater exit capacity. During a fire a lobby stair would be the logical one for evacuating occupants, while a non-lobby stair would be a better one for fire department operations that could block a stair either with men and equipment or with smoke.

Normal stair use and flow

There is considerable opinion and some anecdotal evidence that normal, everyday use of an exit is a positive factor in fire emergencies. Normal use of stairs is therefore worthy of study. Such everyday use can be substantial even in tall office buildings. A recent study found normal use averaging four stair trips per person each day in a 21-storey office building in Ottawa, and even several times greater use in lower office buildings (Johnson and Pauls, 1977).

Without the benefit of such a study for the 58 stairs in which total evacuations were observed, only a rough estimate could be made about normal stair use. Normal stair use was estimated to be:

very frequent	in 6 stairs
frequent	in 14 stairs
medium	in 25 stairs
infrequent	in 9 stairs
very infrequent	in 4 stairs.

This estimate was based on stair condition, location, tenancy, etc. After taking into account the possible effects of other factors, such as population, the cases with infrequent or very infrequent use were compared with cases having medium to very frequent use. Means flows in the thirteen cases with infrequent use averaged 5 per cent lower than those in the other 45 cases. Within four buildings having stairs in each of two groups compared above, the stairs with more frequent normal use carried evacuation populations (per effective stair width) that were 64 per cent larger than in the infrequently used stairs. Thus the difference in population alone would lead one to expect an additional 14 per cent in mean evacuation flow (that is, $1.64^{0.27} = 1.14$).

Discipline or organization and flow

Compared with 44 other cases the fourteen buildings with apparently better fire emergency organizations and better evacuation discipline or training had, on the average, a slightly higher evacuation flow. This is obviously an area where additional research is required to define properly the factors and to determine what correlations there are with performance factors. Of interest is the extent of benefit from periodic training, from supervision during evacuations, and from evacuees having an *esprit de corps*. Related factors include length of time a person has occupied a building, personality, and job status.

Stairwall roughness and flow

Twenty-one stairs with rough or semi-rough textures on walls had mean evacuation flows averaging 7 per cent lower than fourteen smooth-walled stairs. As noted earlier in the discussion of effective stair width this effect might also be taken into account by using a different effective width in relation to measured stair width. One might have to take as much as 400 mm (16 in.) off the measured width to determine the effective width of a stair bounded by rough walls.

Tread dimensions and flow

Stair use observations, particularly those made in greater detail with video records, clearly indicate the importance of adequately large treads. For example, it was determined that 45 per cent of the evacuees descending a particularly steep exit stair, with 230 mm (9 in.) treads, adopted a noticeable twist in body posture or a crabwise gait in order to get adequate footing on the undersized treads. (The stair also had an unusually high level of handrail use; 85 per cent of the evacuees used a handrail.) Note that building codes generally require that exit stair treads have a run, or depth not including a nosing projection, of at least 250 mm (10 in.).

To check what effect tread dimension might have on flow in evacuations the mean flow data, adjusted for possible effects of other factors, were compared between two groups of stairs. Those having treads with runs of less than 250 mm (10 in.) carried evacuation flows averaging about 4 per cent lower than flows on stairs with larger treads.

Unfortunately none of the 58 stairs encountered in the study of evacuation drills in tall office buildings had tread dimensions exceeding 280 mm (11 in.). Thus, aside from providing not-unexpected data regarding movement behaviour on undersized treads, the study findings on flow have little to say about the benefits to flow of larger tread sizes. Other research, however, has shown that the larger treads are justified on grounds of safety and comfort; 300 mm (12 in.) or more should be provided, especially in situations of descending movement—the important direction of travel in most exits (Fitch, Templer, and Corcoran, 1974; Fruin, 1971; Templer, 1974).

Building height and flow

An early study of exit design suggested that 'the rate of movement may tend to diminish after several flights have been traversed', and that 'to allow for this factor it is recommended that for every 10 ft of height in excess of 20 ft above the ground 8 per cent should be added to the calculated number of persons who will use a staircase' to calculate stair width (Joint Committee, 1952).

To check whether observed cases of evacuation in taller buildings had lower flows, as assumed in the 1952 study, 47 flow data, corrected for population differences, were plotted against height traversed by evacuees. There was a small falling trend in flow as height increased. Flow from stairs 54 m (177 ft) high was about 9 per cent lower than for stairs 30 m (98 ft) high. On the basis of this evidence the correction suggested in the 1952 report can be considered excessive if not groundless (because no evidence supporting the contention was given in the 1952 report, see Chapter 13 for additional information on this subject).

Other factors and flow

There is much scope for further speculation and study of how design and occupancy factors might relate to flow or other evacuation variables. Other factors that were considered in the analysis of data from the 58 cases included handrail provision, headroom, length of flights, riser height, direction of stair turn, tread surface, participation of handicapped persons in the evacuations, and any forewarning known to have been given in some of the drills. Incidentally, the last two factors mentioned appeared to have no effect on mean flows. When handicapped people reduced the flow in their immediate vicinity it was countered by increased flow later.

Although the work will be far from straightforward, additional study of existing data from the 58 cases of total evacuation as well as additional data collection in drills should be done to define these factors and suggest relations. Also, a critical search of the literature, particularly literature now beginning to appear describing extensive recent US studies of stair safety, should provide greater insights on some factors (Templer *et al.*, 1978).

Additional study may justify fine-tuning of exit systems using findings about the many factors suggested above. Such fine-tuning is not very important at the current stage of development of egress technology; after all we have been designing exits and predicting egress performance for decades with discrepancies of 100 per cent between our assumptions and what actually happens in real life.

TOTAL EVACUATION TIME OF OFFICE BUILDINGS

If one wants to predict evacuation time accurately it is not a good idea to simply divide the building population by some flow figure. The resulting answer might be correct only in a dimensional sense. To obtain reasonably accurate predictions one should at least consider the exact type of procedure used in the evacuation and the total number of people actually involved. To simplify matters, only the total evacuation procedure will be considered in the following examination of evacuation time of tall office buildings. This will permit comparisons to be made with models developed by other investigators.

In total evacuations all building occupants will attempt, in theory at least, to begin evacuating at the same time (upon perceiving an alarm or receiving an order and deciding that evacuation is the appropriate thing to do). One might therefore ask how long it takes before exits will be operating at their design or mean flow levels. Figure 14.8 provides this information for a representative sample of the 58 cases of total evacuation observed. It shows the growth of flow from exit stairs during 120 seconds elapsing after alarms begin sounding. Evacuations in which people wear or carry outdoor clothing for protection against inclement weather show both a minor delay and a somewhat lower flow—both discussed above.

In Figure 14.8 is shown the time required for flow to build up to one-half of the mean value. This time averaged 41 seconds or 0.68 minute for the no-coats evacuations observed. It is this figure that has been used along with considerations of mean flow to prepare a graph predicting what the total evacuation time will be for a stair having a certain effective width used by a certain number of people.

Figure 14.9 shows the prediction curve for total evacuation time having the equation obtained as follows, using the regression equation for flow as a function of population. The time during which evacuation flow occurs is simply the population per effective width divided by the flow per effective width, that is,

$$\frac{p}{0.206p^{0.27}} \quad \text{in seconds}$$

or

$$\frac{p}{12.36p^{0.27}} \quad \text{in minutes}$$

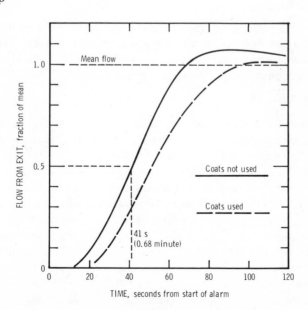

Figure 14.8 Build up of flow from exits in total evacuations of tall office buildings

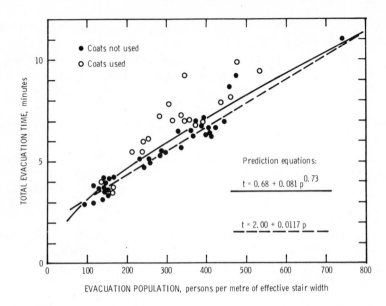

Figure 14.9 Predicted and observed total evacuation times for tall Canadian office buildings

Adding the 0.68 minute starting time one obtains the following equation for total evacuation time in minutes:

$$t = 0.68 + 0.081p^{0.73}$$

where p is the evacuation population per metre of effective stair width.

Because the prediction curve with this equation is almost a straight line, and calculations involving exponentials are somewhat difficult with all but a few calculators, a simpler prediction line and a linear equation are provided;

$$t = 2.00 + 0.0117p$$

In both cases the equations are proposed only for evacuation populations of less than 800 persons per metre of effective stair width, the limiting population that was described by way of an example earlier in the paper.

It must be stressed here that the equations shown were not derived as regressions of the evacuation time data plotted for 56 observed cases. They are predictions based on a knowledge of mean flows and evacuation starting times. It should not be surprising that, on average, there is a good fit. It merely represents internal validity.

The 56 evacuation time data plotted in Figure 14.9 are identified according to use or non-use of coats. With few exceptions, data for cases with coats lie above the prediction curves as would be expected knowing the corresponding flows are lower and the initiation of flow is delayed slightly. With the exception of six cases, to be discussed below, all of the observed evacuation times are within one minute of the prediction curve ($t = 0.68 + 0.081p^{0.73}$). The net error in predicting evacuation times for 50 cases in buildings eight to fifteen storeys high is 0.2 per cent. If more is learned about the evacuation variables listed earlier there is the potential of modifying the prediction equations to take into account many other stair and occupancy factors. The effect of doing this for the coats-factor, for example, should be evident from the locations of coats and non-coats data in Figure 14.9.

Before attempting to refine the prediction of total evacuation time for tall office buildings one should consider the 'law of diminishing returns'. Great effort on improving precision may be an academic exercise. Also to be kept in mind is the fact that criteria for determining acceptability of evacuation times are not well developed.

Building height and total evacuation time

The fact that six cases are not predicted very well by equations in Figure 14.9 is no surprise because these cases are in taller buidlings, eighteen to twenty storeys high. With evacuees spread out over more floors there is simply a reduction in simultaneous demand on stair capacity and thus a model based on capacity flow will underestimate evacuation time. An extreme example would be a tall building with only a few persons on each floor. Here the evacuation time would be governed by evacuees' free speed of descent and the building height. (This condition is covered in early evacuation time prediction graphs by Pauls, 1974, 1975a.)

Also to be kept in mind is the fact that for buildings taller than fifteen storeys other procedures involving selective, partial evacuation are being established (at least in the Canadian office buildings studied). These procedures and some considerations to be taken into account in choosing from a range of evacuation procedures, including use of elevators along with exit stairs, are discussed elsewhere (Pauls 1974, 1975a, 1977b, 1978).

In summary, Figure 14.9 provides a fairly adequate basis for predicting total evacuation times only in office buildings, generally under fifteen storeys, where total evacuation is to be the fire emergency procedure. Of course this procedure must be clearly understood by building occupants. A discussion of the extent to which this is generally true in buildings will be left to other papers.

Large populations and total evacuation time

In Figure 14.10 an addition to the straight prediction line of Figure 14.9 is shown along with prediction lines derived from papers by other investigators (Galbreath, 1969; Melinek and Booth, 1975). In the absence of empirical data the prediction line for populations larger than 800 p/m.eff.—equivalent to 650 persons using a typical 1120 mm (44 in.) stair—has the equation:

$$T = 0.70 + 0.0133p$$

where T is the total evacuation time in minutes, and p is evacuation population per metre of effective stair width.

This is based on the assumption that for such high populations the mean flow will be relatively constant at 1.25 p/s.m.eff. This prediction of total evacuation time is

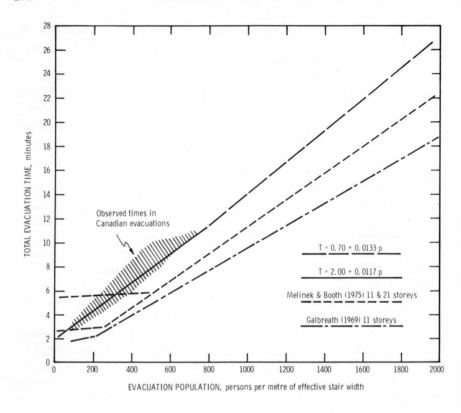

Figure 14.10 Predicted and observed total evacuation times incorporating results from other investigators

more conservative than the predictions of other investigators who had difficulty matching observed and predicted times.

Although there is very little likelihood of ever having the condition, it is interesting to take the case of a 21-storey office building with two standard 1120 mm (44 in.) exit stairs and occupied by the maximum number of people permitted by building and fire codes. Such a building would have 4800 persons (2953 persons per metre of effective stair width–p/m.eff) on its twenty office floors. Using the prediction equation shown, a time of about 40 minutes is predicted to completely evacuate the building by both stairs using the traditional procedure. To understand how unrealistic the condition is we must now turn to an examination of occupant load to see what relation exits between conventional code rules and what has been observed in a number of studies.

Population actually present in office buildings

Throughout the recent Canadian studies of movement of people in buildings particular attention has been paid to accurately assessing the number of people participating in an evacuation drill or in normal egress such as occurs with each event in a grandstand. Early in the study of office building evacuation drills the important discovery was made that buildings are often occupied by fewer people than even management and administrative personnel would estimate for their own buildings or occupancies. Similarly the occupant load figures normally given in codes are not necessarily a good indicator of populations using exits during an egress situation, particularly for tall office buildings.

For office buildings the codes generally require that facilities such as exits and washrooms be designed for a certain number of persons. This number, the occupant load, is established simply by taking the total gross rentable floor area in square feet and dividing by 100. If this calculation is done for the 256 900 m^2 (2 765 000 ft^2) of gross rentable area in the buildings studied one obtains an occupant load of 27 650 persons. In eighteen evacuation drills in the buildings there were only 10 281 persons counted. (In buildings where more than one drill was studied only the drills with larger populations are considered in this calculation.) The mean gross rentable area per person was 26 m^2 (278 ft^2) and not 9.3 m^2 (100 ft^2).

Often figures for population in office buildings are based on the number of workplaces provided. This number could be a good indicator of actual occupancy if for everyone absent there was one visitor. This is not necessarily a good assumption. A very recent Canadian study confirms, for one building at least, that actual occupancy is significantly less than expected if one only counts workspaces (Johnson and Pauls, 1977). This study entailed videotaping the entrance to a 21-storey building with a total of 31 000 m^2 (331 000 ft^2) of gross rentable area on its twenty office floors. Three days of videotaping everyone entering and leaving the building provided many useful data. For example, the peak occupancy during the three typical days was about 1400 persons. Against this can be compared the accomodations officer's estimate of 1700 workplaces and the code occupant load of 3400 persons. Taking the observed peak occupancy, 1400 persons, the actual area per person was 23 m^2 (243 ft^2) which is very close to the figure obtained in analysing all evacuation drill data, 26 m^2 (278 ft^2) per person. An evacuation drill held in this building in 1971, when occupancy was about 2100 persons, is described in Chapter 13.

There was only one building among those studied with a population close to the code occupant load figure and this was only a shortlived phenomenon. This building has a temporarily large occupancy each spring as up to 2500 temporary staff are brought in to work on the annual processing of income tax returns. Here the average area per work station is as little as 6.6 m^2 (73 ft^2) on certain floors at the busiest time of year. Also at this busiest time the average per work station, on all nine above-grade floors, is 8.0 m^2 (86 ft^2). For the majority of the year the average area per work station is 25 m^2 (265 ft^2). At the time of the observed evacuation drill in this building (the one illustrated in Figures 14.4 and 14.5) there were 1486 people present on the above-grade floors with each person averaging approximately 19 m^2 (200 ft^2) of gross rentable area.

The discussion of this subject of occupant load would not be so extensive were it not for the fact that failure to use accurate population figures appears to have introduced significant errors in the data and subsequent models of other investigators. Moreover, if accurate data and models for flow are to be used in future design and regulation applications, then similarly well-founded data should be used for population. If this is not done, any predictions of evacuation time will be almost useless.

SPEED AND DENSITY

One other factor which has an effect upon flow rates, and hence evacuation speeds on stairs, is the density of people on the stairs. It is therefore of value to attempt to identify the relationship between population density, evacuation speed and flow rate so that guidelines for density conditions can be derived.

Density of evacuees on exit stairs has been taken as an independent variable to produce Figures 14.11 and 14.12. Figure 14.11 shows spot measurements of speed and density in total evacuation drills where evacuees did not wear or carry coats. The circles indicate estimates of average speed and density in 21 cases. The linear regression has the equation:

$$s = 1.08 - 0.29d$$

where d is density in p/m^2, and s is the speed in m/s along the slope of the stair (multiply by 0.9 to obtain horizontal component.)

Also plotted is the relation proposed by Fruin (1971) to describe speed and density in non-evacuation situations. Other than commenting that no investigator has data from field conditions for the high-density portion and that the linear regression appears adequate for our purposes, nothing more will be said about this relation so that we can move right on to the flow-density relation shown in Figure 14.12.

Flow and density

Figure 14.12 shows estimates of averages in 21 cases not involving the use of coats. The curve shown is calculated from the regression for speed and density using the well-known flow equation:

$$F = s \cdot d \cdot w$$

where F is flow over the entire stair width, s is horizontal component of speed, d is density, and w is stair width.

A stair width of 1300 mm or 51 in. (that is, an effective width of 1 m) has been used in calculation of the equation for flow as a function of density:

$$f = 1.26d - 0.33d^2$$

where f is flow in p/s.m.eff., and d is density in p/m^2.

The scatter of data to either side of the curve (with this equation) on the graph is partly due to the fact that slightly different curves are needed for different stair widths. The curves shown adequately describe the trend of increasing flow with increasing density up to the density of 2.0 p/m^2. The curve derived from Fruin (1971) does not adequately describe the apparent tendency for flow to fall off at densities exceeding 2.0 p/m^2 (0.19 p/ft^2). In the absence of many data for high-density

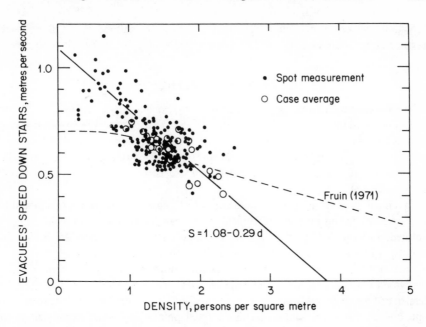

Figure 14.11 Relation between speed and density on stairs in total evacuations

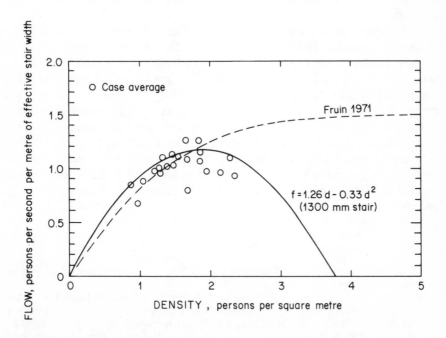

Figure 14.12 Relation between flow and density on stairs in total evacuations

crowd movement down stairs (in evacuations) it is assumed that 2.0 p/m² represents an optimum density for best flow.

SOME CONCLUDING PROPOSALS

There is still much to be learnt about building evacuations and there is still some profound thinking necessary on the objectives for any given evacuation. It is nonetheless of value, at the very least as a way of summarizing this lengthy chapter, to present the optimum evacuation conditions which may be postulated from these findings: *density* is 2.0 p/m² (0.19 p/ft²); *speed* is 0.5 m/s (1.6 ft/s); *flow* is 1.18 p/s.m.eff.

We must keep in mind that these apply to typical stairs with both good and bad features and having normal use typical of the case in the Canadian office buildings studied. In other words, flow could be somewhat higher or lower depending on a variety of factors that were suggested earlier in this paper.

Putting these density, speed and flow conditions into everyday, concrete terms we would have on a typical 1120 mm (44 in.) stair: each evacuee occupying slightly less than two treads; descent of one (office building) storey every fifteen seconds; and one person every second passing a fixed point. Going further we can picture evacuees on such a stair taking a staggered formation; there would be one evacuee directly ahead and four treads distant and another evacuee to one's side and two treads distant. For those readers familiar with the levels of service described by Fruin (1971) these conditions are those of level E, the highest level recommended for stairs.

In an experimental situation, where people disregard conventions that appear to operate in evacuation drills, we can expect that they would reduce by a factor of two the spacings while maintaining the same speed; but this would only be sustained for short distances and short periods of time before a lower density or a lower speed would be adopted. Figure 14.1 a starting point for doubts on current recommendations, shows such a shortlived situation on a 1120 mm (44 in.) stair.

For decades poorly founded traditional practices have been the basis for design and regulation of exits. Empirical evidence against traditional practices was available decades ago. Today the evidence, although still limited, has become too important to be ignored or improperly used. The question is, will such evidence finally be used properly as part of a long-overdue technological approach to exit design and use?

REFERENCES

Fitch, J. M., Templer J., and Corcoran P., (1974). 'The dimensions of stairs', *Scientific American*, **231, No. 4.**

Fruin, J. J. (1971). *Pedestrian Planning and Design* New York: Metropolitan Association of Urban Designers and Environmental Planners.

Galbreath, M. (1969). *Time of Evacuation by Stairs in High Buildings* National Research Council of Canada, Division of Building Research, Fire Research Note No. 8.

Henning, D. N. and Pauls, J. L. (1974) 'Building use studies to solve building regulation problems: some Canadian examples'. CIB 6th Congress, Budapest, Preprint Papers, Vol. 1/1, pp. 26–29.

Johnson, B. M. and Pauls, J. L. (1977). 'Study of normal stair use'. In R. J. Beck, *Health Impacts of the Use, Evaluation and Design of Stairways in Office Buildings.* (Ottawa: Health and Welfare Canada, Health Programs Branch).

Joint Committe (1952). *Fire Grading of Buildings. Part III Personal Safety.* Postwar Building Studies No. 29 (London: HMSO).

London Transport Board (1958). *Second Report of the Operational Team on the Capacity of Footways.* London Transport Board, London, Research Report No. 95.

Melinek, S. J. and Booth, S. (1975). *An Analysis of Evacuation Times and the Movement of Crowds in Buildings.* UK Building Research Establishment, Fire Research Station, current paper.

National Bureau of Standards, (1935). *Design and Construction of Building Exits.* National Bureau of Standards. Department of Commerce, Washington, DC Miscellaneous Publication M. 151.

Pauls, J. L. (1974). 'Building evacuation and other fire safety measures: some research results and their application to building design, operation and regulation'. *Proceedings of Environmental Design Research Association, 5th Annual Conference,* University of Wisconsin. Part 4, pp. 147–168.

Pauls, J. L. (1975a). 'Evacuation and other fire safety measures in high-rise buildings', *Ashrae Transactions,* **81.**

Pauls, J. L. (1977b). 'Management and movement of building occupants in emergencies', *Proceedings of Second Conference, Designing to Survive Severe Hazards,* (Chicago: IIT Research Institute), pp. 103–130.

Pauls, J. L. (1978). 'Movement of people in building evacuations', *Canadian Architect* **May** (also in *Buildings,* **May**).

Pauls, J. L. (1979a). *Building Evacuation, a Review of the Literature,* National Research Council of Canada, Division of Building Research, Building Research Note No. 142.

Pauls, J. L. (1979b). *Means of Egress: Progress Towards a Realistic Technology* (Boston: Society of Fire Protection Engineers, technology report).

Pauls, J. L. (1979c). *Criteria for Exit Stair Width,* National Research Council of Canada, Division of Building Research, Building Research Note No. 143.

Templer, J. A. (1974). 'Stair shape and human movement', unpublished doctoral dissertation, Columbia University.

Templer, J. A., Mullet G. M., Archea J. and Margulis, S. T. (1978). *An Analysis of the Behaviour of Stair Users',* (Washington, DC: National Bureau of Standards), NBSIR 78-1554.

Fires and Human Behaviour
Edited by D. Canter
© 1980 John Wiley & Sons Ltd.

CHAPTER 15

Risk: Beliefs and Attitudes

COLIN H. GREEN
*Duncan of Jordanstone College of Art
and University of Dundee*

INTRODUCTION

There has recently been widespread interest in the problem of ascertaining acceptable levels of risk. In this chapter possible factors influencing attitudes to risk are discussed and some ways of investigating attitudes to risk are suggested. Some experimental work being carried out at Dundee is described.

Attitudes and beliefs occasionally take such a Humpty-Dumpty-like usage, it is worth stating at the start both the meanings in which these terms will be used and those attitudes and beliefs which will be considered. Basically, both 'attitude' and 'belief' will be used in the senses that Fishbein (1967) uses them: an attitude being a 'bi-polar evaluative judgement about something similar in meaning to the economist's strength of preference or utility', and a belief being 'a probability judgement linking something to some attribute . . . '. The purpose of the inquiry dictates which attitudes and beliefs are considered relevant and are therefore to be investigated.

This author's principal interest is in determining what is an acceptable or optimum level of risk in relation to the fire hazard to buildings. Thus, the principal attitude which is relevant to the investigation is the acceptability of any given level of risk. Similarly, the relevant beliefs are those as to the level of risk, and any other beliefs affecting the acceptability of risk.

The phrase 'the acceptability of risk' raises a problem of definition. Rowe (1977), for example, discusses risk and its acceptability without at any point defining a meaning for 'acceptable'. It is of some importance that we know what it is we seek to determine. Pragmatically, what is really of interest is how people will behave in relation to some level of risk. We may want to know, for example, how much people would be willing to pay for some decrease in risk. We are also interested in

This chapter forms part of the work of the Fire Research Station, Building Research Establishment, Department of the Environment and is Crown copyright. It is reproduced here with the permission of Her Majesty's Stationery Office. The work has been undertaken under contract FRO/28/068 for the Fire Research Station.

how people are likely to react if an accident occurs. While there are two basic strands of approach to the relationship between attitude and behaviour (Liska, 1976), here Fishbein's approach will be adopted: that is, that attitudes affect intended behaviour, while actual behaviour depends not only upon attitudes but also upon other factors (Ajzen and Fishbein, 1970). Thus the definition of 'acceptable' that is relevant depends upon in which behaviour we are interested.

Below it will be argued that the term 'risk' is misleading and another term substituted, but for the moment 'risk' will be used in two of the three senses in which it is used in the literature (Rowe, 1977). It may mean either the combination of the probability of an event and the severity of the event, or, and it is to this sense that the term will be restricted in the latter part of this paper, the probability of occurrence of an event. For the third sense, the feature of the situation which gives rise to the possibility of an accident, the term hazard will be used. Thus, in any situation or activity, such as staying at an hotel, there are a number of hazards, such as the possibility of a fire due to an electrical short circuit, which contribute to the total risk.

Taking this analysis a step further, the risk in a situation is a composite of the risks from all the various hazards that exist in that situation. The relevant attributes and beliefs may then concern either the activity itself or each of the hazards that may exist. A belief, say, that those at risk can control the probability of a fire occurring at home and/or the probability of harm occurring to them as a result is a belief in relation to the situation, being at home, and presumably based upon some generalization for all the possible hazards that could lead to fire in the home. Alternatively, it may be that attitudes to risk are held in relation to the risk from individual hazards. If so, then it may be meaningful to define acceptable risk only in relation to a particular hazard and not in relation to the situation as a whole. Thus if beliefs about the hazard of domestic gas explosion differ from those concerning smoking materials, very different risks may be acceptable for each hazard.

A preliminary question is how the acceptable level of risk is affected by an individual accident. Does an acceptable risk imply that if an accident of a given severity occurs then this is an acceptable accident, where an acceptable accident is one where there is no requirement to extend or revise existing safety measures? It is widely assumed that the acceptability of accidents depends only on their frequency and that when an accident occurs we, both as a society and as individuals, consider only whether other similar accidents are likely to occur. However, there is no *a priori* reason to assume that frequency is the only factor which determines the acceptability of accidents. Attitudes may also be influenced by other factors, such as how accidents occur.

If people's beliefs are built up, as seems reasonable, by a knowledge of accidents, then these beliefs may be changed by any one accident. An accident might then be accepted if its occurrence did not change beliefs about the probability of its repetition. On the other hand, an accident might also change people's beliefs about other attributes governing the acceptability of the risk. Any accident which occurs would then be interpreted in the light of those beliefs and may lead to a revision of them. The relevant linkages are given in Figure 15.1.

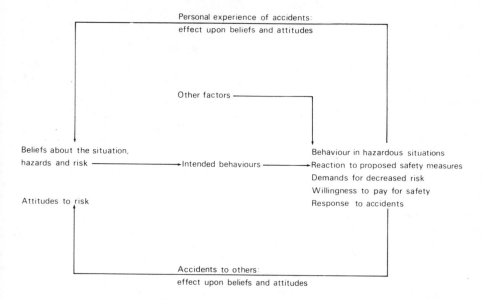

Figure 15.1 Effects of accidents on beliefs and attitudes

It may be argued that this discussion has merely shifted the problem of definition from acceptable risk to accepted accident. While there is some work on the way accidents may be interpreted so as to be tolerable (Walster, 1966; Wortman, 1976), I will suggest below that a rather more complicated explanation is called for.

The model given in Figure 15.1 will be adopted as a framework for the remainder of this paper, though other work in this area has tended to consider either risk and its acceptability, or the interpretation of accidents, in isolation.

While Figure 15.1 joins accidents and risk in something of a chicken and egg conundrum, it is also necessary to consider whether there are any beliefs about risk in the absence of accidents, and whether accidents have any meaning in the absence of expectations about potential accidents. Since much of this work has tended to rest upon unvalidated assumptions, much of the discussions will be taken up with identifying these assumptions and attempting to indicate the form a rigorous methodology would take. I shall start with risk.

PERCEIVED RISK

Elsewhere, a parallel between perceived risk and noise has been drawn (Green and Brown, 1978a). Sound can be measured in decibels by its sound pressure level. However, in order to determine how loud a sound appears, the annoyance it causes, or the long-run hearing damage that will result from it, it is necessary to determine which characteristics of a sound are associated with these consequences and in what manner. It is not sufficient to assume there is a linear relationship between these and the sound pressure level. As Weiner (1920) remarked: ' . . . things do not, in general, run around with their measures stamped on them like the capacity of a freight car; it requires a certain amount of investigation to discover what their measures are . . . '.

Similarly, it is not sufficient to assume that any one statistic of the risk to life, such as the risk of death per man-hour of exposure, accurately reflects what people mean by risk. Rather it is necessary to discover on what basis we make a judgement such as 'A is riskier than B'. If, for example, someone is faced with two situations such as staying at an hotel and staying at home, and asked to decide in which there is the greatest risk, we might expect that person to consider:

(1) the likelihood of a fire
(2) the probability of the various outcomes should an accident occur
(3) the relative seriousness or disutility of the various outcomes.

Thus if an accident does occur, then there is a chance that that person will be killed, that they will receive various degrees of injury but survive, or that they will escape all harm. Clearly, we would expect most people to prefer not to experience an accident at all, and if they are involved in an accident to prefer to escape harm rather than to suffer any degree of injury.

This model illustrates a problem with the use of the term 'risk'; here, risk could either refer to the probability of occurrence of one outcome or the probability of occurrence multiplied by severity. Thus if one is to ask someone what is the risk when staying at an hotel, they may ask 'risk of what?'. If they are told 'the risk of death', then this is to assume that death is the only outcome worth considering. A second problem is that there are quite likely to be social factors operating, particularly if we ask people what risk they would accept. Taking high risks may carry with it some connotations of machismo, and this might influence people to say that they would accept a higher risk than they truly prefer. On the other hand, safety, like law and order, may be something of which we all think we should be in favour though differing in our attitudes towards any measure which it is claimed will increase it. For these reasons, in the experimental work in which I have been involved in Dundee, questions have been phrased in terms of perceived safety. The model described above indicates what factors people might consider but does not say that this is what they do consider. It is therefore useful to examine the assumptions underlying some conjectured measures of safety. For example, the probability of death per man-hour of exposure proposed by Starr (1969) and others (Kletz, 1971) as a measure of safety, assumes that either death is somehow so qualitatively different from all forms of injury that these injuries can be neglected, or that the probability of these injuries is seen as being in a fairly constant ratio to the probability of death itself. In two experiments we have asked samples of respondents to indicate the relative severity to them of suffering a number of different injuries, ranging from not being hurt at all to death. In both studies (Green and Brown, 1978b), it was found that at least for these samples of students, there are two fates worse than death, injuries to which death is preferred, namely brain damage and permanent paralysis from the neck down.

Another series of experiments was conducted, using Stevens' (1975) methodology, to determine ratio scale measures of perceived personal safety and the chance of death should an accident occur. A sample of 70 students from the Departments of Architecture, Mathematics and Social Science were asked to complete ratio scale questions on perceived safety. They were first asked to rank the situations from safest to least safe and to rate that which they had ranked as safest as 100. Then, for the situation they had described as the second safest, to put down a number which represented how many times less safe they considered this situation to be, and so on (Green and Brown, 1978b). On this scale, zero can be defined as the point where there is no chance whatsoever of accident or harm occuring.

Subsequently, part of the sample of respondents who had been used in this experiment were asked, for the same fourteen situations, to complete ratio scale measures of the chance of accident occurrence in these situations, C(A); and the conditional chance of death should an accident occur, C(D|A). In each case, they were asked to rank the situations from greatest chance to least chance. Then, the question asked respondents to indicate the chance of an accident, or death, occurring. Thus if they believed in a situation, that if an accident did occur then death was certain, this would be indicated by stating a 1-in-1 chance. The results are shown in Table 15.1, from which it may be seen that when someone is rock-

Table 15.1 Variation of perceived safety with perceived chance of accident and conditional chance of death should an accident occur

Situation	Perceived safety	Geometric means	
		Perceived chance of an accident occurrence C(A), 1 in:	Perceived conditional chance of death should an accident occur C(D|A), 1 in:
Staying at home (from fire)	163	2640	620
Living within 8 km (5miles) of a major airport	203	15 000	180
Swimming in a swimming pool	212	3530	360
Staying at a hotel (from fire)	262	1760	230
Travelling by train	272	3500	210
Living within 8 km (5 miles) of a nuclear power station	335	4120	130
Crossing the road	361	400	100
Travelling by coach	368	700	180
Living within 8 km (5 miles) of a major chemical plant	376	5760	410
Driving a car	406	180	140
Skiing	512	220	790
Travelling by plane	525	4020	8
Riding a motorcycle	808	60	30
Rock-climbing	1056	100	27
	$n = 70$	$n = 36$	$n = 36$

climbing, it is believed that they have a 1-in-100 chance of an accident, there being a 1-in-27 chance that an accident will result in their death. When the logarithmic means for C(A) and C(D|A) were regressed against the logarithmic means for perceived safety, then:

$$\log \text{Perceived safety} = 3.25 - 0.23 \log C(A), \qquad r^2 = 0.55 \qquad (1)$$

and

$$\log \text{Perceived safety} = 3.11 - 0.26 \log C(D|A), \qquad r^2 = 0.40 \qquad (2)$$

Regressing the means of perceived safety on both C(A) and C(D|A):

$$\log \text{Perceived safety} = 3.54 - 0.186 \log C(A) - 0.193 \log C(D|A),$$

$$r^2 = 0.75 \quad (3)$$

Hence:

$$\text{Perceived safety} = \frac{3500}{(C(A) \times C(D|A))^{0.19}}$$

since the coefficients for C(A) and C(D|A) are not significantly different from each other, although both are significantly different from zero.

Thus both factors are apparently equally considered in assessing perceived safety, and this is what the rational-man model would predict. The fact that perceived safety depends primarily on the risk of death suggests either that the risk of serious injury is ignored or that the chance of serious injury is seen as being in fairly constant ratio to that of death across the different situations.

While the logarithmic means for C(A) and C(D|A) are overall not significantly correlated, there is a somewhat puzzling correlation between them in each of the three situations: driving a car, riding a motorcycle and rock-climbing. In each of these cases, the more likely that respondents believed an accident was to occur the stronger their belief that it would be fatal.

The logarithmic means of perceived safety were compared to the objective statistics of the probability of death per person-hour of exposure. In this regression, only nine of the fourteen situations could be used owing to various problems with the availability of objective statistics. For skiing, for example, the only statistic that was found in the literature was based on one death in one season's skiing. Similarly, no statistics specifically for swimming in a swimming pool were obtained. For nuclear power, it would seem to depend on whose subjective estimates one chooses to believe; and much the same problem applies to chemical plants, although the great variation between plants also adds to the problem. Similarly, the only statistic for the ground risk from aircraft related to Hollywood-Burbank and Los Angeles International Airport (Solomon *et al.*, 1974); the applicability of this figure to the United Kingdom, or any other airports, is doubtful.

However, for those situations where reasonable data are available, regressing logarithmic mean perceived safety on the probability of death per 10^8 person hours of exposure gives:

Perceived safety = $200P(D)^{0.17}$, $r^2 = 0.75$

Again the exponent is significantly different from zero.

The significance of this equation is two-fold. First, for those situations where reasonably reliable estimates of the objective risk of death are available, the respondents appeared to have a surprisingly accurate knowledge of the relative levels of safety in different situations. The situations where such objective statistics are available are also those where we might expect respondents to have the best subjective feel for the safety level. Secondly, the exponent of the power function shows that the respondents underestimated the differences between the levels of safety. This means, for example, that to say, on the basis of objective statistics, that one situation is ten times as safe as another is highly misleading since the subjective difference will be considerably less. A statement of the objective difference might be interpreted, in the lack of other information, as meaning a subjective ten-fold difference.

PRESENT LEVELS OF SAFETY

Another experiment, using eleven-point category scales, had been undertaken to determine which statistics of risk (annual numbers killed, annual risk of death, chance of an accident, conditional chance of death) are correlated with the perceived present level of safety. These experiments provided preliminary indications that it was the risk of an accident, rather than the likelihood of being killed in a year or the annual numbers killed, which is a determinant of perceived safety. Similarly, it was the risk of death rather than the risk of injury that was most strongly correlated with perceived safety (Green and Brown, 1978b).

A number of qualitative attributes of situations were also tested, including whether or not the activity is a voluntary one, and whether the person at risk can control the likelihood of an accident or the outcome should one occur. It was found that present level of safety in the situations examined was correlated with the involuntariness of the activity; the more involuntary the activity, the greater the safety believed presently to exist in that activity.

While it has been hypothesized (Starr, 1969) that the acceptable level of safety in an activity varies with the involuntariness of that activity, to argue that the result given above proves this hypothesis is to fall into the fallacy of equating correlation with causality. Merely because the present levels of safety are higher in involuntary activities than voluntary ones does not mean either that the present levels are necessarily accepted or acceptable, nor that higher levels of safety are expected in some activities rather than in others. Indeed, neither does it mean that any difference in acceptable levels of safety between involuntary and voluntary activities is necessarily because one set is involuntary and the other voluntary. Economic efficiency arguments can be advanced as to why, say, flying by plane should be safer than riding a motorcycle and flying in a large plane safer than flying in a small plane (Green and Brown, 1977).

ACCEPTABLE SAFETY

Given limited resources, a principal problem in setting design standards is: how safe is safe enough? The introduction of reliability engineering methods of design

(Baldwin, 1974; Nelson, 1973), by requiring the explicit statement of some accepted probability of failure, has done no more than expose this problem to public view.

It has been widely hypothesized (Lowrance, 1976; Rowe, 1977; Starr, 1969) that the acceptable levels of safety will vary according to some attributes of the activity in which the risk occurs or the nature of the risk. In particular, the levels of acceptable risk have been hypothesized to be higher in voluntary activities, those in which you may choose whether or not to participate, than in involuntary ones. There is limited experimental evidence (Fischhoff *et al.,* 1976) to support these conjectures.

Unfortunately, discussions on 'acceptable levels of safety' have been obscured by confusing usage of the term 'acceptable'. It may, for example, be simply used in the sense that since this is the level of safety prevailing in some other situation A then that level ought to be good enough for situation B under discussion. Here 'acceptable' is being loosely used in the sense of 'accepted', and the underlying assumptions are, first, that the level of safety in situation A is acceptable by reasons of being that which presently exists: second, that situations A and B are in some sense comparable, and hence what is acceptable for one should be acceptable for the other. This kind of approach has underpinned what has come to be known as the 'revealed-preference approach' to determining acceptable safety (Rowe, 1977), and has been argued to have a number of methodological drawbacks (Brown and Green, 1980).

The economic approach is to say that since there are limited resources, it is necessary to make tradeoffs between various things that are desired in order to achieve the best combination (Kenney and Raiffa, 1976). Thus, in relation to the safety of any one situation, there will come a point where any further increase in the safety of that situation means going without so many other things that an increase in safety is not worth having. Ultimately the value of any increase in safety is a *value judgement* and, as such, not a question of fact but of opinion. In the extreme, if we are considering the safety of a dwelling, it might mean that many people could afford a house only if they went without food. More generally, however, since there are many hazards to life, we might decide that we would be better off if rather than further increase safety in the home, we increase the level of safety in some other situation. This economic approach leads to two further definitions of acceptable safety, one of which is the economic optimum (Brown and Green, 1980).

The point in question is, then, what, if any, is the meaning of an 'acceptable level' of safety? One way to look at this problem, other than to hypothesize meanings, is to consider the implications of the use of an acceptable level of safety in practice. In most situations, including buildings, the use of an acceptable level of safety means that sooner or later we will see some failures. If an acceptable level of safety is defined in terms of a probability of occurrence of death from, say, fires in dwellings, then each year there will be a number of fires and deaths which, if we are successful in designing to this criterion, will average out at the acceptable probability per person. Each fire or series of fires, as events, will then give us information only as to how confident we can be that the acceptable level of risk is not being exceeded. That is, we would not be interested in what happened, only in how often it happens since an acceptable level of risk for dwelling fires treats all dwelling fires as identical events. Furthermore, provided that the frequency, or probability, of occurrence is less than the acceptable level, all such fires must be acceptable. An acceptable level of risk should therefore exclude the possibility that any one fire will result in a requirement to modify the existing stock to reduce or eliminate the possibility of further such fires. If this is not so, then it is difficult to see what, if any, is the use of designing to some acceptable level of risk, unless the idea of an acceptable risk is construed as a gamble on the part of the decision-maker. In the latter case, what is implied is that although any accident may result in requirements for changes, this probability of an accident is sufficiently low for the designer to feel he can live with it. Such a policy is perhaps reasonable where, say, there is a single industrial plant which might go wrong, and the probability is so small that the chance of it happening during the plant's lifetime seems negligible. It is intuitively less sensible in relation to products like homes or cars, where the same probability will result annually in some if not many accidents.

Although most individual accidents are accepted the hypothesis that the only determinant of the acceptance of individual accidents is their frequency does not seem to accord with real life. Any approach using an acceptable risk must account for those situations like Ronan Point (Department of the Environment, 1968) and motorway crossover accidents (Anon, 1970; Mooney, 1977) where, while the risk from that hazard is small relative to the total risk in that activity, an individual accident still does not seem to be accepted. These types of accident suggest either that acceptable levels of safety may need to be set in relation to each hazard within a situation, or that some accidents are by their nature unacceptable in relation to certain situations.

INTERPRETATION OF ACCIDENTS

In any situation, there are a number of sequences of events which may lead to an accident, that event which is considered to have initiated the sequence being generally termed the cause. Thus there are a number of hazards in the home which may result in a fire and possible death thereby.

Beliefs about the present level of safety in a situation, it seems reasonable to suppose, are derived from generalizations about the risks in that situation and their individual characteristics. If these beliefs are true, then they should imply something about the acccidents that are likely to occur in the siutation. To take the simplest and most trivial example, if accidents are believed to be unlikely to occur, then we would not expect them to occur very often. Hence beliefs about the present level of safety and the characteristics of the situation and risks, while being derived from knowledge of accidents, may also imply expectations as to future accidents and the way these accidents will be interpreted. Thus, since a sample of respondents believe that for fire in the home both the risk of a fire and the chance of surviving it are within the control of those at risk, then as a corollary it might be expected that individual home fires would be expected to result from carelessness, a personal failure of control. There is some evidence to suggest that this is believed to be the case (Green and Brown, 1978b; Melinek, Woolley, and Baldwin, 1973). These beliefs and expectations may not necessarily be appropriate but they are consistent.

Beliefs about attributes of a situation that are used to derive a belief about the present level of safety in that situation might therefore be termed a conjecture of safety since they imply that, if true, certain events will occur with some stated likelihood. Any given accident would then be equivalent to an experiment which would either support or negate this conjecture of safety.

Strictly, used in this way, there are two conjectures; first as to the present level of safety. Secondly, assuming that the acceptability of any given level of safety is dependent upon some beliefs about the attributes of the situation or risk, then the acceptability of a given level of safety depends upon a further conjecture as to these attributes. Any accident or accidents might then negate either of these two conjectures: a situation might come to be perceived as less safe than was believed and hence the revised belief as to its safety indicate an unacceptable level of safety. Alternatively, while the level of safety might not be seen as being less than was previously believed, revised beliefs of the nature of the situation may now demand a higher level of safety.

This hypothesis as to the nature of the relation between beliefs about and attitude towards risk and the interpretation of accident leads to a series of questions: first, what is the formal structure of these conjectures. Second, what events will be interpreted as refuting one or other of these conjectures; and what will happen if a conjecture is interpreted as being refuted. In regard to the second question, will an accident change beliefs about the situation as a whole, or only about the particular hazard in that situation which has given rise to the accident in question? It is this question which is most relevant in the light of the case of Ronan Point cited earlier. It may be that there are different acceptable levels of safety for each class of hazards in a situation; alternatively, there may be some hazards which are just not acceptable in a particular situation. To suggest that there may be conjectures of safety is to argue that people do not make abstract probability judgements as to the likelihood of an accident, but inferential statements of the form: 'if A then the probability of an accident is P'.

If the above theory on conjectures of safety is correct, studies of responses to accidents should reveal a process of comparison of an event to an hypothesis, revealing the formal structure of these hypotheses. Responses to accidents may be examined in a number of ways, including the examination of media coverage and subjective studies of public attitudes to major accidents. However, while such studies may be useful in generating hypotheses (Drabeck and Quarantelli, 1967), they are only to some varying degree reflections or interpretations of the beliefs and reactions of individual members of the population.

Similarly, examination of the development of the tort of negligence is a useful source of hypotheses in so far as it is considered that this formal process is an analogue of the process by which individuals or society assess accidents. One of the more interesting characteristics of this tort is that, as a branch of common law founded on case law, it has shown a steady evolution over the years. For example, until the late nineteenth century in Great Britain, an employee was held to have voluntarily assumed all the risks of employment and could not recover any damages resulting from an accident from his/her employer. The position was similar in other areas but has now evolved to a markedly different position.

Although the study of the tort of negligence is complex (Atiyah, 1970; Heuston, 1973; Prosser and Smith, 1967) there are a number of principles which underlie a judgement in any particular case. The first point to note is that even though an individual may have suffered harm as a consequence of the actions of another, this is not in the general case sufficient to establish liability. Before liability can be established it is necessary to demonstrate that the defendant had a 'duty of care' to the plaintiff, that is, a social contract, and that they were required to take into account the possible consequences of their action to that individual. It can be argued that the extent of these duties of care represent an evolving consensus as to the form of behaviour which is equitable under the social contract. Secondly, even where there exists a duty of care, it is limited to those consequences which are 'reasonably foreseeable'. An accident can be so unlikely to result from a particular course of action that the defendent had no duty to take that remote possibility into account, and guard against it. The mere fact that an accident happened is not proof that it could and should have been foreseen. Thus, in tort, a person is not liable for all his/her actions.

It should not be assumed that the judicial process of tort is ncessarily identical to that which an individual will use in assessing an accident in another context. Heider (1958), has hypothesized that the attribution of responsibility exists at five levels, only one of which matches that with which tort is concerned. Similarly, Jones and Thibaut (1958) have suggested that there are a variety of concepts which may be used in personal perception. However, the principles of the tort of negligence provide at the very least a framework for assessing the adequacy of studies of actual and hypothetical accidents.

Less formalized studies of accidents can be undertaken either in relation to actual accidents or to hypothetical accidents. There is a fair body of studies of hypothetical accidents based on attribution theory (Crinklaw and Vidmar, 1971; Mackillop and Porovac, 1975; Shaver, 1970; Walster, 1966), but these studies have not included all the factors that a court would review and many are subject to other methodological criticisms (Vidmar and Crinklaw, 1974). For example, they generally fail to consider the varying degree of foresight present in each case. Similarly, as in tort, causality need not necessarily imply culpability as Reismand and Halperin have pointed out (unpublished, cited in Vidmar and Crinklaw, 1974). Further, negligence under tort is fundamentally about 'ought' and duties, where 'duty is a notional pattern of conduct' (Heuston, 1973). Tort recognizes a complex hierarchy of duties; an employer, for example, is generally liable for the torts of his/her employees. Thus the fact that one person is held responsible for an accident may impose liability on another. This kind of transfer should be taken into account in any study of responses to accidents. In the first study, undertaken by Walster (1966), for example there were varying consequences of a parked car running away down a hill, in the worst case crashing through a shop window and injuring the occupants, the person liable for the damages might not have been the car owner, but the dealer who sold the car. The latter might have been held liable on the basis that the car she or he sold should not have hidden defects. being reasonably fit for the purpose intended; in the case in point, that the car should not have had defective brakes.

The one detailed study of responses to an actual accident that seems to exist is that of Bucher (1957) on the Port Elizabeth air crashes. While this study started after the first crash, and was continued after the second and third crashes, no analysis of any systemic changes of attitudes or beliefs is reported. This is particularly regrettable since three types of perception are reported: 'no problem', 'violation of expectations' and 'emergent features'. The degree to which these, especially the latter two, are alternative or successive views is of some interest. The 'no problem' group defined a crash as an accident and considered that accidents will happen. Of the other two groups, Bucher defines the 'violation of expectation' response as '. . . the disaster violates all conceptions of the usual and acceptable. Features which would be expected to be present in the situation according to prior conceptions are not only lacking, but the features which are present are directly contrary to expectations.' An 'emergent features' response, on the other hand, is described as being characteristically ' . . . the situation is initially labelled in accordance with prior conceptions, but there are aspects to the particular case which shed a different light on previous responses. A new and unique situation exists, one which requires special evaluations.' The conjecture of safety hypothesis implies that these responses are potentially successive, changing from 'no problem' via 'violations of expectations' to 'emergent features', but no data are given to indicate whether or not this was the case.

Once an individual had decided there was a problem, Bucher describes the process whereby respondents sought to explain the accidents as follows:

> In order to organise their thinking about the situation, the respondents had to see in what way their previous knowledge applied to this particular case . . . when the respondent was able to label the situation as being basically of a certain kind, he could proceed to collect and evaluate the facts of the case . . . the situation was scrutinized for significant features. There are myriads of facts available to these respondents. In order to make any sense out of them at all, they had to be weighed against some reference points. It was on the basis of the analogues aroused that significant facts were picked out and evaluated.

The two distinct forms of analogue that Bucher reports each lead to a distinct definition of the cause of the crashes and what could be done to prevent them. One school of thought visualized aeroplanes as inherently fallible machines and these respondents draw the conclusion that the airport should be moved. The other group conceived of machines as requiring care to operate properly; this group concluded that improvements were necessary in servicing machines.

In an interesting parallel with tort, Bucher also reported that the laying of responsibility must be clearly distinguished from blame. Who is held responsible, Bucher reports, results directly from the analogy employed and was laid at the feet of those with power to alleviate the conditions underlying the crashes: 'It was not instrumentality in causing the crashes which determined responsibility but ability to do something to prevent their recurrence'. But someone could be responsible without being to blame; agents of responsibility were only held to blame only when respondents were convinced that (a) those responsible had violated basic moral standards and values; and (b) that they would not of their own volition take action which would remedy the situation.

Bucher's study has been discussed in some depth as it indicates some further problems in examining response to accidents. One accident is not necessarily representative of the whole universe of accidents, and therefore the response to any one accident cannot be generalized to all other accidents. In so far as Bucher's study suggests that accidents are compared to expected accidents, where expectations are defined not only in terms of expected frequency but by other attributes as well, it is the differences that seem to be significant. If this is so, then a criterion of acceptable safety is to be seen as being not only an expectation of frequencies but also of the characteristics of likely accidents, and it is only those accidents which are seen as conforming in both respects that will be accepted.

CONCLUSION

To use reliability engineering approaches to design requires both a measure of safety and a definition of the acceptable probability of failure. No measure of safety can be arbitrarily assumed as appropriate; what people mean by safety and what characteristics of a situation are used to assess its safety must first be determined in order to derive an appropriate measure of safety. Nor equally can an 'acceptable' level of safety be assumed to be an intrinsically meaningful term. In practice, the use of an acceptable level of safety as a design criterion presupposes that some accidents at some frequency will be accepted if and when they occur. While there is some evidence that beliefs about the level of safety presently existing and the nature of the situation are reflected in expectations about the types of accident that will occur in that situation, there is little present knowledge as to whether it is merely the frequency of accidents that determine whether they are accepted, or whether they are assessed in light of more complex expectations. Until the way individual accidents are interpreted in the light of expectations, and the conditions under which they lead to the review of beliefs about the nature of the risk and attitudes towards it are clarified, the application of an acceptable level of safety is insufficiently defined to be a reliable design criterion for fire safety design. It will not be such a criterion until we have defined what it is we hope to predict, and for which we use this label of acceptable safety.

REFERENCES

Ajzen, I. and Fishbein, M. (1970). 'The prediction of behaviour from attitudinal and normative variables', *Journal of Experimental Social Psychology*, 6, 466–487.
Anon (1970). 'Motorway crash decisions', *The Economist*, 22 **August**, p. 21.
Atiyah, P. S. (1970). *Accidents, Compensation and the Law.* (London: Weidenfeld and Nicolson).
Baldwin, R. (1974). 'The analysis of fire safety', *Accident Analysis and Prevention*, 6, 205–222.
Brown, R. A. and Green, C. H. (1980). 'Precepts of safety assessment', *Journal of the Operational Research Society.*

Bucher, R. (1957). 'Blame and hostility in disaster', *American Journal of Sociology,* **62**, 467–475.

Crinklaw, L. D. and Vidmar, N. (1971). 'Inferential sets, loss of control and attribution of responsibility for an accident', Research Bulletin No. 203, University of Western Ontario.

Department of the Environment (1968). *Report of the Inquiry into the Collapse of Flats at Ronan Point, Canning Town.* (London: HMSO).

Drabeck, T. and Quarantelli, E. L. (1967). 'Scapegoats, villains and disasters', *Trans-Action,* **4**, 12–17.

Fishbein, M. (1967). 'Attitude and prediction of behaviour', in M. Fishbein (ed.). *Readings in Attitude Theory and Measurement.* (New York: Wiley).

Fischhoff, B., Slovic, P., Lichtenstein, S. and Read, S. (1976). *How Safe is Safe Enough? A Psychometric Study of Attitudes Towards Technological Risk and Benefits.* (Eugene Ore.: Report 76-1, Decision Research).

Green, C. H. and Brown, R. A. (1977). 'The use of quantitative risk criteria in hazard analysis', *Journal of Occupational Accidents,* **1**, 85–94.

Green, C. H. and Brown, R. A. (1978a). 'Counting lives', *Journal of Occupational Accidents,* **2**, 55–70.

Green, C. H. and Brown, R. A. (1978b). *Life Safety: What is it and how much is it worth?* Building Research Esatablishment Current Paper CP 52/78, Borehamwood.

Heider, F. (1958). *Psychology of Interpersonal Relations,* (New York: Wiley).

Heston, R. F. V. (1973). *Salmond on the Law of Torts, 16th Edition.* (London: Sweet and Maxwell).

Jones, E. E. and Thibaut, J. W. (1958). 'Interaction goals as bases of inference in interpersonal perception'. In R. Tagiuri and L. Petrullo (eds.) *Person Perception and Interpersonal Behaviour* (Stanford: Stanford University Press).

Keeney, R. and Raiffa, H. (1976). *Decisions with Multiple Objectives.* (New York: Wiley).

Kletz, T. A. (1971). *Hazard Analysis—A Quantitative Approach to Safety,* Symposium Series No. 34. (London: Institute of Chemical Engineers).

Liska, A. E. (1976). *The Consistency Controversy.* (Cambridge, Mass.: Halsted).

Lowrance, W. W. (1976). *Of Acceptable Risk.* (Los Altos: Kaufmann).

Mackillop, J. and Poravac, E. J. (1975). 'Judgements on responsibility for an accident', *Journal of Personality,* **93 (2)**, 248–265.

Melinek, S. J., Woolley, S. K. D., and Baldwin, R. (1973). *Analysis of a questionnaire on attitudes,* Building Research Establishment Fire Research Note 962, Fire Research Station, Borehamwood.

Mooney, G. H. (1977). *The Valuation of Human Life.* (London: Macmillan).

Nelson, H. E. (1973). 'Goal orientated systems approach to building fire safety', in *Planning and Design of Tall Buildings.* (New York: American Society of Civil Engineers), Volume 1B, ASCE.

Prosser, W. L. and Smith, Y. B. (1967). *Cases and Materials on Torts, 4th Edition.* (Brooklyn: Foundation Press).

Rowe, W. D. (1977). *An Anatomy of Risk,* (New York: Wiley).

Shaver, K. G. (1970). 'Defensive attribution: Effects of severity and relevance on responsibility assigned for an accident', *Journal of Personality and Social Psychology,* **14**, 101–113.

Solomon, K. A. *et al.* (1974). *Airplane Crash Risk to Ground Population.* (Los Angeles: School of Engineering and Applied Science, University of California), Report No. UCLA-Eng. 7424.

Starr, C. (1969). 'Social benefit *vs.* technological risk', *Science,* **165**, 1232–1238.

Stevens, S. S. (1975). *Psychophysics: Introducing its Perceptual, Neural and Social Prospects.* (New York: Wiley).

Vidmar, N. and Crinklaw, C. D. (1974). 'Attributing responsibility for an accident:

a methodological and conceptual critique', *Canadian Journal of Behavioral Science*, **6 (2)**, 112-130.

Walster, E. (1966). 'Assignment of responsibility for an accident', *Journal of Personality and Social Psychology*, **3 (1)**, 73–79.

Weiner, N. (1920). 'A new theory of measurement, a study in the logic of mathematics', *Proceedings of the London Mathematics Society*, **Series 2, 19,** 181–205.

Wortman, C. B. (1976). 'Causal attributions and personal control', In J. H. Harvey, W. J. Ickes and R. F. Kidd (eds.) *New Directions in Attribution Research, Volume 1.* (Hillside, New York: Lawrence Erlbaum Associates).

Fires and Human Behaviour
Edited by D. Canter
© 1980 John Wiley & Sons Ltd.

CHAPTER 16

Modelling Fire Safety and Risk

ERIC W. MARCHANT
Department of Fire Safety Engineering
University of Edinburgh

INTRODUCTION

This contribution is a brief summary of some of the models used to represent the complex real fire situation. Because fire *is* complex, many models are used in research, investigation and fire safety engineering. Models may represent small parts of the combustion process or attempt to represent the total fire safety components in a building design. Fault trees are an example of analytical models which have been developed by systems analysts and are now being applied to fire safety problems. These models help the fire safety engineer to identify weaknesses in the relationships between people, buildings and possible fires, so that the greatest increase in safety is achieved for the least amount of effort and resources.

In this chapter a brief description of the types of model is given first, followed by a section on the use of fire safety information in models. A third section deals with the processing of information in the model. The fourth section deals with the application of some of the concepts of systems as applied to fire safety.

TYPES OF MODEL

Fire safety is a complex subject which can be divided and subdivided into many areas, each of which is considered to be important both in its own right and as a component of the whole subject. For example, a quantitative understanding of the combustion characteristics of wood is essential knowledge for the prediction, or analysis, of the behaviour of real fires. A knowledge of this behaviour is needed to be able to assess the threat any fire might pose to the human occupants of a building or the potential loss of inanimate contents or the degree of damage the building is likely to sustain. The study of fire safety is complicated further because fire is a phenomenon of rare occurrence for a particular building and even more rare for the individual occupant. This suggests that the majority of individuals in any population will not have a regular experience of fire and therefore it can be difficult to predict how they will behave when involved in a fire situation.

Burning produces total loss. The protection of people, their property and buildings from the effects of fire is a necessary and continuous activity. The techniques used to achieve the required standards of protection are defined by the responsible authorities (for example, the Home Office, Department of the Environment and the insurance companies). These authorities are advised by research and development staff.

The study of fire phenomena is difficult because of its complexity and a major component of this complexity is the fact that all fires present a dynamic situation for which it is difficult to set appropriate limits to any study. However, several research programmes, internationally, are devoted to the study and analysis of this complex dynamic problems. To help in this study, models of various components and divisions of fire safety have been developed and these include the use of psychological, physical, analogue and mathematical concepts, all attempting to gain some information to improve the understanding and prediction of the complex interactions of fire behaviour and the reaction to fire of people, property and buildings.

Many real situations are complex combinations of objects, environment, people, plants and animals, all capable of a range of responses depending on a given stimulus. The success of some models of systems can be measured, usually, against the performance of the real system; for example, a mathematical expression representing the time an object will take to drop 2 m (6.5 ft) can be tested readily. In studies of fire safety very few models can be tested so readily. In an endeavour to study real situations closely models of many types have been developed to represent such real situations.

INTERNAL AND EXTERNAL MODELS

Gregory (1966) suggests that no branch of study can be pursued without the use of some type of model and he defines 'internal' and 'external' models. The former is 'patterns in the head' whilst the latter may be 'any kind of information-bearing or information-yielding "substance"' including three-dimensional solid objects, graphic models, word pictures, symbolic statements, analogues, mathematical expressions and computer programs. They may contain more or less information than a real situation, but will be successful if they enable the 'operator' to make the correct decision. The complete success of any model cannot be guaranteed as the information input required by the model may be incomplete and the manipulation of available information may not be related accurately to the real situation. However, with a subject as complex and as imprecise as fire safety, any method which simulates human or fire behaviour with a close estimate to reality will be useful.

The essential components of a model of any system have been summarized as six points by Glen and Evans (1979):

(1) Formulate the problem.
(2) Collect relevant information.
(3) Check the solution.
(4) Implement the solution.
(5) Monitor the performance of the system.
(6) Take any necessary corrective action.

In their analysis of the building design process Lang and Burnette (1974) considered that three basic actions need to be carried out: (1) analysis: (2) synthesis; and (3) evaluation. In broad terms, analysis requires the function(s) of a building to be separated and defined and would include a definition of the activities to be pursued in a building. A statement on the objectives of all environmental control systems would be required also. Within this statement, fire safety objectives could be defined clearly—one objective may be that not more than five people should be killed by fire in the lifetime of the building.

Synthesis would include a study of all relevant fire safety documentation which would indicate the basic level of fire safety standards which are acceptable traditionally for a particular type of building. The evaluation activity is the assessment of adequacy of the selected system to meet the initial objectives. Naturally, detailed evaluation techniques may be required to help in this assessment, especially if the protection system selected first is found to be 'too good' or 'too bad' for the mitigation of the identified risk. Figure 16.1 (after Brill, 1974), summarizes the character of the analysis, synthesis and evaluation sequence emphasizing the need for evaluative feedback which may cause changes in the earlier decisions or standards. The possibilities of integrating building design and fire safety design has been developed further by Marchant (1978). The system described in Figure 16.1 is based on the general concepts for evaluating buildings on a performance basis. For fire safety, a sequential model has been developed by Rasbash (1977) which follows a similar pattern. Although the model has twenty steps, the main sequence can be described by seven components. Each of these components requires a positive decision to be made, that is, a data processing step, before continuing the sequence.

(1) Estimate the probability of a fire being caused   (fire prevention).

(2) Estimate the courses of fire behaviour   (manage fire).

(3) Estimate the production of harmful agents by fire   (manage the

(4) Estimate direct hurt and damage   exposed risk).

(5) Estimate total expectation of loss and harm by fire   (fire accommodation).

(6) Estimate effects of changes (in protection or hazard)   (fire safety

(7) Define acceptable methods of achieving objectives   management).

The input information required to enable these estimates to be made includes definitions of hazard and objectives; the identification and quantification of

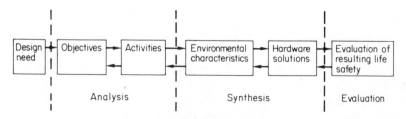

Figure 16.1 Design sequence

combustible materials; sources of ignition, conditions for fire spread, and the causal agents (human failure, mechanical and electrical forces, natural forces). A knowledge is required also of the means available for controlling a fire, and the harmful products of combustion and techniques for their control. This model aims to relate all components of fire safety into a sequential system but a depth of analysis is required to produce information useful as input information for the decision-making parts of the model.

TYPES OF INFORMATION USED IN MODELLING

Studies of fire safety problems are being pursued at many different levels ranging from international politics to the molecular structure of materials. The type of information needed to build models varies with the problem under consideration and the information needed is not always matched by the information available.

Numerical information

In direct terms, numerical information would appear to be the most useful but it is only as good as the model which generated the numbers. Fire statistics in some countries provide a broad data base useful for gross estimates and predictions of loss, of both life and property, but tell little about the small fires not reported to the relevant authority or the 'near-miss' fire or explosion.

Experimental results should be used with care as the results of an 'attitude to danger' survey is only representative of the sample of people used to model society; the fire test on a full-size structural member produces comparative results only, and its behaviour in a real building fire may be quite different to the test performance. Part of the reason for differences is that is usual to test beams, for example, as simply supported structural components. Such components in real buildings may become part of a three-dimensional reinforced concrete framework. In a building fire, beams in concrete frames can be subjected to additional stresses due to longitudinal compressive forces and torsional forces which do not exist in the standard fire test. Similarly, smoke produced from a tiny specimen under controlled experimental conditions may bear no relation to its gross performance when used as a component material in some saleable and usable product.

Observational information

This type of information may be the most reliable as it is gained from real situations, but it relates usually to one situation only. Studies of people movement (cf. Pauls and Jones, Chapters 13 and 14 of this volume) provide data which can be used directly in models of escape route design assuming that an appropriate relationship between normal and emergency behaviour can be found.

Real, experimental, fires can be set in buildings and their course of development monitored. Such tests have been instrumental in providing empirical formulae which express the probable severity of a fire in terms of the geometry of the building and the amount of fuel available. From this type of experiment mathematical models of fire development have been developed.

A third type of observational information is the elaborate case study of a fire and its reconstruction as a sequence of words and pictures, (see Lerup, Cronrath, and Liu, Chapter 10 of this volume), or as a model (reduced scale) replica (Sykes, 1977). The study of such case studies can be rewarding when common problems or deficiences in building design or the care of occupants becomes apparent. In the latter type of study, the reduced-scale real building model, if the reconstruction is successful then a great impact can be made on the teaching of fire safety to the building professions as the actual mode of smoke and fire spread through the building can be observed in laboratory conditions.

Experience information

Because the contribution to safety of the related components of fire safety cannot be quantified readily, the help of experts is used to work these components, and assign relative values, the resulting numbers being used in some models. The models of fire loss for basic premium rating and the various percentages used for additions and deductions by insurance companies are examples of expert agreement which have a long history. The making of Codes and some Regulations for fire safety in buildings include value judgements based on the experience of the members of Code committees. The same system has been used in a recent objective evaluation scheme for fire safety in health care facilities. Bannister (1977) has drawn attention to the possible weaknesses in using experts to make value judgements by suggesting that the views of the expert are not remaining as advice to policy-makers but the basis of policy. He considers also that it is wrong to entrust the assessment of risk to people of one discipline when the problem is multidisciplinary as each discipline can have an influence on the identification, quantification and mitigation of risk. However, the agreed opinion of experts on the contribution of a fire safety technique to an improvement in the overall level of fire safety should be regarded as a useful expedient in the absence of valid experimental or observational data.

Hypothetical information

In some parts of fire safety a fire development sequence which contains some assumptions may give some reality to a model of a subsystem of fire safety. Escape route design is one such subsystem. Here a common element is elapsed time. The amount of time available for safe escape is a complex function of at least 51 variables (Marchant, 1976) and an attempt was made to rank these variables with respect to their influence on the elapsed time components of escape. The values relate to the occupants under direct threat from the fire, so different values may be appropriate to occupants in other parts of a building. Similarly, the values for fire safety components reached by experience will vary with each change in the value-risk relationship.

Hypothetical information, such as random numbers, can be useful in computer simulations of fire behaviour and occupant behaviour in buildings, either where no experimental or observational information is available, or to analyse a particular configuration of building in terms of the most likely choice of escape paths. Observational and experimental information can provide a basis for the analysis of probable environmental conditions and behaviour for well-defined fire situations.

Naturally, hypothetical information can be used to build theoretical models, and because such information is from the result of some study of some fire safety components, the models will have some validity at least for isolating common problems, that is problems which are found to be important in many potential fire situations and not confined to a particular case.

PROCESSING INPUT INFORMATION

The models of information processing can be divided into the 'internal' and 'external' processes (Gregory, 1966), which may be described as psychological and technical processes; there is, however, no clear division between these processes and systems, engineering techniques are being developed to integrate both processes with respect to fire safety, although as components of the total fire safety problem they are sufficiently discrete to warrant separate discussion.

Psychological processing

Human response to variations in ambient stimuli (for example, heat, light and sound) are well documented but information on human response to fire dangers has not been collected systematically over a long period, and of course experiments are not encouraged. However, Wood (1972) (and see also Chapter 6 of this volume), Bryan (1977), and Breaux (see Chapter 8) have made significant contributions to the understanding of the gross behaviour of people in fires. From this work there is some evidence to support the notion that familiarity with a building, especially a knowledge of its spaces are important to escape attitudes.

A positive attitude to escape problems was manifest in the emergency evacuation of Robinson's store in Singapore (Smith, 1974) where, despite poor communications, more than 500 people were guided to safety by staff members. A second attitude, a non-positive reaction to escape, was taken at a fire (Whitaker, 1972) in Japan. Here, staff in a seventh floor night club did not know the location of all emergency arrangements. As a consequence 22 people jumped from windows to their death, and 96 died from the effects of smoke from a fire burning on the third and fourth floors.

The concepts of space that are held in the mind vary from person to person, as only spaces, or objects, that hold special significance for a person will be recorded properly. Lynch (1966) showed that a 'religious person', for instance, will have a distorted map of a city where the landmarks are churches. Canter (1977) reports that different users of the same space have different concepts of how that space will be used. Stea (1974) discusses the problems of cognitive mapping of buildings and refers to a case where, in normal working conditions, hospital workers were given a choice of two paths, both of the same measured length. The results showed that the route passing outside the building was given a subjective length that was twice the other.

A cognitive map contains spatial relationships to spaces outside immediate perception. Such maps, even distorted images, take time to build and will be related to the interest of the individual and his, or her, ability to form maps. It could be proposed that people of low intelligence using unfamiliar buildings would be more at risk in a fire than intelligent people in a familiar building. However, people may not be conscious of potential danger or may discount the possibility of threat to life completely. It is clear therefore that deep study of the cognitive maps and mental images that people have of the buildings which they use and their response to safety objects (fire extinguishers, etc.) and signals (alarms and emergency messages) is required, as their own and others' lives could be 'at risk' if buildings have complex spatial relationships which are difficult to learn.

Acceptable risk

A companion problem to cognitive mapping is the appreciation or realization of risk. The importance of the problem is emphasized by Rasbash (1977) who lists the principal causes of fire (and explosion) as (a) human failure; (b) failure of mechanical and electrical forces under human control; and (c) natural forces. A perusal of fire statistics will confirm this list. This suggests that whatever risk exits, its reduction could be helped by increasing the quantity and quality of safety evacuation for the population at large. Safety training could be emphasized for people pursuing activities within society where a malfunction would cause an unacceptable loss to the community. Fires which have caused such reaction include the Flixborough explosion in 1974, the Summerland fire of 1973, and the Beverly Hills Supper Club Fire in Kentucky in 1977.

Starr (1969) has considered the problems of acceptable risk and finds that for voluntary activities the risk is balanced by the individual in terms of reward or convenience. For example, in the mining industry a coal miner may accept a risk (of accident) four times greater than a stone quarry worker and his cash reward is 60 per cent greater. An acceptable level of risk, measured in deaths per person per year, is thought to be between 10^{-2} and 10^{-3}, the level of the risk of death from disease. The risk of death from fire appears to be acceptable at 10^{-5} times per person per year (Martin, 1976). If the likelihood of dying in fire became greater, then greater public anxiety would be generated assuming that such a general trend became general knowledge in a nation. Acceptability depends also on the number of people likely to die in one fire, that is, multiple fatality fires. Multiple fatality fires quickly arouse public anxiety, especially if the fires occur in public authority controlled buildings. Such fires are rare but can still precipitate hasty legislation because the concepts of public safety carried by people can be distorted easily. So much so, that a single fire causing eleven deaths can result in legislation affecting all commercial buildings, whereas eleven deaths caused separately makes little impact on the population. Involuntary risks, those which are imposed on society (nuclear power stations, for example) need to be 1000 times safer for the same benefit given to society as voluntary risks. In fire safety terms this suggests that a theatre should be safer than a dwelling (perhaps 100 times safer) and a power station 1000 times safer than a dwelling.

Technical processing

In industry and commerce the concept of acceptable risk is related to the risks of the process, or activity, the severity of the potential risk, and the cost of insurance for the particular loss potential. Haller (1977) has described the scale of risk with its effect of a system (Table 16.1).

Depending on the nature of the process, the risk to the human life associated with the system could follow the same order of potential loss. In an attempt to measure the life risk, the concept of fatal accident frequency ratings (FAFR) has been developed (Kletz, 1972). FAFR for any industry or any specific activity can be calculated if the number of deaths for each lifetime of exposure (10^8 hours of exposure to particular risks) are known. The FAFR, once calculated, can be used as a target, or objective, for the improvement of the safety of a system. Some typical FAFR's are tabulated in Table 16.2 below. This approach to the attainment of fire

Table 16.1 Risk and its effect on a system

Risk (loss)	Effect on system
Catastrophic Large	Destroys the system
Medium	Forces particular goals and expectations to change
Small Nuisance	Forces a change in particular processes

Table 16.2 Fatal accident frequency rates (FAFRs)

Industry/ activity	FAFR
(1) Professional boxers	7000
(2) Rock-climbing	4000
(3) Motorcycling	660
(4) Travelling by air	240
(5) Construction workers	67
(6) Travelling by car	57
(7) Coal mining	40
(8) Steel industry	8
(9) Chemical industry	3.5
(10) Staying at home	3

safety levels depends on the ability of the analyst to identify the causes of fatal accidents and to assign failure probabilities to all components of the system. A simple fact, that should not be ignored, is that in industries processing dangerous materials (such as, petrochemicals or plastics) a good safety record enhances worker morale.

SYSTEMS APPROACHES TO FIRE SAFETY

Malfunctions of a system which cause accidents are likely to originate with human error. Meister (1971), discussing errors in man–machine systems, states that error may be revealed as: (a) failure to perform a required action; (b) incorrect performance of a required action; and (c) the execution of an action at the wrong time. Such events lead to a conclusion that about 40 per cent of systems failures can be attributed to human error. Generally, a more detailed study of human errors and system malfunction would allow the identification of the apparent weaknesses in a system.

Fire death scenarios

The technical processing of factual information for fire safety analysis can be approached in a number of ways. Clarke and Ottoson (1976) have proposed fire death scenarios as a means of identifying those areas of fire safety where investigation should be given a high priority. They use the standard statistics available and rank the type of loss, the occupancy class of the building, the ignited agent, and the ignition source.

Figure 16.2 indicates the elements of the fire scenario and the specific loss paths. In the example, deaths were the main concern. The majority of deaths occur in residences and of these the most common item first ignited was furnishing, and the

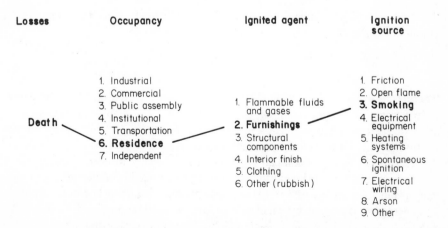

Figure 16.2 Elements of fire scenario and specific loss paths

most common source was smoking. In this case only 378 (7 X 6 X 9) causes of death are possible, but each cause can be classified by elements of fire safety, frequency of occurrence and many other ways. The objective of this scenario is to assess the likelihood of death from the various combinations of the elements so that research and legislative effort can be directed against the most common elements causing death. The scenario approach, a reverse fire sequence, that is, starting an analysis with the final result, loss (either people or money), and identifying the most likely cause of ignition to cause death, is extendable to other areas of fire loss. The basic input for this scenario method is national fire statistics, or if a large corporation is involved, then its own fire records could be useful for developing such scenarios and the records could show the 'near-miss' fires. A knowledge of these would be useful when assessing potential hazards.

LIFE SAFETY MODELS

Other types of scenario used for the study of behaviour of people and fire include the sequential analysis of actual fires (Lerup, 1977 and see also Chapter 10). This detailed 'architectural' approach endeavours to collect information on the operation of the building under normal conditions and under emergency conditions. A detailed time sequence of escape movement, smoke spread and fire spread is required, and the analysis of the mapping can lead to improvements to common problems from the analytical model of words and pictures. One important aspect of this model is that it uses a common time base for the growth of fire and the reaction and action of people. It could be fruitful to study those identified situations where things nearly went wrong. This is called the critical incident method of analysis (Singleton, 1976). In some contexts this technique is useful as there is less likely to be a distortion of the truth because no actual event takes place.

Escape route analysis

Time is a common element in the approach which attempts to relate the 51 components of escape route design (Marchant, 1976):

$$\frac{Tp + Ta + Trs}{Tf} \leqslant 1$$

Where Tp = elapsed time from ignition to perceive that a fire exists; Ta = elapsed time from perception to the beginning of safety action; Trs = elapsed time from initiation of safety action to reach a place of relative safety; Tf = elapsed time from ignition for the fire to develop untenable environmental conditions.
The factors in the expression can be developed, especially the denominator which can be extended into: (a) conditions produced by smoke in compartment of origin and then in adjacent spaces; and (b) the same divisions for flame and heat spread.
    The model developed by Stahl (1978) also uses a time base as does the 'code' approach developed by van Bogaert (1976). The former uses time units in which events take place with a particular probability (such as, which direction will escaper

choose) and has the objective of identifying those fixed planning arrangements of buildings which influence escape time. The latter assumes a target time for complete evacuation and ensures that all occupants can remain safe for any portion of the time spent inside the building waiting to escape, providing that the total time is, for schools, not longer than 4 minutes. A similar predictive method, to arrive at adjacent escape routes, has been proposed (Marchant, 1976) which assumes simultaneous warning and evacuation of all occupants. Time block analysis considers the relationship between the number of people flowing at any section in the escape route for each unit of elapsed time and the capacity of the route The technique can be used to show which sections of an escape route could be overloaded at definite time intervals after the beginning of evacuation. This would indicate a redesign of the geometry of the route or a change in evacuation pattern.

In an endeavour to relate components of threat to human occupants of buildings with safety components, a study was made in which they were identified and related simply (Figure 16.3). A further step converted the relationship chart to a simple algorithm (Figure 16.4). This type of approach has been developed by others (National Academy of Science, 1975) into probabilistic networks for assessing fire growth potential between workspaces in an office building. The input for this probabilistic model is based upon simplified probabilistic predictions of each component of fire growth, the thermal and geometrical properties of the boundaries to the space. The model is capable of extension to provide probabilities of fire spread beyond the room of origin. This model did not contain specific input on fire resistance of the construction surrounding the fire. However, the L-curve (Nelson, 1974; Wilson, 1977) for the walls or other barriers defines the probable limit to flame movement by comparing potential fire spread with the flame-stopping characteristics of the barrier. Models exist also which help to predict smoke movement through buildings and can therefore help in the design of an appropriate smoke control system. These models include three-dimensional scale models (Thomas *et al.,* 1963), mathematical equations (Morgan and Marshall, 1975; Morgan, Marshall and Goldstone, 1976) and computer programmes (Shannon, 1975; Shaw, Sander, and Tamura, 1974; Wakamutsu, 1971).

Whole building evaluation

An evaluative model for health care facilities (National Bureau of Standards, 1977) is a comparative model. The comparison is between the relevant requirements of the NFPA Life Safety Code (1976) and the fire safety and risk encountered in any existing or projected building. The objective of this evaluation system is to determine the equivalent safety of a building when evaluated against the accepted construction standards, for particular occupancies, set out in the NFPA Life Safety Code (1976).

The risk comprises five principal components: (1) patient mobility; (2) patient density; (3) fire zone location; (4) attendant:patient ratio; and (5) average age of patients. These components were chosen as representing some easily measured factors which affect the safety of the patients directly, health care buildings being

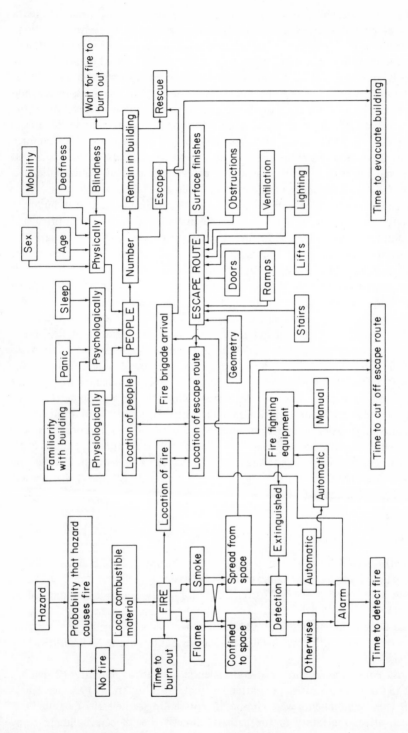

Figure 16.3 Relationship chart for people, fire and escape routes

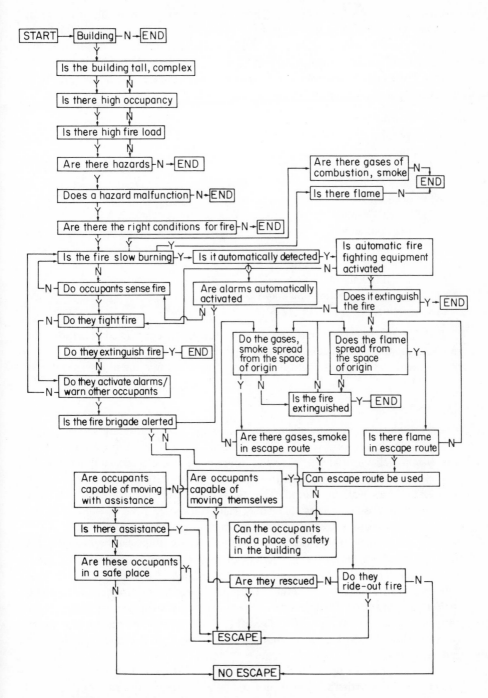

Figure 16.4 Simple algorithm for people, fire and escape route

regarded as having a high life risk. These components are given values and when combined yield an occupancy risk factor, which are checked against the general level of safety given by various features of the building, all of which are surveyed easily. There are thirteen safety features, each of which is given a value appropriate to its quality. For example non-combustible barriers have a greater fire safety value than combustible barriers. The thirteen safety features are:

(1) Construction
(2) Interior finish (corridors)
(3) Interior finish (rooms)
(4) Corridor partition walls
(5) Doors to corridor
(6) Corridor lengths
(7) Vertical openings
(8) Hazardous areas
(9) Smoke control
(10) Emergency movement routes
(11) Manual fire alarm
(12) Automatic detection and alarm
(13) Sprinklers

Each of the thirteen parameters is given values. In defined combinations the resulting numbers are checked against the risk factors and Code requirements for containment and extinguishment of a fire and for people movement (escape) safety.

Fault tree analysis

The development of probabilistic models for components of fire safety was brought into worldwide exposure in 1971 (General Services Administration, 1971). The systems reliability approach to fire safety components was established and the concept of decision trees or fault trees to model fire safety components and their interrelationships was introduced. The use of fault trees (Martin, 1976) to make systematic analyses of events that could cause unwanted fires or explosions, has developed and decision trees are available which define all the component parts of fire safety and their relationships. Although all relationships can be included, the best that can be achieved in the analysis of total fire safety is a qualitative interactive model of the fire safety system (NFPA undated). To achieve a quantitative fault tree, that is one that can define the probability of the occurrence of the event, be it a component or the objective of the fire safety system, a great deal of reliability data is required as input. Woolliscroft (1977) has proposed the use of a simple quantitative fault tree to help discuss the relative merits of fire detection and compartmentation.

A limitation in the use of fault trees (Martin, 1976) is that they can handle only discrete events and therefore may not model adequately a fire situation which is continuously varying. Fortunately, this problem may be solved by accepting that any fault tree must be no more detailed than the basic functional unit of the real

system. Values of the variables within each unit would be unique input for a particular calculation. Similarly, not all possible components will be needed to represent
a particular fire situation so that the omission or admission of particular components can be made easily. When probabilities of occurrence of particular events or
components in a logic tree are known, it is a simple arithmetic step to calculate the
probability of failure of the objective or conversely to decide which components
need to be made safer so that the objective can be met. A simple example is given in
Figure 16.5.

Assume that Figure 16.5 represents a few components of a fire safety system
for a shop. If A is an OR gate and B is an AND gate, then the probability of no
escape is 0.001 due to the smoke control, emergency lighting and door subsystem,
but the route is more likely to be blocked. This means that Pf for safe escape is
0.201, but by definition it must be 0.01. Solution—move the rubbish! If both A
and B are OR gates, then any one event could cause a malfunction. The probability
of no escape increases to 0.55. Solution—improve all subsystems or change objective.
From this example it may be clear that the use of fault trees is a simple yet powerful
modelling technique for applying to fire safety problems.

Fire insurance

Finally, a brief note on the attitude of insurance companies to quantitative models
of fire problems. Traditionally the insurance company's fire underwriter was not
concerned with the risk of loss of life in a building fire but with the value of the
building and its contents. However, the situation now is that discounts on premiums
can be given for 'good housekeeping', simple layout and for the provision of fire-
fighting equipment—all of which are fire safety components which enhance life
safety.

Assessment techniques are based on loss experience of which the majority of
fires will be recorded in insurance loss statistics. Depending on the loss performance
of each class of insured premises, the basic rate of premium may vary. Loss experience is a good guide to the risk inherent in a particular trade or process, but for a
new process or an unusual building, experience is of little use. So methods of
assessing risk and the mitigating factors have been developed. Purt (1974) describes

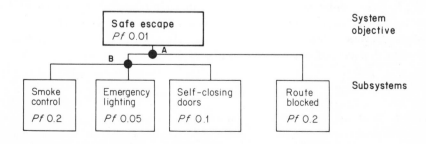

Figure 16.5 Hypothetical partial—fault tree

a simple model which helps in decisions about the level of protection. The two major risks, the building and its contents, are evaluated by two expressions:

$$GR = \frac{(QmC + Qi)BL}{WRi}$$

and

$$IR = HDF$$

where GR is the building risk
          Qm is the mobile fire load
          C is a combustibility index
          Qi is the stationary fire load
          B is a size factor
          L is extinguishing delay
          W is a fire-resistant feature
          Ri is a risk reduction factor based on combustibility

          IR is the content risk
          H is a value for danger to humans
          D is a value for danger to property
          F represents the danger of smoke formation

Figure 16.6 indicates the use of the results of this type of computation.

A Swiss evaluation scheme (Gretner, 1973) was developed also because of the lack of a system of uniform risk appraisal, especially for new risks. The basic expression used is:

$$B = \frac{P}{M} = \frac{P}{N \cdot S \cdot F}$$

where B is a fire risk
          P is total factor of potential risks
          M is total factor of protective measures
          N is a factor for normal measures
          S is a factor for special measures (say a Halon system)
          F is a fire-resistance factor

Each of these broad factors can be broken down into smaller units and quantities assigned to each from a straightforward survey and report. The allowable fire risk B is given an agreed value by the local authority. All combinations and variations in the protection methods can be made to achieve the target safety level.

CONCLUSIONS

These notes on the modelling of fire safety have been confined to a brief review of those techniques developed to improve the understanding of the people–building–fire relationship. A deeper understanding of the behaviour of these three principal components of building fires will give the building designer guidance on the

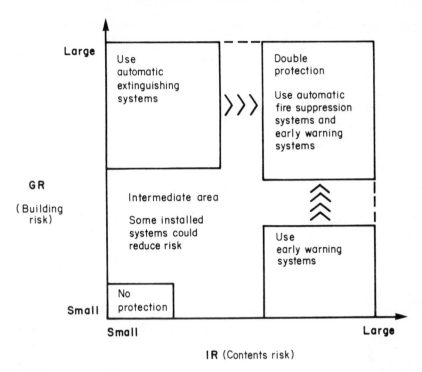

Figure 16.6 Qualitative relationship between risks and fire protection measures

problems which may be created by using particular design features or particular spatial geometries.

At present many simple combinations of the components of fire safety are understood (for example fire duration and severity are influenced directly by component geometry, window openings and the quantity and distribution of fuel). However, little knowledge is defined on the relationships between cognitive maps of buildings, fire safety education, compartmentation, the smoke release characteristics of combustible materials, and communication systems in complex buildings. The use of models, especially those containing interactive networks, may be useful in improving our knowledge of such complex relationships.

Many national governments are concerned about fire safety standards as various building types have been identified as having too high an incidence of fatal fires. In the Unites States domestic homes and health care buildings are both recognized as high life risk buildings. In the United Kingdom, hotels have been over the last six years the subject of intense upgrading activity (that is building work and improved management aimed at increasing the level of fire safety). This work on hotels is expected to spread to all countries in the European Economic Community. In all upgrading work, or in the assessment of projected buildings, some acceptable level of safety must be worked towards using a very flexible approach in the choice of fire safety techniques to give the expected standard of safety. The whole building evaluation scheme described briefly above gives a framework of an equivalent safety system of appraisal. This shows the various choices of fire safety technology to be evaluated against a common standard thereby enabling the most cost-effective measures to be taken.

When fault trees analysis techniques are developed fully, they will give building designers the opportunity to work towards definite fire safety objectives (perhaps an annual likelihood of death by fire and/or the chance of losing property or goods to a particular value). The designer should then have complete flexibility in the choice of fire safety components because each component will have a quantifiable contribution to safety and if, after an initial choice of component, the risk is still too high, then further choices can be made to make the *actual* risk match the *acceptable* risk.

Finally, the use of models in fire safety is essential to the development and understanding of the subject and to the proper practical application of the results of research and testing.

APPENDIX: SOME COMPONENTS OF AN ESCAPE ROUTE SYSTEM

Table 16.3 shows some of the components of an escape route system is an attempt to identify their contribution to the safety of an escaper in relation to the time for

Table 16.3 Components of an escape route system

Component		Tp	Ta	Time Trs	Ts	Tf
Detection systems:	Gas	1				1
	Smoke	2				2
	Heat	3				3
	Human	1	2			4
Alarm systems:	Audible	1	2			
	Visual	1	2			
	Tactile	2	3			
	Olfactory	2	3			
People factors:	Sensory:					
	Hearing	1	2	3	3	
	Vision	1	1	2	2	
	Touch	4	4	4	3	
	Smell	2				
	Psychological function	1	2			4
	Physical function			1	1	
	Physiological function		2	2	2	
	Social habits	2	2	2	3	2
	Number		3	1	1	4
	Type of group		3	1	1	
	Distribution of people	4	4	1	2	
	Density			1	1	
	Mobility			1	1	
	Type of escape possible			1	1	
	Training	3	2	1	1	2

Table 16.3 Components of an escape route system *(Continued)*

Component		Time				
		Tp	Ta	Trs	Ts	Tf
	Management		2	2	2	2
	Time of day	2	2	3	3	2
	Season of year			3	3	4
Building factors:	Size of room		3	2		3
	Shape of room	3	3	2		3
	Smoke load	2	2	2	3	1
	Fire load	2	2	2	3	1
	Sources of ignition	2	2	2	3	1
	Exit location		3	1		
	Exit size			1		
	Escape route size			2	1	
	Direction finding				1	
	Stair geometry				1	
	Lifts				2	
	Refuges				2	
	Maintenance			3	2	2
	Secondary power supply			3	2	
Fire control systems:	Fire resistance			3	2	3
	Active suppression			2	3	2
	Smoke control			3	2	1
	Emergency lighting			3	2	
	Communication systems	3	2	2	3	
Fire brigade:	Attendance time			3	2	3
	Access to building			3	2	2
	Rescue			2	1	
	Care of escapers				3	
	Firefighting					2

Sources: Building Standards (Scotland) (Consolidation) Regulations (1971). Milinskii, A. I. (1966); Caravaty, R. D. and Winslow, W. F. (1970); Ferguson, R. S. (1972); Melinek, S. J. and Booth, S. (1975).

safe escape where Tp is the elapsed time from ignition to perceive that a fire exists; Ta is the elapsed time from perception to the beginning of safety action; Trs is the elapsed time from initiation of safety action to reach a place of relative safety; Ts is the elapsed time from reaching a place of relative safety to reaching a refuge or open air; and Tf is the elapsed time from ignition for the fire to develop untenable environmental conditions. In each case the relative importance of each component is ranked between 1 and 5, 1 being the most important influence. The list is not exhaustive, nor would all components need consideration in each building. A careful consideration of the interactions between components is necessary once some facts about each component are known.

REFERENCES

Bannister, J. E. (1977). 'Managerial aspects of risk analysis—The contribution of the expert to disasters' *Foresight*, **III, September,** 15.
van Bogaert, (1976). *Prospective dans la Construction Scolaire.* (Louvain: Vander).

Brill, M. (1974). 'Evaluating buildings on a performance basis' in Lang *et al.*, (eds.) *Designing for Human Behaviour: Architecture and the Behavioural Sciences,* (Stroudsburg, Pa.: Dewdon, Hutchinson and Ross, Inc.). p. 316.

Bryan, J. (1977). 'Smoke as a determinant of human behaviour in fire situations (project people)'. University of Maryland June.

Building Standards (Scotland) (Consolidation) Regulations (1971). (Edinburgh: HMSO).

Canter, D. (1977). *Psychology of Place.* (London: Architectural Press Ltd).

Caravaty, R. D., and Winslow, W. F. (1970). 'A new approach to fire codes', *Architectural and Engineering News,* **March,** 22–25.

Clarke, F. B., and Ottoson, J. (1976). 'Fire death scenarios and fire safety planning', *Fire Journal,* **May,** 20.

Ferguson, R. S. (1972). 'User-need studies to improve building codes', Technical Paper No 368, Division of Building Research, National Council of Canada.

General Services Administration (1971). *Proceedings: Reconvened International Conference on Fire Safety in High Rise Buildings.* (Public Buildings Service, General Services Administration). (Washington DC).

Glen, J., and Evans, J. O. (1979). 'Management techniques relevant to fire safety', in Fire Safety Management, Edinburgh University.

Gregory, S. A. (1966). 'Models in practical design'. In S. A. Gregory *The Design Method* (London: Butterworths), Chapter 17.

Gretner, M. (1973). 'Assessment of fire risks and the deduction of safety measures'. Fire Prevention Service for Industry and Trade (VKF).

Haller, M.G. (1977). 'The aim of risk management'. In *Approaches to Risk Management.* (London: Keith Shipton Developments Ltd.).

Kletz, T. A. (1972). *Hazard Analysis—A Quantitative Approach to Safety.* Scottish Fire/Safety Conference Report No 35 1971–72.

Lang, J., and Burnette, B. C. (1974). 'A model of the designing process', in Lang *et al.,* (eds.) *Designing for Human Behavior: Architecture and the Behavioral Sciences* (Stroudsburg, Pa.: Dewdon Hutchinson and Ross, Inc.) p. 43.

Lerup, L. (1977). *People in Fires: A Manual for Mapping.* (Berkeley: University of California).

Lerup, L., Cronrath, D., and Liu, J. K. C. (1977). *Human Behavior in Institutional Fires and its Design Implications.* Berkeley: University of California.

Lerup, L., Greenwood, D., and Burke, J. S. (1976). *Mapping of Recurrent Behaviour Patterns in Institutional Buildings under Fire: Ten Case Studies of Nursing Facilities.* Berkeley: University of California.

Lynch, K. (1966). *The Image of the City.* (Cambridge, Mass.: The MIT Press).

Marchant, E. W. (1976). 'Some aspects of human behaviour and escape route design'. 5th Int. Fire Protection Seminar, Karlsruhe, September.

Marchant, E. W. (1978). 'Integration of fire safety design', *Public Service and Local Government,* **February,** 8, (2), 42–44.

Martin, J. (1976). 'Systems performance: Human factors and systems failures—Unit 7/8—Engineering Reliability Techniques'. (Milton Keynes: Open University Press).

Meister, D. (1971). 'Human error in man-machine systems'. In D. Meister, *Human Factors: Theory and Practice.* (New York: John Wiley & Sons).

Melinek, S. J., and Booth, S. (1975). *An Analysis of Evacuation Times and the Movement of Crowds in Buildings.* Building Research Establishment Current Paper 96/75. Department of the Environment.

Milinski, A. T. (1966). 'Principles for regulating the evacuation of public buildings', in *Fire Prevention and Firefighting Symposium.* Trans. from the Russian.

Morgan, H. P., and Marshall, N. R. (1975). *Smoke Hazards in Covered, Multi-level Shopping Malls: An Experimentally Based Theory for Smoke Production.* Building

Research Establishment Current Paper 48/75. (Borehamwood: Fire Research Station).

Morgan, H. P., Marshall, N. R., and Goldstone, B. M. (1976). 'Smoke hazards in covered multi-level shopping malls: Some studies using a model 2-storey wall'. Building Research Establishment Current Paper 45/76. (Borehamwood: Fire Research Station).

National Academy of Sciences (1975). Task Group T-57 'Program for developing and implementing a new approach to designing for fire safety in buildings'. Technical Report No 67, Washington.

National Bureau of Standards (1977). *Manual for the Fire Zone Fire Safety Evaluation Worksheet for Health Care Facilities.* (Washington DC: Center for Fire Research).

National Fire Prevention Association. (Undated). *Decision Tree for Fire Safety.* (Boston, Mass.: NFPA).

Needham, A. M. (1966). 'A practical design: An oil burner for large water tube boilers'. In S. A. Gregory, *The Design Method.* Chapter 10 (London: Butterworths).

Nelson, H. E. (1974). 'The application of systems analysis to building fire safety design'. (Washington DC: General Services Admin.).

Purt, G. A. (1974). 'The evaluation of fire risk as a basis for planning automatic fire protection systems'. *Fire Technology* **July,** 291–300.

Rasbash, D. J. (1977). 'The definition and evaluation of fire safety', *Fire Prevention Science and Technology,* **16,** 17–22.

Shannon, J. M. A. (1975). 'Computer analysis of the movement and control of smoke in buildings with mechanical and natural ventilation'. Paper 9 in CIB Symposium on the Control of Smoke Movement in Building Fires. *Proceedings Vol 1,* pp. 99 (Borehamwood: Fire Research Station).

Shaw, C. Y., Sander, D. M. and Tamura, G. T. (1974). 'A Fortran IV program to simulate stair-shaft pressurisation system in multistorey building'. Division of Building Research, National Research Council, Canada, DBR Computer Program No 38.

Singleton, W. T. (1976). *Human Aspects of Safety.* (London: Keith Shipton Developments Ltd.).

Smith, P. B. (1974). 'Nine dead in lift at Robinson's Department Store, Singapore', *Fire,* **66, 823,** 399–400.

Stahl, F. I. (1978). Final Report on the 'BFIRES/Version' Computer simulation of emergency egress behavior during fires: calibration and analysis. (Washington DC: Dept. Health Education and Welfare).

Starr, C. (1969). 'Social benefit versus technological risk'. *Science,* **165,** American Association of the Advancement of Science.

Stea, D. (1974). 'Architecture in the head: cognitive mapping'. In Lang *et al., Designing for Human Behaviour: Architecture and the Behavioural Sciences,* (Stroudsburg, Pa.: Dewdon, Hutchinson and Ross, Inc.). p. 157.

Sykes, F. (1977). Private communication about investigation at Leeds Polytechnic, England.

Thomas, P. H. *et al.,* (1963). *Investigations in the Flow of Hot Gases in Roof Venting.* Fire Research Technical Paper, No 7. (London: HMSO).

Wakamatsu, T. (1971). *Calculation of Smoke Movement in Buildings.* (Tokyo: Building Research Institute), Research Paper No 46.

Whitaker, E. H. (1972). '118 die in smoke from fire three floors below', *Fire,* **65, (807),** 165, 167.

Wilson, R. (1977). 'The L-curve: evaluating fire protection tradeoffs', *Specifying Engineer,* **May,** 110.

Wood, P. G. (1972). *The Behaviour of People in Fires.* Fire Research Note No 953.
(Borehamwood: Fire Research Station).
Woolliscroft, M. (1977). 'Detection *vs.* compartmentation−hospitals and residential
homes', *Fire Surveyor,* **6, (4),** 20−23.

Fires and Human Behaviour
Edited by D. Canter
© 1980 John Wiley & Sons Ltd.

CHAPTER 17

Communications Strategies for Fire Loss Reduction

RICHARD R. STROTHER AND
LAURA BAKER BUCHBINDER
United States Fire Administration, Washington

INTRODUCTION

Well-planned, targeted public fire education programmes have been effective in reducing fire loss in local communities. This paper focuses on strategies for transferring technological fire protection solutions to users and changing human behaviour related to fire and fire safety. A review of effective education strategies is followed by a description of how these strategies are being used by the US Fire Administration to achieve fire loss reduction within communities.

Fire is a major problem within the United States, causing 7500 deaths, over 300 000 injuries and S12 billion dollars in property loss each year (United States Fire Administration, 1978). It is responsible for more loss of life and property than all natural disasters combined, and the second most frequent cause of accidental death in the home.

USFA FIRE LOSS REDUCTION STRATEGY

Solutions to fire problems generally fall into three categories: behavioural, structural and suppression. In all cases, the solutions must be disseminated to the intended user in a form that will promote the probability of adoption. The Office of Planning and Education's strategy for reducing fire loss through public fire education is simple in concept. If fire is, in fact, a local problem, then OPE should:

(1) Involve local fire educators in determining public fire education needs and resources.

(2) Provide the technical assistance and support needed to initiate public education programme development at both the state and local level.

(3) Develop field usable products (tools and techniques) to aid local fire educators in planning, implementing and evaluating their own public fire education programmes.

In 1971 the National Commission on Fire Prevention and Control was established to investigate the United States fire problem and recommend measures for
reducing fire losses. The Commission's report (Bland, 1973) stated: 'Among the
many measures that can be taken to reduce fire losses, perhaps none is more important than educating people about fire'.

The Unites States Fire Administration (USFA) (formerly the National Fire Prevention and Control Administration) was established by the Federal Fire Prevention
and Control Act of 1974. The Office of Planning and Education (OPE) was given
the mission to reduce fire loss through planning and implementation of public fire
education programmes.

TARGETING AND LOCAL INVOLVEMENT KEY FACTORS

To develop its strategies, the OPE first identified fifteen fire education programmes
which were successful in reducing fire losses, and studied the basis of their success,
(Strother, 1975). Our examination indicated that two factors were associated with
loss reduction: (a) targeting education programmes at local fire problems; and (b)
involving the community in programme development and implementation. The
research indicated that for an individual to change unsafe behaviour, the problem
must be perceived as local, immediate and personally relevant. In addition, delivery
or reinforcement of the prevention message by a community leader increased the
probability of acceptance.

The Louisiana study: local 'contactors' used

The importance of targeting the education programme as a local problem and
involving the community in its solution was illustrated in 1973 in Beauregard Parish,
Louisiana, by the US Forest Service and sociologists from Louisiana State University (Doolittle and Welch, 1974). The education programme to reduce intentional
burning of woods in rural areas used local opinion leaders to deliver the education
message.

This programme focused on the communications networks in rural communities.
Although Beauregard Parish had been the consistent target of mass media and other
types of messages designed to change existing fire attitudes and practices, previous
programmes had not been effective. The project report (Doolittle and Welch, 1974)
states: 'The Southern Region of the United States has had a 38.1 per cent increase
in the number of fires and a 66.2 per cent increase in the number of acres burned
from 1961 to 1967'.

The problem, according to the report, was not one of fire prevention message
reception. Rather, it was one of selective perception. The woods-burner did not
perceive that the messages being sent via TV, radio, newspaper or billboards applied
to him. His selective interpretation of fire information recived via the mass media
was apparently rationalized in terms of economic and other motives which lie
behind deliberate firesetting activities in rural communities.

To change the traditional firesetter's behaviour patterns, a prevention programme was designed which used influential local people and opinion leaders to
reinforce the prevention messages in each community. A 55 per cent reduction in
set forest fires in Beauregard Parish was realized. At the start of the fire prevention
programme, the ten-year average of forest fires in the county was 800 per year.
Within the county, Beauregard Parish accounted for 22 per cent of those fires. Five
years after the start of the programme, the incidence in Beauregard Parish had
decreased to 11 per cent of the county average (Doolittle and Welch, 1974).

Boston, Massachusetts community arson prevention programme

Community involvement may occur through the direct involvement of individuals
living in a high-risk area. Such was the case in the Symphony Road neighbourhood
in Boston, Massachusetts. This once elegant neighbourhood had experienced a series
of fires in which six people had died. It was not, however, the tenants who were
causing the Symphony Road fires; the fires were a reflection of a deeper problem—a
problem common to scores of inner city neighbourhoods throughout America. The
housing had reached the bottom of an economic cycle. Faced with building and
health code violations, high tax bills, and maintenance problems aggravated by
years of neglect, the landlords sought to renovate their properties at insurance and
public expense. Arson for profit was the cause of the Symphony Road fires.

It was through the efforts of a neighbourhood group, the Symphony Tenants
Organizing Project (STOP) that a six-million dollar arson ring was discovered and 33
people were indicted. The group provided State Attorney General, Francis Bellotti,
with documentary proof of a pattern of real estate transactions and inflated insurance values preceding the fires. Information supplied by STOP provided the basis
for the more detailed investigations conducted by private insurance investigators
that resulted in convictions.

THE NEED FOR A CLEAR AND SPECIFIC EDUCATIONAL MESSAGE

In 1975, a study of the motivational psychology related to fire prevention behaviour, (Strother and Kahn, 1974), funded by the National Fire Protection Association, provided insight on how to communicate fire safety information to the
public. The study, which provided the basis of NFPA's successful 'Learn Not to
Burn' media fire education programme, stressed that fire prevention messages must
be explicit and must show the desired behaviour in the context where the action
should occur.

Many people interviewed in the study expressed a strong underlying fear and
concern about fire. The contradiction of apathy in the face of strong emotional
response appeared to be caused by feelings of inability to cope with vague and
complex fire safety procedures which the people are told to follow. Under threat or
tension people will seek ways to reduce that tension: Some will seek new information in order to act effectively; others will block out the threat which is creating
the tension. Therefore, to be effective, fire safety information must be directed

toward specific, positive motives which already exist within the target audience. Messages must be explicit and they must not overload the receiver. They must show the context in which the desired behaviour will occur and clearly explain specific steps to be taken.

MATCHING THE EDUCATIONAL MESSAGE TO THE TARGET AUDIENCE

The importance of specificity cannot be overly stressed. Various age groups, socio-economic sectors, and geographic regions are prone to particular types of accidents involving fire. The hazards encountered in the urban core are not the same as those in the suburban or rural areas. Clothing and heating needs in the North lead to different hazards from those in the South. People from impoverished backgrounds will react differently to perceived dangers than those raised in the more affluent middle class. For all these reasons, school curricula and other public education campaigns must be designed not only with the fire problem in mind, but with an awareness of the precise audience to whom the programme will be directed. Those community public fire education programmes which have resulted in loss reduction have been effective because they were targeted at specific local fire problems and have involved the people of the community. The desired change in behaviour is clearly shown (for example, 'If your clothing ignites, stop, drop and roll'.). The messages have been relevant to the local problems and to the specific target audience to be reached.

SMOKE DETECTOR CAMPAIGN

Public fire education can be used to disseminate technical innovations as well as fire safety information. Our intensive smoke detector campaign was based on a two-step communication strategy. Technical papers and research on smoke detectors from the National Bureau of Standards was translated and packaged for public consumption in six targeted publications. First, smoke detector facts were broadcast through the media and informational leaflets. The second step made use of community leaders, senior citizens and members of women's clubs to go from house to house to reinforce the messages and assist the homeowner in properly installing the detector and developing an escape plan. At the same time, the homes visited were 'inspected' for other fire hazards. In one California town using this programme fire loss dropped 46 per cent.

THE FIVE-STEP PLANNING PROCESS

Systematic planning of public fire education programmes has proven to be essential to achieving a measurable impact. To this end, a detailed planning process has been developed by OPE (Public Education Office, 1978).

(1) Identification—asks the local fire educator to analyse his local fire rates to determine times, locations, causes, victims, etc. to build a detailed profile of the local problems.

(2) Selection—calls for a review of resources, and manpower, a survey of the intended audience and a selection of educational objectives.

(3) Design—requires that the fire educator focus on his target audience in planning the programme content, message and format.

(4) Implementation—involves training of volunteers, distribution of materials, and monitoring activities.

(5) Evaluation—calls for an assessment of the impact the programme has had on
   (a) education: changing awareness knowledge and behaviour;
   (b) organizations: changing allocation of manpower funds and mission;
   (c) risk reduction: removing hazards and leaving proper behaviours;
   (d) loss reduction: measurable changes in death, injury, incidence or loss rates.

The Five-Step Planning Process has proven to be a valuable tool in planning and focusing fire education efforts.

PUBLIC EDUCATION ASSISTANCE PROGRAM (PEAP)

Through interactions with leaders in the fire education field, OPE learned that fire educators need guidance, materials and training to be effective in educating their communities about fire safety. Because the state seemed to be the logical jurisdictional entity for focusing on the needs, the Public Education Assistance Program (PEAP) was established.

The purpose of PEAP is three-fold: To improve the state's capacity to provide (a) leadership: (b) information and materials; and (c) technical assistance in planning, implementing and evaluating public fire education programmes to communities within their jurisdiction.

To meet these goals, PEAP has three components: administrative, resource system, and technical assistance. The objective of the admistrative component is to make the public education programme an integral part of the state fire service organization. The resource system component provides fire educators with fire education programme materials from centrally located resource centres. Building the local capacity to plan, implement and evaluate effective public education programmes is the objective of the technical assistance component. While the three components are distinct, each contributes to the overall effectiveness of the programme.

Common objectives and individual accomplishments

While all states participating in PEAP share common objectives, the unique needs and resources of each state lead to different types of programmes, organizations and accomplishments. California has adopted a regional resource system concept so that educators can locate existing programmes and materials from neighbouring communities, and adapted fire education planning techniques to use data available from the California Fire Incident Reporting System (CFIRS). In this way, educators identify fire problems using local data and then target education strategies to solve that problem.

In Delaware, both civilians and fire fighters are certified as Fire Safety Instructors after completion of special training at the Delaware State Fire School.

Illinois educators have developed a 40-hour course which is now required for state certification as a Public Education Officer.

In Oregon, ten communities have each received $500 grants to identify local fire problems and develop local programmes to solve those problems. The results of these programmes are shared with other Oregon educators through a statewide newsletter and conferences.

Each of the four pilot states conducted extensive preplanning to include a year-long planning phase. Each state has hosted statewide public fire education conferences. Regional meetings, steering committees, and newsletters are among the other ways being used to get local input into statewide fire education planning.

The philosophy of the PEAP approach to states has been summed up by Connecticut Fire Administrator Bill Porter: 'Mutual aid is a way of life for the fire service. We are planning ways to apply the concept of mutual aid to public eduation programmes. Our final state plan will tell us how to take advantage of the ideas, materials and expertise of educators throughout the state.'

Measures of success

Evaluating the impact of statewide fire eduction programmes is central to the PEAP concept. Participating states and OPE both evaluate programme impact; documentation of programme accomplishments and pitfalls helps states plan public education programmes. In addition, evaluation results will help guide other states. 'We are reluctant about formally evaluating our programme at first,' says one fire educator. 'Now we see evaluation as a part of overall programme management. The information is useful for monitoring ongoing activities and for future planning.'

RESOURCE EXCHANGE BULLETIN

OPE serves as a national 'switchboard' or clearinghouse for fire education resources. Information on existing programmes is collected by interviewing fire educators. Descriptions of programmes are then assembled and published in a 'Resource Exchange Bulletin' (REB). The name, address and telephone number of the local fire educator who developed the programme is included. Readers of the REB are encouraged to contact the resource person directly for more information and programme reprints. The REB has become a valuable tool for fire eduators by providing visibility for local programmes, current programme information, and a forum for presenting and discussing fire education issues.

Public fire education—present and future

State and local fire educators were actively involved in OPE's programme from its inception. The programmes they developed and their experience in community public fire education provided a pool of resources to be used by both OPE staff and those entering the fire education field. Local input and advice was crucial in shaping a federal programme that responded to local needs and supported state and local efforts.

Public fire education is now recognized as a fire loss reduction technique within the United States. The USFA is convinced that continued support of state and local public fire education efforts will result in decreased fire incidence, property and life loss within the United States.

REFERENCES

Bland, R. E. (1973). Chairman, *America Burning*. Report of the National Commission on Fire Prevention and Control.

Doolittle, M. L. and Welch, G. D. (1974). 'Personal contact pays off', *Journal of Forestry*, **August**.

Public Education Office (1978). *Public Education Assistance Program Guidelines*, National Fire Prevention and Control Administration, and *Public Fire Education Planning: A Five-Step Process*, United States Fire Administration.

Strother, R. R. and Kahn, C. (1974). *A Study of Motivational Psychology Related to Fire Preventive Behavior in Children and Adults*. National Fire Protection Association.

Strother, R. R. (1975). *Review of Successful Programs in Fire Safety Education*. National Fire Prevention and Control Administration.

United States Fire Administration (1978). *Fire in the United States*.

Fires and Human Behaviour
Edited by D. Canter
© 1980 John Wiley & Sons Ltd.

CHAPTER 18

Experiencing Fires

CHARLES CLISBY

THE AGONY AND THE WOE OF IT

A child cries loudly in the night, its mother sharply coughs,
For death has made an early start, at kind and creed it scoffs.
Lost race against a ticking clock to infant's cot is run,
Loved one is gathered up, the last maternal act is done.
The infant and its mother soon unconscious of their doom,
Await the gasping Firemen in a corner of the room.

Experiencing fires leaves a mark on a man. The mark left on me after experiencing fires for over three decades in our capital city is indelible. I nurture a savage resentment in my breast. The heart bursts with pride when a life is saved—the heart breaks when one must walk with care so as not to further violate the wretched remains of a life lost.

Fire is a phenomenon of challenging magnitude, a test of experience, a fight to be fought, an engagement to be won and quickly too if life and property are to be saved. Destruction wrought by fire can shock when viewed in the clear cold light of day.

It was a privilege of rank that I left the scene when all persons were accounted for and when smoke changed from ruby-black through lightening browns, a sandy-white mixture, penultimately to near white. Pure white indicates steam is being given off. At that final stage of fire fighting, the fireman himself should be allowed the triumph due to him, soaked through to the skin, sweat squelching in his boots, exhausted, heart thumping, gulping fresh air, beetroot-complexioned below the soot and grime where any part of his face or neck has been exposed to the effects of fire.

It was also my duty to return when all was cool, close down the incident, carefully phrase the messages and arrange a damping-down watching brief, often completed in roofless conditions over many days and nights in all kinds of weather.

Still probing about purposefully was the fire prevention officer who, like a referee—certainly not like a linesman—had observed at first hand the scene from start to finish, seat of fire, direction of travel, reactions, intensities and effects.

It is a matter of fact that loss of life by fire cannot be compared to that lost in other public spheres such as on our roads. Yet I am struck with another fact, which is that there is more public concern over one life lost by fire than by any number on the roads, except where groups of people expire together. Regional television, national radio and local newspapers feature a life lost at a fire but never feature a single death in a road accident except where the victim is a person considered important in the news value sense.

The fireman has accountability. He is expected to save life, no matter whose it is; failure to do so is widely reported. James Hunt, the internationally famous racing driver, when commenting on death on the racing circuit, singled out death by fire in the following manner: 'The thing about fire is that it is a particularly nasty way to die. Frankly, I admit I am frightened to die that way.' All of the human senses detect the presence or approach of fire. Firemen are familiar with these symptoms, meeting with them constantly. In my time I have, on seven separate occasions, assisted at breaking, or personally broke the news where a firemen has been killed at a fire. The book *Red Watch* by Gordon Honeycombe (1976) deals with one of those seven occasions. It should be read for the day-by-day account it gives of what the fireman does for a living. A great many firemen suffer injury from fire or are subject to ill-health arising from cumulative doses of exposure due to the effects of fire. That is why the fireman's efforts to save life are near frantic. He abandons safety procedures. Frankly, I admit it, the ticking clock is his enemy too.

The fireman wins or loses in his task of saving life and property in an amount directly proportional to what has gone on in planning in fire cases—who actually did the planning and on what basic thinking the planning was initiated, that is, the true purpose of the planning, prominent or underlying, overt or covert. The main procrastinating arguments have been firstly to await the outcome of international and later European deliberations of experts, none of whom are firemen; the French representative is a bridge-builder, the Italian a railway employee. Most work behind desks for the government concerned. Politicians also appear to be involved, for point-scoring is evident in reports. The fireman has to fight for recognition. He is considered an outsider. He does not, for instance, slot neatly into what he sees as the callous philosophy of writing off hundreds of thousands of living people in a nuclear war. The fire service itself is written off during a nuclear war. Its prime function is to provide uncontaminated water necessary to the survival of those chosen to be saved. I can see firemen sneaking water to the abandoned and being shot for doing so, but to the fireman, the loss of even one life is a disaster of importance.

By now the aims of my share in this worthwhile book should be abundantly clear. My intention is to disturb complacency. The reader is judge. When the fireman fails, what must be fully appreciated is that people other than he, people who have not experienced fires or who do not consider experiencing fires as necessary to their expertise, are to an ever increasing degree becoming involved to the exclusion of the fireman in setting the scene.

COMPLICATING FACTORS

I lay and smoked and wondered,
After we got back last night,
Child at fire had turned quite black.
Never seen a black turn white!
How is that explained away?
Racialists could put me right.
Whispered little prayer for it.
Thanked our God that I'm not bright.

In line with the vivid impressions I have gained over 30 years and in context with this unique book, it is important that the true nature of the man who responds to a call of fire is completely understood. I offer that what I have said so far, and what I have yet to say, will prove conclusively that the fireman should participate fully in any planning to save life and property from fire, administrate and apply the outcome, and above all have the power of veto over all other interests. Experiencing fires counts, and the fireman should not be blinded with science, nor should his straightforward approach to fire problems be disregarded as unqualified.

To the observing fireman, fireproof buildings are no safeguard at all for the people who work or reside in them. Fireproofing is a fatuous criterion. There has been an obsession over the years to limit destruction by fire in buildings which has resulted in a blurring of preferences—fire loss versus loss of life. The human species is certainly not fireproof or even fire-resisting, nor are the *usual* items humans take into or install in buildings. Easy means of escape should rate far and above fire protection except for the escape route. Requirements in fire-fighting media suggest delay in making an escape. It ignores the point that there is a fire service, and if attendance times and initial attendances are wanting those deficiencies should be put right. People, like God, act in strange ways.

Some people, who we will call black, obdurately walk unhurriedly in the path of fire until they are consumed by its extraordinary speed of travel or by that which precedes fire, the invisible products of the visible. Then there are those who we will call white, who dash away from the smallest flame or smell of smoke, felling all who get in their way. Naturally many shades of grey exist, yet all colours possess chameleon-like properties. Predicting reaction is wholly worthless. Beware of sampling, ours is a multicoloured society.

Firemen hold suspect all buildings, structures and engineering products. These marvels never seem to match up to design factors, and even when design factors match up to a fire situation, complicating factors intervene to thwart fire-fighting and rescue. A small fire exposes structural steelwork found to be inadequately clad in concrete, a floor or roof gives suddenly; material of smaller size or of less factors of safety than that specified appears to be involved. It could perhaps be that factors of safety are ill-conceived in respect of fire.

Then there are the continuing arguments between the fireman and the various departments over the soundness in fire of structures in which high alumina cement are incorporated or where calcium chloride, corrosive to steel, has been used for quick drying of concrete. The fireman 'treads warily', heeding the warning from the Demolition Contractors' Association, based on their experience.

Ever since officialdom gave way by setting aside Christopher Wren's plan for rebuilding London following the Great Fire, both city and metropolis have followed a worrying course. The situation recovered in part for a while when the fireman's voice was heard. For this reason the fireman, in the presence of disaster, works on the premise that people involved do not act in a manner prescribed so as to save them. That is why the fireman wins more times than he loses. In 1976–7 many more persons were saved from fire than lost their lives to fire. To quote a case in point: not so long ago, no building was erected above the height attained by the tallest ladder in the brigade. In this precaution was embodied the result of actually experiencing fires, but officialdom gave way again. Today the tall building is London's chief headache (let alone eyesore). The plaudits of the whole world went out to those planners who opened up views of Westminster and St Paul's Cathedral at local street level, views swathed through tall buildings. All over London, however, tall buildings are springing up, denying whole sectors of London long-range views of those magnificent edifices.

For example, the man who died on the thirteenth floor of a block of residential flats came out on to a rear balcony to escape the blaze. The brigade's turntable ladder (reaching as it can no more than ten floors) could not get to the rear. Attempted rescue by hook-ladder from the balcony below was also thwarted; a hook-ladder could not be manoeuvred without difficulty up the stairway in time.

At Shene Street in Shoreditch (and then at Glasgow hardly a month later) numbers of people were prevented by iron-barred windows from making their escape from fire. Crime prevention had superseded prevention of loss of life from fire. One wonders at the centuries of lack of composite policy. Let us look further into this perplexing conundrum. The aged and infirm were expected to run inordinate distances to smoke-stop doors. They were even expected to drop to the safety of a balcony below, until the fire service brought the matter to public notice by ridicule. Admittedly, anybody can make a mistake. It is indifference to error which is so distressing, the unaccountability which is so frustrating.

One hears of fire prevention officers being caught up in a welter of meetings and letter-writing with say, factory inspectors, while work of a dangerous nature continues on an overtime basis. Employees of a fire authority may not criticize the authority or another department of the authority to draw public attention to deficiency. Nor may they openly criticize local authorities, although a coroner's inquest sets the embargo aside.

Civil servants can have their own pet consideration incorporated in regulations, giving themselves power and enhancing their job descriptions, or can place those financially profitable resources in the hands of counterparts in the service of a local authority. The wrong man is empowered to make decisions and take action. I argued with a local authority dangerous structures surveyor that a 18 m (60 ft) high wall affected by fire would fall into the River Thames. He said it would not. It did.

DOES THE FIREMAN DO ENOUGH?

We are weary, sad at heart,
Never had a chance from start,
Far too late we got the call,
Gave us wrong address and all.

Broken in was large front door
Giving vent to fire in store.
By the time we did arrive
No-one there could be alive.

One was found at head of stairs,
All the rest lay dead in pairs.
Can you say to those who try,
'Have you never reasoned why?'

It is in the old buildings that we lose life and yet these products of Victoriana somehow seem to gain an exemption to almost every new act of parliament. The fireman's expertise has been tested to the limit in coming to terms with them and arranging fire safety improvements in them. The fireman is utterly dismayed when lives are lost at a fire in needless circumstances but, where the victim did not avail himself of obvious means of escape, loss of life appears to the fireman to be an incredible tragedy. At Shene Street escape could have been achieved by going down into the basement. Fire was upstairs as well as blocking the way of escape to the street. The way left open was via stallboard doors, but the victims were aware the stallboards doors were padlocked on the outside. How they must have searched their minds for a way out! Would anyone hear them if they went down into the basement and hammered on the doors? Was the Fire Brigade coming? The street was derelict, due for demolition. The iron-barred windows cruelly defied the victims. How were they to get out? Which way to safety? Fire never got into the basement. Five people died! Too often, people are faced with a similar dilemma. Should people have to face a dilemma few are capable of unravelling? Better salvation! No way except to safety!

Victims are often well aware of the location of means of escape having been resident in or worked in familiar surroundings over many years. The ideal surely must be to eliminate the need to remember when under stress. Certainly opposing circumstances arise which deny escape to the unfortunate. Would-be rescuers, themselves ignorant of the enormity of their crime, burst open doors and smash windows at a fire in a frenzy equalling the last moments of those they have caused to be trapped. Some people, native to favourable climes, and unused to space heating, place the source of their undoing between themselves and means of escape.

Short-term residents at hotels and boarding establishments are often blissfully unaware of the whereabouts of means of escape. Safe within a well-fitting door, they dash themselves from the window of their room onto the pavement or sharp railings below as the fireman fights his way along the passage outside their room or erects his ladders. At one hotel fire in London the brigade rescued over 50 persons using every type of ladder carried; one young fireman rescued a man from the sixth floor by hook ladder. Conversely, an Italian, trapped with his family by a serious fire in a West London hotel room, soaked bedsheets in water in the washbasin, forced the sheets into cracks around the ill-fitting door, and left the taps running so that water cascaded on to the steaming carpet. That family was saved by firemen. Many people died in adjoining rooms and rooms above and below. Italians are considered an excitable race. Beware of categorizing people.

There is a greater language problem than ever before in our cosmopolitan cites. Vandalized public telephone boxes are a symptom of our times. A London fire brigade survey disclosed that some 60 per cent of boxes in poor areas were vandalized, so boxes were not available in areas where the private telephone is an expensive luxury. Post Office repair crews turn out immediately it is reported that a box is out of order, but far too few reports are received. While it is an offence to vandalize a public telephone box, I have always considered it a worse crime, morally, not to report boxes vandalized or out of order. The repair crew can usually reconnect the emergency side of communications even if they cannot restore the full service.

The verse following was prompted by possibly the saddest incident in my experience.

> 'I ran from box to box', he wept.
> 'A knowledge of their whereabouts
> In case of need in mind I kept.
> That one would work I had my doubts
> And so I noted nearest three.
> The thief not only all had robbed,
> He took from me my family.'

The war brought an end to street fire alarms. Bombs destroyed the network of underground cables. False alarms were numerous but vandalized alarm boxes were unheard of. No fireman ever complained about false alarms. Stories abounded of the ignorant poor, dare I say the immigrant, waiting at boxes and jumping up on fire engines, pointing the way they could not speak. Boxes were physically checked twice in each twenty-four hours by firemen. To install a land-line type of alarm box system is beyond the pocket of most fire authorities, but what of radio-operated boxes as used in the United States, which could be sited in areas where vandalized public telephone boxes are commonplace. Is there a regular data collection to define such areas, or should the fireman inspect telephone boxes at night and whenever shops and other telephone-equipped establishments are closed? (I have often wondered whether firemen should be nomadic.) Should concrete rafts be set in place where analysis discloses life is seasonally lost. Should temporary fire stations be set up as required, outposts, detachments? If only . . . is not good enough.

Few fire stations are sited correctly, that is, originally the majority were sited next to or near town halls when local authorities ran fire brigades. When the counties or cities were amalgamated or enlarged recently, how could the majority be sited so as to serve properly larger areas? To correct the position would, in the short term, also be beyond the pocket of fire authorities. Moreover, it could prove unwise, for changes in road patterns, residential or industrial complexes are certain to take place, misplacing the new stations. Bear in mind that a change of political party invariably results in a change of environmental policy.

Should appliances in seasonably less busy areas of a brigade be moved into areas where fires are expected? Should a fire station ever be left empty? John Horner, once General Secretary of the Fire Brigades Union, was a man of courage. Echoing his sentiments, I wrote to the editor of a fire service journal:

Dear Editor,
What the hell, above day-time inspections, is the British fire service doing as a corporate body about hotels or any other place where large numbers of persons are collected together for roasting en masse?

Time is rapidly running out, winter has just about set in bringing with it the seasonable spate of fires, deaths and serious injuries. The fireman sleeps safely in his fire station at night while the guests sleep unsafely in many hotels and the like. It is two decades since John Horner said 'do away with beds in fire stations and earn professional status'.

We all know that employees appointed to stay awake in hotels have become infected by the sleeping watch of fire brigades, this in 1973 and after a dreadful record of fatalities in hotels and places of public entertainment.

Should we not be out and about at nights checking that a wakeful watch and patrol is being carried out at such places, recording hazard and reporting infringement of regulations?

To hell with complaints from hoteliers and guests about noise and inconvenience, or to hell with them in reality.

Charles Clisby

Sir Charles Cunningham stated in his 1971 review of the fire service:

If economies are going to be achieved, then there are clearly grounds for some re-evaluation of the job of the fireman. This could be justified by reference to the additional skills needed for this work.

(a) powers of observation with regard to the structure of buildings factory processes and chemicals used, etc.

(b) Skill in dealing with supervisors and work people in offices and factories with tact and commonsense.

(c) Skill in understanding the basic requirements of legislation that is often quite complicated.

It should be noted that these, in some measure, all apply to the basic grade of fireman, as well as to more senior ranks, where additional skills again are required.

This work provides 'job enrichment' and 'job enlargement' for firemen. In addition, it brings them into regular contact with the public. It is a service which feels to a degree *isolated* and also seems to feel *misunderstood* and perhaps *undervalued* by the community at large, this may not be the least important result of this experiment.

In this way economics would be practised, productivity. An army of inspectors would not be needed. Charges could be levied for tuition.

In these days of paging signals that can penetrate into every nook and cranny of the largest city, of duty systems which provide ample time for rest off-duty, there is no excuse for a sleeping watch or free time except for emergencies at weekends and on public holidays. It is also important that more senior ranks are available at their desks for more than two or three days a week. There is a necessity to enliven management of fire brigades and put an end to restrictive practices originated by the representative bodies. Evening classes and weekend schools are called for and the fireman should be tutor as well as unexpected visitor at all hours, especially early ones.

Since Sir Charles made that perceptive statement, the conclusion being evergreen and certainly topical, the London fire brigade in all ranks has carried out the following work in a field of human endeavour of Churchillian character.

SAVING LIFE BY FIRE PREVENTION

A survey in 1975 disclosed that:

Besides the massive undertaking of the London Fire Brigade's Prevention Branch, fire prevention work undertaken at fire stations includes the following:
(a) Inner London fire stations have responsibility for
  (i) Annual inspection of Section 20 premises under the London Building Acts (some 3000 premises).
  (ii) Reinspection of the hotels (acting as agent for the Council's Architect (some 500 premises).
  (iii) Testing and annual inspection of dry and wet rising mains and firemen's lifts (some 3000).
  (iv) Fire drills in schools, hospitals and like institutions.
  (v) School visits and lecturing of children.
  (vi) Instruction of hospital nursing staff on fire routines and use of hand-held appliances.
  (vii) Talks and lectures to outside organizations, youth organization training.
(b) Outer London fire stations have responsibility for all the above except (a) (i) and (ii) but assist inner stations in their work in these respects.
In addition, various private fire protection installations are tested by

all stations on request, street hydrants are inspected and the general public are encouraged to 'call in' at any fire station for advice on matters of fire prevention while visits to and exercises at premises incorporating a high life risk are a part of training in firefighting and rescue carried out on a routine basis.

THE POWER OF VETO

I simply do not understand
The Architect with smile so bland,
Who tells me what I loud demand
Is not required by law of land.

It was two members of parliament, Anthony Greenwood and Bob Boothby, members of a governmental committee reviewing fire legislation, who said the fireman should decide, not the man who deals with sanitation.

Being aware of what has gone on where people other than firemen have had the last word, those in the fire service who have experienced fires and devoted a large part of their lives successfully to fire prevention are deeply concerned over present legislation.

Michael Doherty, head of fire prevention in London speaks out his worries on every possible occasion as do his counterparts in fire brigades nationwide. I am in good company. The building regulations in Britain are concerned primarily with structures and the materials from which they are constructed. There are at present only limited provisions for obtaining means of escape in case of fire. These provisions relate to three types of premises as detailed in regulation E23 which accepts the British Standards code of practice provisions as deeming to satisfy the regulations for:

(1) Flats and masionettes
(2) Offices
(3) Shops and departmental stores

These codes, in addition to building matters, give advice to owners/occupiers/management or staff. Proposals are in hand in Britain to extend the provisions of Building Regulations considerably in the area of escape routes in case of fire. I am not very happy about that. If extended as proposed they would have the effect of placing responsibility for determining means of escape for all future developments in the hands of the building authority and the Department of the Environment, not the fire authority. Thus, the 'statutory bar' would be extended to all premises when dealt with by the fire brigade for certification. This means that if a local authority has approved the arrangements at plan stage the fire brigade may not make requirements for alteration, *even if the means of escape are inadequate.*

I believe the answer lies in the Fire Precautions Act, Section 12, wherein the Secretary of State may make regulations for this purpose. The Home Office, rightly I say, would be the ultimate authority and the fire brigade would be the arbiter on such matters, both at planning stage and when certification is required. Hardly any class of building will escape the Fire Precautions Act when all premises covered are designated.

An easy way to achieve this commonsense aim would be for the Home Office to do what the Department of the Environment has done in building regulations, Section 23, that is to accept the existing Home Office codes as satisfying provisions of their regulations if made under Section 12 of the Fire Precautions Act.

Consequently, the question would be answered of what could be done in the future to take account of people's actions in fires. Consider once more an example of confusion in preferences. The very nature of the Fire Precautions Act allows the extension of its application to most classes of premises except single portable dwellings. The eventual extension of the Act will, I am convinced, be the most effective factor in controlling people's actions in fire situations. Under the Building Regulations, a door is not an element of structure, so in a private dwelling there is no requirement for any door to be fire-resisting unless there is a second occupied floor. In a conventional house therefore, with fire-resisting partitions separating the living areas from the hall and staircase, the doors are an inherent weakness. Similarly, the modern style of building incorporating an open staircase direct from an open-plan living area, exposes the upper floors of a house to any fire which breaks out in a lounge, dining room or kitchen. A London fire brigade investigation into fatal fires 1969–75 showed that half the fires which broke out in dwellings broke out in a lounge, dining room or kitchen. The number of fatal victims in dwellings as against other places was in the ration of 4:1.

The work of the fire service in this country in fire prevention has brought about fire safety standards second to none in the world. Our record of saving lives at fires is heroic. Our fire service colleges are accepted as the ultimate. Officers attending courses come from all over the world. What is taught is written into legislation abroad.

Who, then, are these impertinent people, unaccountable people who have never experienced fires, who could not stay the very course they have set for us to run in saving life, whose preference is buildings, not life?

REFERENCE

Honeycombe G. (1976). *Red Watch*. (London: Hutchinson).

Index

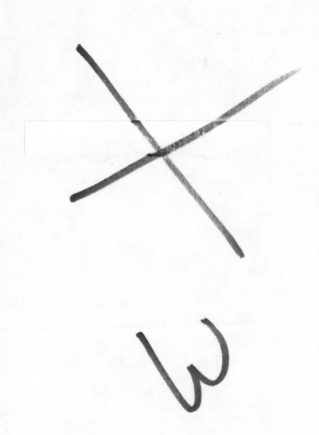